Entrepreneurship for Engineers

Entrepreneurship for Engineers

Kenji Uchino

CRC Press is an imprint of the
Taylor & Francis Group, an **informa** business

CRC Press
Taylor & Francis Group
6000 Broken Sound Parkway NW, Suite 300
Boca Raton, FL 33487-2742

Printed in the United States of America on acid-free paper
10 9 8 7 6 5 4 3 2 1

International Standard Book Number: 978-1-4398-0063-8 (Paperback)

Library of Congress Cataloging-in-Publication Data

Uchino, Kenji, 1950-
Entrepreneurship for engineers / Kenji Uchino.
p. cm.
Includes bibliographical references and index.
ISBN 978-1-4398-0063-8 (pbk. : alk. paper)
1. Entrepreneurship. 2. New business enterprises. I. Title.

HB615.U35 2010
620.0068'4--dc22 2009030735

Visit the Taylor & Francis Web site at
http://www.taylorandfrancis.com

and the CRC Press Web site at
http://www.crcpress.com

Contents

Preface

There is a general trend, regardless of country, toward an increase in entrepreneurial enterprises. A young engineer who invents or develops a new technology or comes across a new idea tries to set up his or her own company in order to commercialize the invention. From a historical viewpoint, the large corporation period ended in the twentieth century in advanced countries, and I would like to call this new twenty-first century the "high-tech entrepreneur" age, when small business corporations take the lead in Industrial Evolution (not Revolution!). This is analogous to the historical transition from huge dinosaurs to small mammals.

However, many young researchers hesitate to start their own company. This textbook is designed to answer frequently asked questions from these ambitious entrepreneurs. Topics will include:

1. How to start up a company
2. How to create product lines
3. How to get venture capital funding
4. How to write successful federal funding proposals
5. How to manage finances in a small firm, in particular, cash flow
6. How to keep a university position while also operating your own company (a conflict of interest)

Although many are ambitious enough to start their own companies, only a few ventures can survive and succeed in business. Why? Lack of fortune, perhaps? Yes, that is partially true. But, I do not believe it is merely due to lack of fortune, since few fortune-tellers are millionaires. I believe there is a systematic preparation process to be followed in order to build a great reputation in the commercial marketplace (i.e., to create a best seller). This is the motivation for this book: *Entrepreneurship for Engineers*.

I have been a professor for 35 years and a research-center director. During this time, I have also spent 19 years as a corporate executive in the roles of president, vice president, and lab director, and have been closely involved with the commercialization of over 60 ferroelectric devices with various international firms. I have been a professor, vice president, research and development (R&D) center deputy director, and a standing auditor at several universities and private companies in both Japan and the United States. I am a company executive *and* (not or) a university professor, simultaneously. I find this important for commercializing research immediately into real-world products.

This textbook is particularly suitable for graduate and undergraduate students majoring in engineering and also working engineers who are interested in starting up their own companies using their technological expertise. This book is also appropriate for those who are just starting or are about to start a company, as well as MBA students who are interested in high-tech companies.

I recently finished an MBA program and noticed two points which differ from the typical coursework for high-tech entrepreneurs:

1. Most of the scenarios for management courses start from a supposition that: "You are employed as a middle manager in a 1000-employee company. How will you work under the corporate executives and above your subordinates?" However, an entrepreneurial firm such as my company, MMech, has only a couple of partner–employees with similar corporate rank. Regular human resources management in a large firm may not be applicable to a start-up.
2. Most entrepreneurial classes in MBA curricula teach "How to start up a restaurant or an apparel shop," or another general small business. The July 2008 issue of the journal *Inc.* contained an article titled "How to Launch a Successful Start-Up." It introduced four young companies: a sweets shop, a photo shop (conversion from old photos to digital format), a sweater business, and an event Web site business. These were very attractive from a business model viewpoint. However, high-tech start-ups are quite different. High-tech start-ups introduce new technology to the market, and therefore their needs and structures are different from companies that replicate other similar businesses.

Based on the above points, I have emphasized two concepts in this book. First, the fundamentals of global economics, accounting, finance, and quantitative business analysis are provided in simple language for engineers as survival skills, because ordinarily engineers lack this knowledge. Second, that high-tech entrepreneurial success is based on (1) the business plan, (2) successful proposal writing, (3) SWOT analysis, (4) intellectual property, and (5) engineering ethics. I selected only important and basic items to help you understand how to start-up and operate a company by commercializing a new invention. The emphasis is on practical and utilitarian applications rather than on abstruse matter that a student may need to learn to pass an MBA exam. Various actual examples are provided which can be revised for your use. I also wish to describe my personal philosophy on how to develop best-selling devices, especially in the area of smart materials and structures, as examples. This sort of strategic "business and administration" viewpoint is, I believe, as important as the practical technological development of devices to a junior engineer, who is just starting or is going to start his or her own entrepreneurial company.

Chapter 1 discusses my concept of "Industrial Evolution," which is different from Industrial Revolution. High-tech entrepreneurs may take the industrial lead position in the twenty-first century. In this book, a fictitious entrepreneur, Dr. Barb Shay, and her company, Micro Motor, Inc. (MMI), are introduced, so that the case studies using her company can be easily understood. You can imagine that MMI is a hybridized firm of multiple existing corporations.

In Chapter 2, "Best-Selling Devices," three creativities for business success are introduced, based on the ideas of Sony's former president Dr. Morita. We will start from technological creativity. Of course, the engineering reader who wishes to start his or her own company is usually required to have (or will already have) his or her own invention. I will suggest how to cultivate your technological creativity in this chapter. Then, I will describe the key point, *product planning creativity*, detailing how to turn a primitive idea into a real product. *Marketing creativity* is considered briefly, taken up again in detail in Chapter 8. This development sequence (seed-push) is typical in an entrepreneur company, but the sequence may be just the opposite in a big company; that is, market research first, followed by technological development, and product planning (need-pull). Chapter 2 provides you with a quick overview of the context of this textbook.

Chapter 3 explains "Corporation Start-Up." Team-setting and legal procedures will be reviewed. Chapter 4, "Business Plan," is essential for finding investors. It describes how to write an attractive and practical business plan. More than 50% of start-up companies will become bankrupt in less than 4 years, according to U.S. statistics. Chapter 5, "Corporate Capital and Funds," covers information essential for an entrepreneur owner to collect the necessary funding, though many engineers may not be interested in it. I will introduce how to apply for Small Business Innovation Research (SBIR) programs in this Chapter, which is a unique "high-tech" start-up opportunity. Chapter 6, "Corporate Operation," describes some financial and accounting basics needed to operate a corporation on a daily basis.

Chapter 7, "Quantitative Business Analysis," may be easy for the typical engineer. Fundamental mathematical skills will be introduced with an Excel "Solver" tool used to analyze linear programming, PERT, and game theory problems. Chapter 8, "Marketing Strategy," explains the famous four Ps of marketing: product, price, promotion, and place, which will be new to typical engineers. Chapter 9, "Intellectual Properties," will discuss how to protect your new idea from the competitors. Chapter 10, "Human Resources," describes how to find suitable engineers for your start-up company, including dealing with visa applications for international employees.

Even though we write a business plan in Chapter 4, Chapter 11, "Business Strategy," is devoted to deeply analyzing the original strategy 1 or 2 years after start-up. This is to evaluate business success in order to modify or change the original plan, if necessary, before catastrophes such as bankruptcy occur. Chapter 12, "Corporate Ethics," is a sort of appendix from my viewpoint. Ethics differ widely according to education history, country, culture, and religion. There are no unique absolute ethics in the world. The epilogue chapter discusses the future of your company, through your desire for success: Chapter 13, "Now It's Your Turn." After the summary of this book's content, the rest of the chapter relates to business globalization.

Although I have extensive international business experience, I had not received systematic training in business and administration until my education at SFU. This textbook is essentially my MBA thesis: each chapter represents a course taken at St. Francis University (SFU) over a 4-year period. Actually, I have been a professor in the morning, senior vice president in the afternoon, and a graduate student at night in these 4 years.

First, I acknowledge the following professors at SFU, without whom I could not have completed this book.

Dr. Randy Frye, MBA Program Director
Dr. Mahabub Islam
Dr. Jim Logue
Dr. Dennis P. McIlnay
Dr. John D. McGinnis
Dr. John S. Miko
Mr. Robert A. Low
Mr. Robert E. Griffin
Mr. Adam Conrad
Dr. Larry Rager

Secondly, I would like to thank the following colleagues, who reviewed this book and provided helpful modifications and comments:

Dr. Michael Strauss, President of HME
Dr. Alfredo Vazquez Carazo, President of Micromechatronics, Inc.

Dr. Seung-Ho Park, R&D Director of Micromechatronics Inc.
Dr. Gareth Knowles, President of QorTek

Last, as I will discuss in Chapter 3, one of the most important factors for an entrepreneur is understanding of his or her family. For that, this book is dedicated with my deepest love to my wife, Michiko Uchino, who permitted and encouraged my new business.

Kenji Uchino
Pennsylvania State University
State College, PA

About the Author

Kenji Uchino, a pioneer in the development of piezoelectric actuators, is the director of the International Center for Actuators and Transducers (ICAT), and a professor of electrical engineering at Pennsylvania State University. Since 2004, he has been senior vice president of Micromechatronics Inc. He has been a university professor for 35 years and a company executive for 19 years.

After being awarded a PhD (per CMS) from the Tokyo Institute of Technology, Dr. Uchino became a research associate in the physical electronics department. In 1985, he joined the Sophia University in Japan as an associate professor in physics. In 1991, he moved to Pennsylvania State University. He was also involved with the Space Shuttle Utilizing Committee of the National Space Development Agency (NASDA) in Japan between 1986 and 1988, and was the vice president of NF Electronic Instruments in the United States, between 1992 and 1994. He has been a consultant to more than 90 Japanese, U.S., and European industries to commercialize piezoelectric actuators and electro optic devices. He is the chairman of the Smart Actuator/Sensor Study Committee, partly sponsored by the Japanese Ministry of International Trade and Industry (MITI). He was also the executive associate editor for the *Journal of Advanced Performance Materials* (Kluwer Academic) and the associate editor for the *Journal of Intelligent Material Systems and Structures* (Technomic), and the *Japanese Journal of Applied Physics*. He has also served as an administrative committee member for IEEE, Ultrasonics, Ferroelectrics, the Frequency Control Society, and as a secretary for the American Ceramic Society, Electronics Division.

Dr. Uchino was made a fellow of the American Ceramic Society in 1997. He has been a member of IEEE for 20 years. He received the SPIE Smart Product Implementation (SPIE) Award in 2007, the R&D 100 Award in 2007, the American Society of Mechanical Engineers Adaptive Structures Prize in 2005, the Outstanding Research Award from the Penn State Engineering Society in 1996, and the Best Paper Award from the Japanese Society of Oil/Air Pressure Control in 1987.

Dr. Uchino's research interests are in solid state physics—especially dielectrics, ferroelectrics, and piezoelectrics, including basic research on materials, device designing, and fabrication processes, as well as development of solid state actuators and displays for precision positioners, ultrasonic motors, and projection type TVs. He has authored 550 papers, 55 books, and 26 patents in the piezoelectric actuator field. He currently teaches three graduate courses: Micromechatronics, Ferroelectric Devices, and Applications of the Finite Element Method for Smart Structures. His business interests are, of course, in commercializing his inventions worldwide. He received his MBA at Saint Francis University in Pennsylvania, and is establishing his entrepreneurship teaching materials for young researchers and engineers. He has taught How to Develop Best-Selling Devices for seven years, and plans to teach Entrepreneurship for Engineers using this textbook.

In addition to his academic career, Dr. Uchino is an honorary member of KERAMOS (the National Professional Ceramic Engineering Fraternity) and in 1989, he received the Best Movie Memorial Award at the Japan Scientific Movie Festival as the director and producer of a series of educational videotapes on "Dynamical Optical Observation of Ferroelectric Domains" and "Ceramic Actuators."

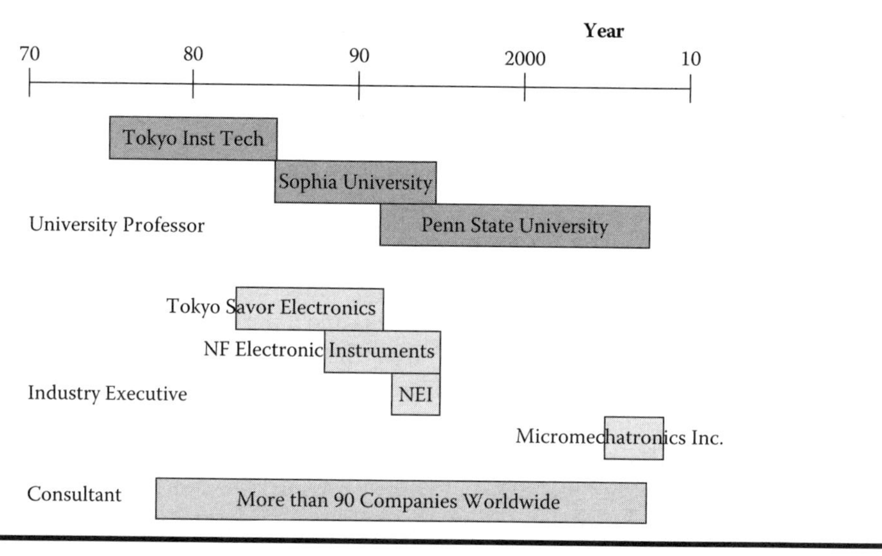

Figure P1 Personal history of Kenji Uchino.

Chapter 1

Industrial Evolution—Why Become a Small High-Tech Entrepreneur?

1.1 Necessity of New Industrial Viewpoints

Most university students try to find positions in large companies because they think small companies are not stable, and the salaries they offer may be lower. This is partially true, but not completely true for the so-called *high-tech entrepreneur*. Figure 1.1 [1] compares major electric equipment manufacturers and electronic component manufacturers in Japan, in terms of total sales, employees, and profit. Hitachi, Toshiba, and Mitsubishi Electric are world-famous giant companies, while Rhom, TDK, and Murata are component companies with sales and employees roughly one-tenth the size. During the severe economic recession in Japan during the late 1990s, the three big companies had only $1.5 billion profit, which is not terribly bad. However, the smaller company group (they are not "small corporations," just smaller than the former category) created twice the profit from one-tenth of the sales. If you are an engineer, which of the two categories would you choose for your career? Certainly, high net profits per employee will be reflected in bonus and salary payments.

Business administration classes in the United States usually advise students to create best-selling devices to make significant profits for a firm, leading to growth in terms of stock, employees, property, plant, equipment, and so forth. In addition, students sometimes learn about mergers and acquistions (M&A), which increase corporation size. However, is firm growth really the goal of a company? Recent legal problems disclosed by large firms such as Enron and WorldCom suggest there may be a *threshold of company size*. Corporate control mechanisms may not function smoothly above this threshold, leading to *moral hazard*, i.e., an increase in reckless or immoral behavior because members of the firm believe they will be saved when things go wrong.

In the 1980s, when the Japanese economy was booming, a giant firm, NEC, decided the firm should divide into nine smaller firms or divisions. The Japanese managers understood the concept of a threshold size of the company, contrasting with the U.S. managers' eagerness for M&A.

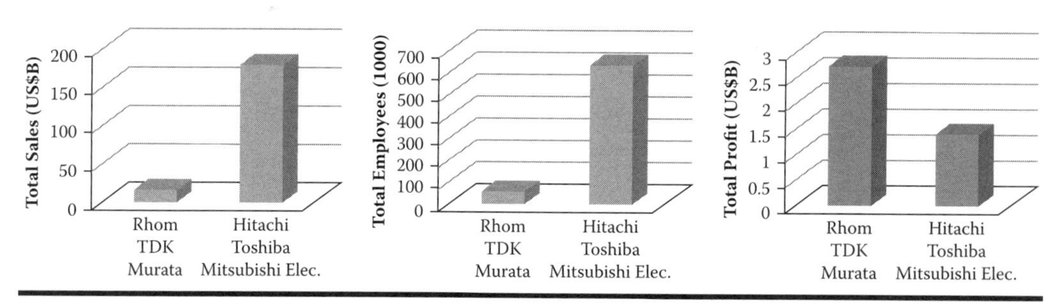

Figure 1.1 Comparison between major electric equipment manufacturers and electronic component manufacturers in Japan, in terms of total sales, employees, and profit. (Data from Data Stream, Japan, 1998)

As an introduction to this chapter, I describe my opinion on "Industrial Evolution," that is, that small high-tech firms will lead industries in advanced (matured) countries such as the United States and Japan in the twenty-first century. Corporation size suitable in advanced countries is discussed from historical, biomimetic, and organization viewpoints. Incidentally, "small business," as defined by the U.S. Small Business Administration [2], is a firm with 500 or fewer employees and with an annual revenue under $5 million. Therefore, small businesses encompass a large variety of companies.

1.1.1 Culture Transition

The earliest civilization in the world is believed to have started in mainland China around the Yangtze River more than 20,000 years ago. Independently, civilizations started in India around the Indus and Ganges Rivers, and in the Middle East (Mesopotamia) around the Tigris and Euphrates Rivers. Around 3000 years ago, these cultures motivated a new civilization in Greece. Rome took the initiative politically and economically and became the world power around the transition from BC to AD. Rome's place was taken over by Spain, Portugal, and the United Kingdom during the Middle Ages. They invaded many countries and expanded their colonial territories. With their economic resources, these European countries also cultivated sophisticated arts and sciences, culminating in the Renaissance in the fifteenth century. Refer to Figure 1.2 for the geographical positions of these countries and areas.

After World War II, the United States became powerful because most of the industrial world was decimated while the industrial capacity of the United States remained mostly unscathed. The United States was able to supply the world with mass production of automobiles and electric devices such as refrigerators, washing machines, and TVs in the 1950s. However, in less than 30 years, Japanese manufacturing industries (in which the triple best sellers are the "3Cs"—car, cooler, and color TV) caught up to U.S. technologies, making Japan's corporations the most profitable. This economic dynasty lasted only 10 years. The current economic dynasties are South Korea and Taiwan with their production factories in mainland China, which are fabricating digital electronic devices (laptop computers, digital cameras, cellular phones, displays, and printers). "Made-in-China" products currently exceed 80% in the United States.

As a private consultant, I recommend U.S. and Japanese companies work in the "Oriental Triangle," that is, Tokyo–Taipei–Shanghai, which is presently the most profitable combination of product manufacturing. Some Japanese companies opened factories in mainland China, but many have not operated smoothly. There are two major reasons for this failure: (1) language barriers, and (2) cultural differences. When SARS flu spread through mainland China in 2000, one Japanese company demanded that all executives in its Chinese subsidiary return to Japan. During their absence, the subsidiary became uncontrollable because of lack of directorship. If this Japanese company had chosen a Taiwanese partner as an agent, the language and culture barriers would have

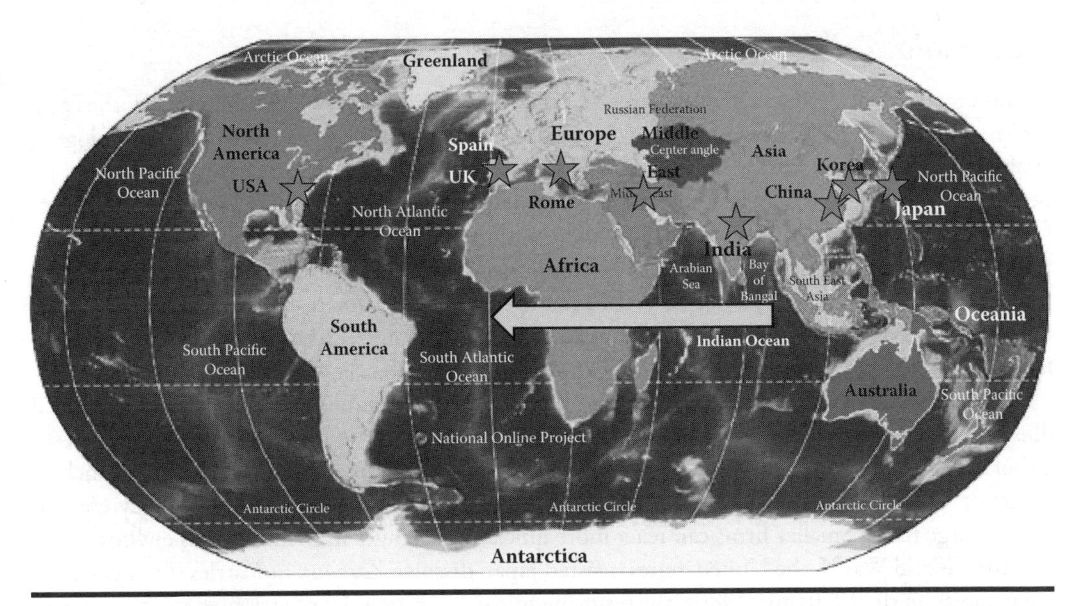

Figure 1.2 Historical civilization of key countries/areas in the world. The position transition follows from East to West.

been solved first. The Taiwanese did not leave the mainland during the SAAS incident (a working culture difference), leading to the continuous strong control of their factory.

I believe that the cultural, political, and economic key generally is to *transfer from East to West.* This is analogous to the sun's motion; sunrise is earliest in Eastern territories. I also believe that China has just started activities as the first runner of the second round. The economic and political power of mainland China is inevitably increasing because they have the highest product manufacturing capabilities. I do not know whether India will become an economic world power in the next decades, but I do know that the United States will not be a powerful manufacturing country from now on. It is obviously difficult to mass produce goods in the United States due to high labor costs.

A symbolic event happened in my hometown of State College, Pennsylvania, in early 2005: Murata Manufacturing Company closed its factory. Murata is the world's largest manufacturer of capacitors. This factory supplied 80% of the multilayer capacitors used in North America. Though demand is still rising due to the increasing electronics market, price reductions are also significant due to decreasing prices of electronics. The total sales (quantity times price) have not changed much due to market saturation. This factory closing was incredible because it was almost automatic: there was no manpower in the fabrication line. It had multiple tape-casting systems that could continuously produce multilayer capacitors without direct labor. So, why was it closed? The former president of Murata U.S. kindly explained it to me: "The manufacturing cost was already minimized, i.e., almost equal to the raw material cost. However, labor for the product packaging, purchase order paperwork and the raw material and product delivery is still expensive in State College or anywhere in the U.S." As a result, the factory was transferred to Asia, close to the raw material suppliers and assembly factories. Keeping a factory in the United States is almost impossible nowadays, with the exception of software companies, which do not have significant delivery costs like hardware companies.

It is time for the United States (and maybe Japan) to seriously consider how to transform corporate or industrial structures. My solution: Industrial Evolution, that is, a paradigm shift from large mass-production firms to small high-tech enterprises. Mass production has already been taken over by Korea, Taiwan, and China.

1.1.2 Biological Evolution

As most readers know, life started in seawater. According to a discovery by W. Schopf [3], the earliest known life on earth seems to be microbial algae around 3.5 billion years ago. Then, biological evolution created various underwater organisms, including fish. Coelacanth is the famous fish link between fish and amphibians; it has four legs, which evolved from fins. The evolution sequence from fish to amphibian to reptile generally increased the body size. The final form of these classic reptiles is the dinosaur, some as large as 30 m long. Even primitive mammals, such as killer whales (the largest water animals), do not exceed 10 m. There may be limits to the size an animal can be on the earth. For example, nerve signals propagate between 1 and 120 m/s. If an animal gets too large, the brain may not know what is happening in other parts of its body until it is too late.

Humans, modest-sized mammals with large brains, evolved, conquering other animals. Figure 1.3 illustrates biological evolution, that is, the animal transition over time. There appears to be a threshold of the animal size for survival. The dinosaur bears a resemblance to a giant company such as Enron or WorldCom. Control mechanisms, like nerves in animals, do not work quickly enough in such large firms. Smaller firms can react more quickly to changes in their business environment.

After World War II, the United States insisted Japan dissolve *Zaibatsu* industries (financial conglomerates). In the 1970s and 1980s, the resulting midsized companies contributed to the Japanese miracle of economic and technological success. However, the economic recession in the 1990s accelerated restructuring of Japanese industries; that is, consolidation back to giant conglomerates. The major reason was to reduce the manpower and research budget by eliminating the overlapped sections and divisions. In the cement group, Chichibu Cement and Onoda Cement merged first to become Chichibu-Onoda Cement. Then, Chichibu-Onoda and Nihon Cement merged several years ago to become Taiheiyo Cement, which is now the largest cement company in Japan. The merger of Mitsubishi Kasei Corporation and Mitsubishi Petrochemical resulted in Mitsubishi Chemical in 1994. This restructuring helped the Japanese economy recover. However, I can easily imagine future Enron-type problems in these giant companies. It appears that Japanese industries unfortunately are following U.S. industries with a lag time of around 15 to 20 years. Interestingly, Korean economy/industries seem to follow Japanese practices with a 10- to 15-year lag.

1.1.3 Management Structure

Figure 1.4 illustrates the difference between the corporate management structures of American and Japanese industries [4]. The American structure resembles a regatta, where only the coxswain (president

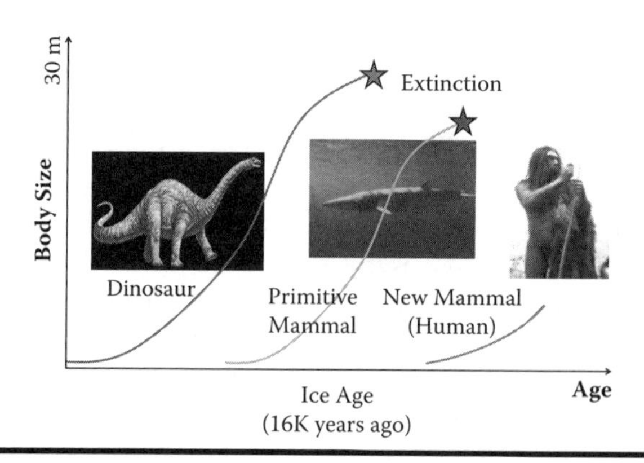

Figure 1.3 Biological evolution: animal transition as a function of age. Note that there seems to be a threshold to animal size for survival.

of the company) sitting in the back knows the destination, and the oarsmen (employees) just row synchronously to the cox's command. The oarsmen do not communicate with each other in general, which is a suitable management structure for "lone wolf"-type engineers. If the president does not command correctly, the regatta will quickly pile up. Similarly, the American economy changes significantly depending on the U.S. president's control. Further, American managers seem to prefer more power in a larger company, which is illustrated in Figure 1.5a. This may be called a "whale" or "brontosaurus" type.

In contrast, Japanese company structure resembles a *Mikoshi* (portable shrine), where all carriers (employees) know the destination of the *Mikoshi*. Walking about, staggering left and right, the *Mikoshi* does not go straight to the destination (major shrine), but reaches the destination, even if the flagman (president) standing on the *Mikoshi* falls down during travel. The employees behave as a team with a patriotic loyalty to the *Mikoshi*, not to a person or a president. In a similar fashion, even if the Japanese prime minister changes, the Japanese economy will not change much. In contrast to the whale type for the American managers, the Japanese managers prefer a "sardine"-type structure (Figure 1.5b). Loose coupling by medium- and small-sized companies creates power similar to a big whale. But, unlike a whale, a group of sardines can change shape adaptively according to an enemy's presence. Based on each sardine's patriotic and synchronized intention, these companies can make a *Keiretsu* (industry family tree).

A discussion similar to the aforementioned is found in an article titled, "The Fourth Economic Crisis" by N. Makino [7] in which he uses an analogy to *kabuki* and musical theaters. The *kabuki* attracts the audience by one or two key actors (there are no actresses in *kabuki*; only male actors), which resembles American industries, such as Mr. Iacocca when he was at Chrysler. In contrast,

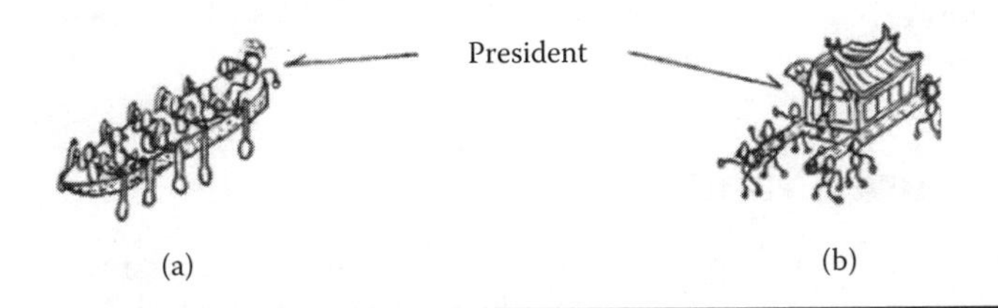

Figure 1.4 Management structure difference between the United States (a) and Japan (b).

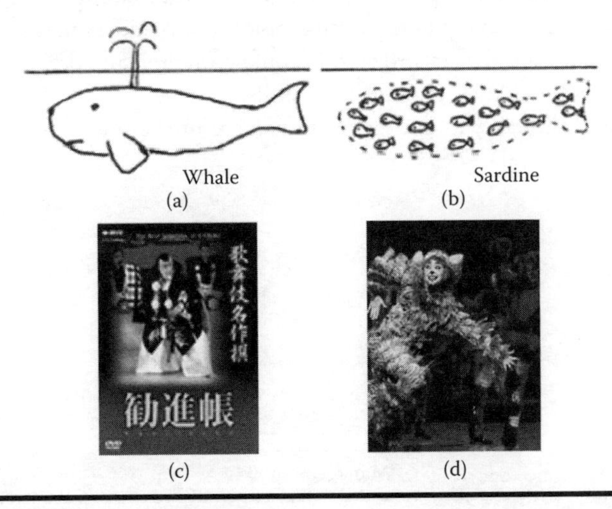

Figure 1.5 Difference between the United States (a and c) and Japan (b and d) in terms of the company structure that managers desire.

the musical is an assembly of many minor actors and actresses, which is close to Japanese industry's structure. Exemplified by the long-running musical *Cats*, the teamwork is brilliant, but not many audiences remember the names of the actors and actresses (see Figure 1.5c and d).

When the United States was taking the lead in mass production during the 1960s and 1970s, the "regatta" or "whale" control seemed to be the best way. However, if firm sizes become smaller in the future, like high-tech entrepreneurial companies, by shifting mass-production lines to Asian production centers such as Korea, Taiwan, and China, *Mikoshi*, or sardine-style management, could be employed in matured industrial countries. This is the paradigm shift I am suggesting in terms of the corporate structure and management concept.

In conclusion, I suggest the ideal style and structure for large product manufacturing firms in countries such as the United States and Japan (maybe 10 years from now). High-tech entrepreneurship will be the key to understanding Industrial Evolution, which may

1. Keep firms small- to medium-sized for strict corporate control
2. Keep small-scale production using differentiated technology
3. Outsource large-scale production using partnerships with developing countries' companies

Basically, small, active entrepreneurial companies are a suitable solution for developed (matured) countries such as the United States or Japan, with tight partnership with developing countries' large production companies. I am not claiming that this paradigm may be applicable to service industries such as banks and insurance companies, though service industries are outsourcing back-office functions, including call centers and accounting. The focus here will be on high-tech manufacturing industries.

1.2 Entrepreneurial Mind

Timmons and Spinelli indicated in their textbook [6] that the successful entrepreneur needs to have both *creative aptitude* and *management skills*. Figure 1.6 depicts the four categories of business people on these two scales: *entrepreneur*, *inventor*, *manager/administrator*, and *promoter*. Many graduate students with science and engineering backgrounds may be eligible to be inventors, while graduate students from management and administration departments may become managers or administrators. However, note that in order to become entrepreneurs or start their own companies, engineers should improve their management skills.

The following tests will help you decide where you are in the four categories and what skills you need to enhance.

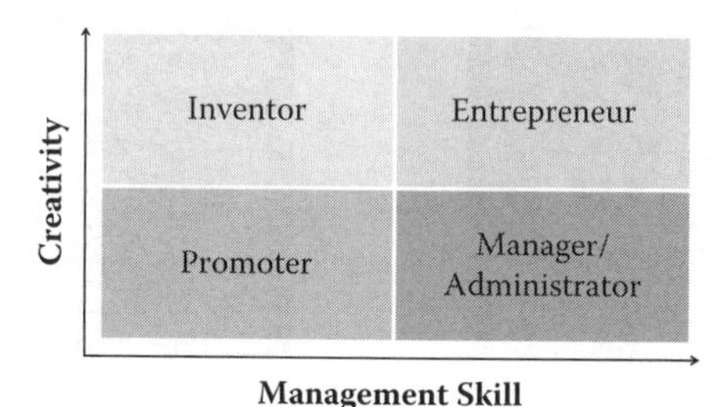

Figure 1.6 The four categories of business people: entrepreneur, inventor, manager/administrator, and promoter.

1.2.1 Creativity Test

The personality and aptitude of researchers are, of course, important factors to the entrepreneur. Suppose that you are an engineering student, or a young researcher in a manufacturing company; are you confident to continue your career path? Or, are you feeling a slump in your research and development (R&D) activity, and are you more likely to be comfortable in a management position? Even I am still wavering between the two desires—to be an engineer by believing in my creativity or a manager by expanding my production capability. Example Problem 1.1 will help in assessing your creative aptitude.

Example Problem 1.1

Figure 1.7 on this page is a test picture with text randomly cited from an academic journal.

1. First, familiarize yourself with the contents and picture of the academic article in Figure 1.7 on next page for 60 seconds. Do not peek at the following page, which includes the Solutions. Remember, this is your own personality check, before reading this textbook further.
   After studying the picture for 60 seconds, you may turn the page.
   Allowance: 60 seconds
   Stop after 60 seconds, and move to the next page.

Kenji Uchino is an enthusiastic explainer of the mysteries of miniaturization who manages to not only explain complex subjects, but to get his listeners caught up in the excitement of discovery. In the classroom, he introduces students to the joys of ferroelectricity. In industry, he has collaborated in the development of dozens of specialized patented devices.

Figure 1.7 Picture to accompany the Creativity Test.

2. Second, answer True or False for the following sentences *without looking at the previous page.*
   a. His name is Ohuchi.
   b. His article is printed on p. 15 of an academic journal.
   c. He has a moustache.
   d. He wears a dotted tie.
3. Third, score 1 point for each correct answer according to the solution box on page 9.

Comments

	Your Score	Aptitude
Really recognized	4	You can be a good engineer.
Some guesses	2–3	You are fit to be a manager/sales engineer.
No idea	0	Abandon your dream to be an engineer.

People with engineering aptitude generally remember the written content first. If you failed to answer questions (a) and (b) correctly, then your aptitude in this direction is lacking. Recognizing the moustache is also expected of engineering types, because it directly belongs to the person. However, remembering the tie is relatively rare. You can see it only when you try to. This attention to detail is one of the most important aptitudes to cultivate for innovative creativity.

This can be used to think of unconventional questions for a job interviewee. The following are two I have used:

1. You climbed up a staircase a couple of minutes ago. How many stairs did you climb?
2. You must have seen the pedestrian traffic signal just before entering the company entrance. Do you remember an illustration of a walking man lit up in blue? Was he walking toward the left or toward the right?

For the second question, most of the interviewees recognize the illustration, but the answers differ remarkably. When the answer is "I don't remember," we usually find the candidate unsuitable. Even when the answer is correct, "left," it may be a guess with a probability of 50%. This candidate may be hired for a management position. Only when the correct answer arises from a confident memory will we hire the candidate as a professional engineer.

I occasionally use a similar test for hiring suitable people for the university research center or the company. Once you become a corporate officer or a manager of your new venture who is responsible for hiring research and engineering employees, you may need to create similar test questions.

Solutions

(a) F (b) T (c) T (d) F

His name is Uchino, not Ohuchi. An engineer is probably trained to read the page of the references. Page 15 should be remembered. With a normal careful observation, most engineers should easily recognize baldness, moustache, beard, or glasses on the face. The design of necktie or jacket is rarely remembered. But the person who correctly recognizes subtle details will be really precious, and may discover new things. By the way, did you notice whether or not Uchino has a handkerchief in his breast pocket?

1.2.2 Entrepreneurial Mind Test

Timmons and Spinelli [6] reported six dominant themes of desirable attitudes and behaviors for the entrepreneurial mindset:

1. Commitment and determination
2. Leadership
3. Opportunity obsession
4. Tolerance of risk, ambiguity, and uncertainty
5. Creativity, self-reliance, and adaptability
6. Motivation to excel

Example Problem 1.2

You can check your entrepreneurial mindset using the following questionnaire:

1. Are you willing to sacrifice your personal issues for the company's sake? Y or N
2. Are you persistent in solving problems? Y or N
3. Do you like teamwork, rather than being a "lone wolf"? Y or N
4. Do you wish to share the wealth with all the people who helped you? Y or N
5. Are you familiar with customers' needs and specifications? Y or N
6. Are you obsessed with your product performance/design improvement? Y or N
7. Can you tolerate minimum risk for your firm, including financial debt? Y or N
8. Do you have confidence in enduring and in resolving dilemmas or uncertainties? Y or N
9. Are you a nonconventional and open-minded thinker? Y or N
10. Are you a quick learner and chance-taker without fear of failure? Y or N
11. Do you have perspective and a sense of humor? Y or N
12. Are you aware of your weaknesses and strengths? Y or N

Questions 1 and 2 are concerned with commitment and determination; 3 and 4 with leadership; 5 and 6 with opportunity obsession; 7 and 8 with tolerance of risk, ambiguity, and uncertainty; 9 and 10 with creativity, self-reliance, and adaptability; and 11 and 12 are related to motivation to excel.

If you answered Yes to all questions, you are confident of your skills to start your new firm. The number of No's is proportional to the attitudes you lack as an entrepreneur. If you answered No to Question 3 or 4, you may be a lone-wolf researcher who is better suited to working in a large firm. If you answered No to Question 5, you may be suitable to be a university scientist, not a corporate developer. If you did not answer Question 7 in the affirmative, you should not start your own company. The company founder, owner, or officer (CEO) should take joint liability/responsibility and be a cosignatory to the company's loans. Question 9 relates directly to Example Problem 1.1. Creativity requires open-minded thinking—You can see it only when you try to. This is one of the most important attitudes for a high-tech entrepreneur.

There are three steps to starting your own company: (1) collecting money (financial resources), (2) collecting people (human resources), and (3) keeping team harmony with a solid strategy (management development). Financial resources include bank loans and venture capital; the entrepreneur needs to recruit money lenders with self-reliance, enthusiasm, motivation, and a solid business plan, in addition to risk-taking willingness. The entrepreneur needs to persuade colleagues who work with him or her as partners to be as enthusiastic as they are. The entrepreneur also needs a solid strategy in order to convince subordinates to work for this new start-up. Establishing the initial corporate culture for working together to solve problems for new product development is very important for the founder. Rather than the *autocratic* and *custodial* management styles, the small business manager should take the *participative* and *collegial* management styles, to keep harmony.

The details of the aforementioned description will be elaborated in Chapter 10, Section 10.4.

1.3 Background of the Case Study Used in This Textbook

A fictitious company, Micro Motor Inc. (MMI) located in a fictitious town, College Park, Pennsylvania, is used as the case study consistently in this textbook. Dr Barb Shay, a graduate from the fictional State University of Pennsylvania (SUP), founded this company after accumulating a couple of patents on piezoelectric ultrasonic motors. Barb has a business and financial partner, Mr. Lenny Chu, president of Cheng Kung Corporation, Taiwan, who is also president of MMI. Cheng Kung has two factories—one in Taiwan and the other in Thailand—that manufacture piezoelectric actuators. A major customer of MMI is Saito Industries in Japan, a manufacturer of camera modules for cellular phones. Barb believes in her company's technology development capability, while Lenny has an ambition to expand his market in the world with R&D support from MMI. Figure 1.8 lists the key players in this case study.

Figure 1.8 Characters in the case studies for this book.

1.3.1 Background: Piezoelectric Multilayer Actuators

Piezoelectricity in a material was discovered by the famous brothers Pierre and Jacques Curie, who first examined the effect in quartz crystals in 1880. By applying stress on a quartz crystal, the crystal exhibited electric charge or voltage (direct piezoelectric effect). As a converse effect, subjecting the material to an electric field causes it to stretch or squeeze (induces strain). Refer to Figure 1.9 [5].

The *Titanic* shipwreck in 1912 spurred on piezoelectric technology development for underwater transducer applications. World War I accelerated technology development in order to search for German U–boats in deep water. Dr. Langevin, a professor at the Industrial College of Physics and Chemistry in Paris, who had colleagues including Albert Einstein, Pierre Curie, and Ernest Rutherford, started experiments on ultrasonic signal transmission into seawater, in collaboration with the French Navy. Using multiple small, single-crystal quartz pellets, sandwiched by metal plates (original Langevin-type transducer), he succeeded in receiving the returned ultrasonic signal from deep seawater in 1917. During World War II, a superior material, barium titanate ($BaTiO_3$) ceramic was discovered, followed by lead zirconate titanate (PZT). PZT-based ceramics are still the dominant piezoelectric materials at this time.

Piezoelectric ceramics can be deformed under an applied electric field. This is the basis for actuators for moving objects. Because the efficiency of piezoelectric devices is much higher than the conventional electromagnetic motors and they can be miniaturized, applications are expanding dramatically, particularly for cellular phones.

One of the initial problems with piezoelectric actuators was their rather high drive voltage (300 to 1000 V). In order to achieve a low driving voltage (3 V drive is required for cellular phones), I invented a piezoelectric ceramic multilayer structure for actuator applications in 1978. Figure 1.10 shows the structure of the multilayer actuator. By using a thinner layer of piezoelectric

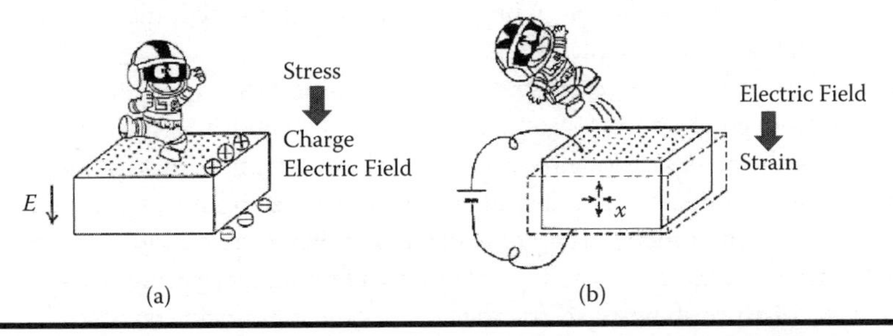

Figure 1.9 (a) Direct piezoelectric effect, and (b) converse piezoelectric effect.

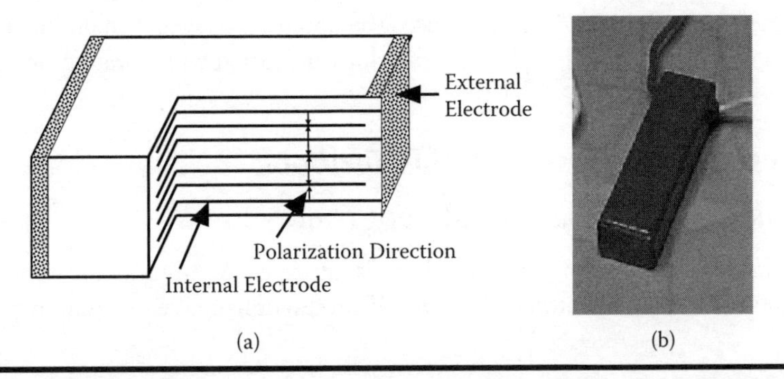

Figure 1.10 (a) Structure of a piezoelectric multilayer actuator, and (b) its appearance.

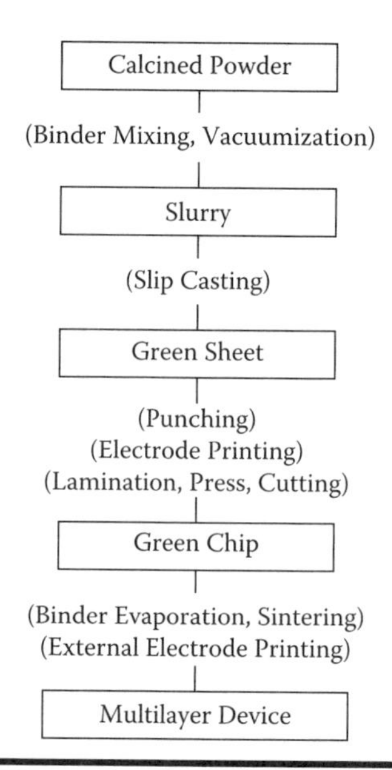

Figure 1.11 Tape-casting fabrication process of a multilayer ceramic actuator.

ceramic, the drive voltage of the device can be proportionally reduced, because the electrode gap is smaller. Note that the electrode configuration that provides the electric field directions opposite each other between adjacent layers is known as an *inter-digital* electrode pattern.

There are two techniques for making multilayered ceramic devices: the *cut-and-bond method* and the *tape-casting method*. The cut-and-bond method does not require expensive facilities, but does require a lot of manpower/labor. Workers cut a bulk piezo-ceramic slab into thin plates, electrode them, and bond multiple plates together.

Alternatively, tape-casting requires an expensive fabrication facility, but almost no labor. This is suitable for mass production and can produce more than 100,000 pieces per month. Figure 1.11 shows a flowchart of the manufacturing process of multilayer ceramic actuators. *Green sheets*, flexible films with piezoceramic powders in a binder, are prepared in two steps: slip preparation of the ceramic powder and a doctor blade process. The *slip* is made by mixing the ceramic powder with solvent, deflocculant, binder, and plasticizer. The slip is cast into a film under a special straight blade, called a *doctor blade*, whose distance above the carrier determines the film thickness. After drying, the green sheet has the elastic flexibility of synthetic leather. The green sheet is then cut into an appropriate size, and internal electrodes are printed using silver, paladium or platinum ink. Tens to hundreds of such layers are then laminated and pressed using a hot press. After cutting into small chips, the green bodies are sintered at around 1200°C in a furnace. The sintered chips are then polished and externally electroded, lead wires are attached, and finally the chips are coated with a waterproof spray.

1.3.2 Topics to Be Discussed in This Textbook

1.3.2.1 Break-Even Analysis, Investment Theory

The introduction of a tape-casting facility, which is an expensive automation production system costing $300,000 per set, is recommended only if production exceeds 1 million pieces per year.

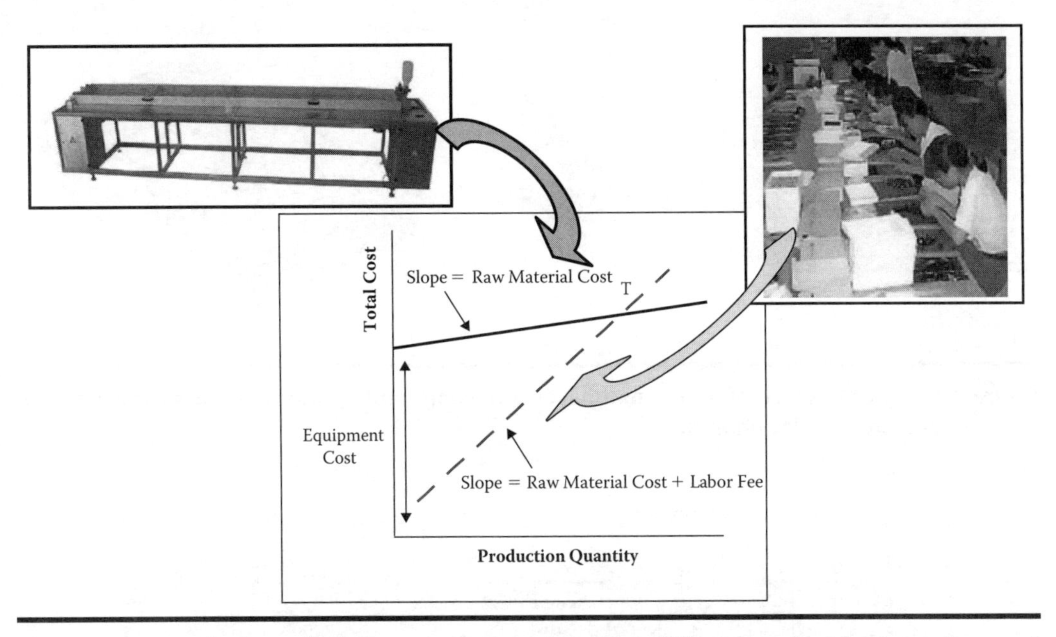

Figure 1.12 Total cost calculation comparison for the multilayer product between tape-cast equipment automatic production and a cut-and-paste manual production.

Otherwise, the conventional cut-and-bond method may be employed by hiring several manufacturing workers.

The basic of this calculation is depicted in Figure 1.12. For a tape-casting system, the labor fee is negligible but the initial investment is very high. For the cut-and-bond method, the labor costs are proportional to the product quantity; even the equipment cost becomes negligible. Because of the difference in slope, we can find an intersection (point T in Figure 1.12). This product quantity is the threshold above which the equipment installation starts to provide a profit benefit. This quantity is about 1 million pieces in multilayer actuator production. We will adopt the *break-even analysis* for this case study in Chapter 6 and learn the fundamentals of manufacturing costs (direct material and direct labor costs, variable and fixed costs) and investment theory.

1.3.2.2 Research Contract

In collaboration with Saito Industries in Japan, MMI developed a new multilayer ultrasonic motor for a zoom/focus mechanism of next generation cellular phone cameras (see Figure 1.13).

You will learn how to structure a research contract in Chapter 5, and about sharing intellectual property with a partner company in Chapter 9.

1.3.2.3 Production Planning

In April 200X, MMI provides 1000 multilayer actuators to Saito Industries for initial tests. Once these tests are finished, much larger production will be required starting in January of the next year. Production is expected to reach more than 10 million pieces per year. MMI needs to develop a production plan for supplying actuators to Saito.

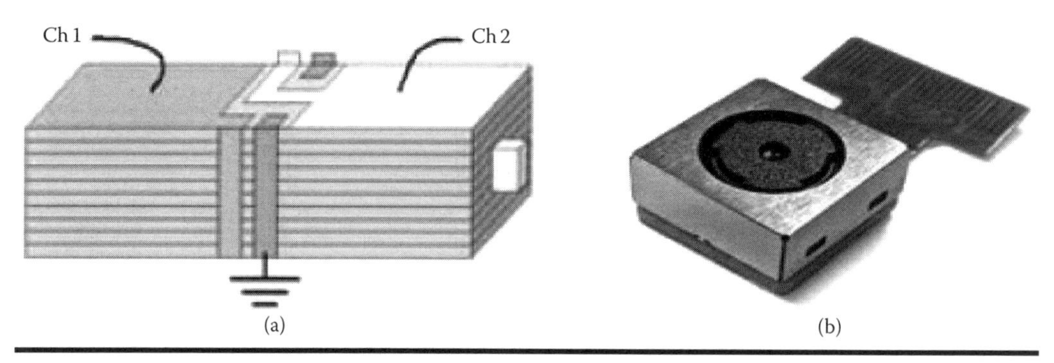

Figure 1.13 (a) Structure of a new multilayer ultrasonic motor, and (b) its assembled zoom/focus module for cellular phones.

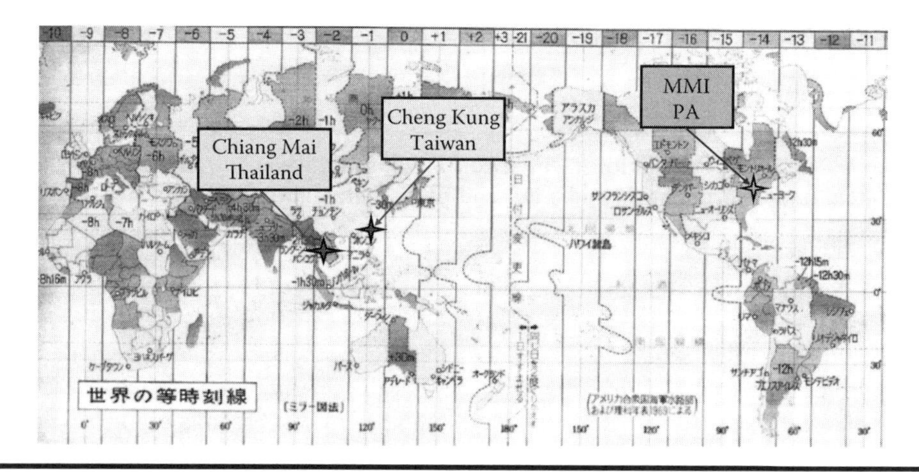

Figure 1.14 MMI multilayer actuator manufacturing factory locations: MMI in Pennsylvania, Cheng Kung Corporation in Taiwan, and Chiang Mai factory in Thailand.

The customer's demands include (a) supply to be increased exponentially in 2 years, and (b) unit price to be drastically decreased with quantity. MMI has three manufacturing options: MMI in Pennsylvania, a partner company; Cheng Kung Corporation in Taiwan; and Chiang Mai factory in Thailand. Labor cost is highest in Pennsylvania, lower in Taiwan, and much lower in Thailand. Taking into account transportation distances, which are equivalent to manufacturing delays, MMI will distribute the manufacturing load to these three places in order to minimize the manufacturing cost. Manufacturing locations of these three facilities are shown in Figure 1.14.

The production planning of the multilayer actuators is made using linear programming methods for minimizing the production cost in Chapter 7, Section 7.1

1.3.2.4 Cash-Flow Analysis

Financial analysis can then be performed, in terms of monthly cash flow. Note that the bankruptcy risks of small high-tech firms increase with increasing contract amount. (Refer to Chapter 6, Section 1.4, and Chapter 11.)

1.3.2.5 Strategic Business Plan

MMI faces a decision: Should we shift to a mass-production company? The Saito Industries contract is an attractive opportunity to expand MMI dramatically. MMI may exhibit weaknesses in manpower and financial resources to set up a factory even though MMI has strength in developing this product. The monthly operating costs for setting up large-scale production are more than 10 times the present small-scale costs. This may initiate a serious cash flow problem, leading to the bankruptcy of MMI. Even if MMI starts production successfully, competitors may enter the market with similar or alternative products in a short time. This is a big threat. You will learn strategic management theory, or SWOT (strength-weakness-opportunity-threat) analysis, for decision making in Chapter 11, Section 11.2.

Although this case study content is fictional, I believe it is similar to the scenario encountered by most high-tech entrepreneurs.

Chapter Summary

1.1 Historical, cultural, political, and economic concerns transfer from East to West, in general. The peak period of U.S. and Japanese production has already passed.

1.2 Biologically, evolution theory suggests that animals above a particular size threshold are disadvantaged and become extinct.

1.3 The ideal style and structure of the firm for product manufacturing after the matured status of the firm (giant size) in the United States and Japan may be high-tech entrepreneurship, the key to understanding Industrial Evolution, which may
   1. Keep firms small- to medium-sized for strict corporate control.
   2. Keep small-scale production using differentiated technology.
   3. Outsource large-scale production using partnerships with developing countries' companies.

1.4 The six dominant themes of desirable attitude and behaviors for the entrepreneur from a consensus [6] are
   1. Commitment and determination
   2. Leadership
   3. Opportunity obsession
   4. Tolerance of risk, ambiguity, and uncertainty
   5. Creativity, self-reliance, and adaptability
   6. Motivation to excel

1.5 Topics discussed in this textbook:
   1. Fundamentals in accounting: break-even analysis, manufacturing cost
   2. Fundamentals in finance: present and future value, cash flow analysis, investment theory
   3. Research contracting: Small Business Innovation Research (SBIR), Small Business Technology Transfer Program (STTR)
   4. Intellectual property: patent, trademark
   5. Conflict of interest
   6. Product planning: linear programming, PERT
   7. Marketing strategy: product, price, promotion, and place
   8. SWOT analysis
   9. Strategic business planning
   10. Corporate ethics

Practical Exercise Problems

P1.1 Business Concept Questions

This exercise will help you write a business plan for your venture. Refer to the MMI example and write up your own plan.

P1.1.1 Function/Feature of Product

What is the important and distinct function or feature of your product—cost, design, quality, capabilities—in comparison with competitive products? Include any photographs or sketches to visualize your idea.

MMI Example

R&D Division

MMI developed a micromotor called *metal tube type* consisting of a metal hollow cylinder and two PZT rectangular plates (see Figure P1.1a). When we drive one of the PZT plates, Plate *X*, a bending vibration is excited along the *x* axis. However, because of an asymmetrical mass (Plate *Y*), another hybridized bending mode is excited with a phase lag along the *y* axis, leading to an elliptical locus in a clockwise direction, much like a Hula-Hoop motion. The rotor of this motor is a cylindrical rod with a pair of stainless ferrules pressed with a spring. The assembly is shown in Figure P1.1b. The metal cylinder motor is 2.4 mm in diameter and 12 mm in length and is driven at 62.1 kHz in both rotation directions. A no-load speed of 1800 rpm and an output torque up to 1.8 mN·m are obtained for rotation in both directions under an applied rms voltage of 80 V. A rather high maximum efficiency of about 28% for this small motor is a noteworthy feature. Figure P1.1c shows a newly developed self-oscillation circuit for driving the micromotor.

The key features of MMI micromotors are

- World's smallest motors (no competitive products)
- Low manufacturing cost competitive with the low cost conventional electromagnetic motors

In collaboration with Saito Industries in Japan, MMI developed the world's smallest camera with both optical zooming and autofocusing mechanisms for cellular phone applications, as shown in Figure P1.2. Two microultrasonic motors with 2.4 mm diameter and 14 mm length are installed to control zooming and focusing lenses independently in conjunction with screw mechanisms.

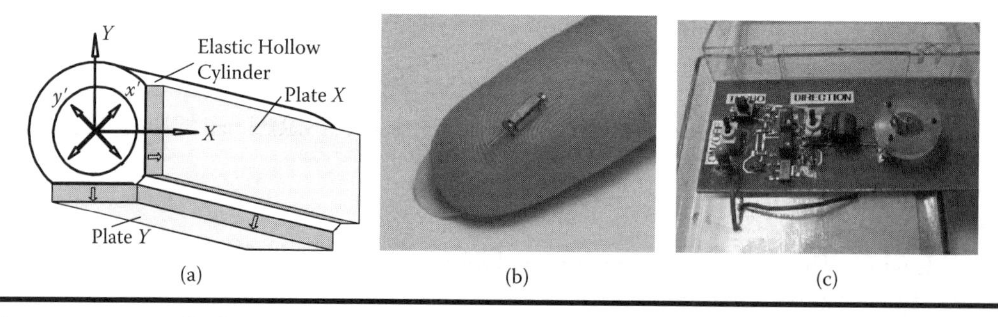

Figure P1.1 Metal tube motor using a metal tube and two rectangular PZT plates. (a) Schematic structure, (b) world's smallest motor (1.5 mm diameter), and (c) drive circuit on a board.

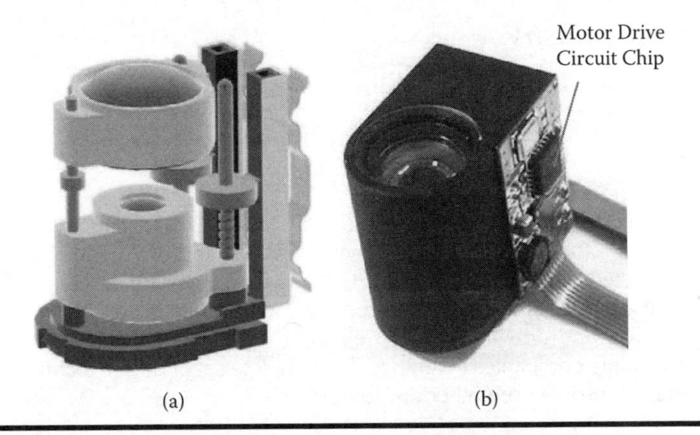

Figure P1.2 Camera auto zooming/focusing mechanism with two metal tube ultrasonic motors in Saito's cellular phone. (a) Schematic illustration, and (b) prototype product.

The large "wobbling" motion at the middle point of the metal tube motor can be transferred to a rotor to rotate the screw rod, which provides up–down motion of the lens unit. The drive circuit in Figure P1.1c could be integrated into one chip at the side of this camera (3 × 3 mm square black chip).

Sales Division

MMI has 13 partner companies in the world, through which they import various piezoelectric devices and distribute them to North America, Asia, and Europe. "All-in-one" is their policy; that is, the customer can find any devices (materials, components, devices, drive circuits, and designing software) relating to the piezoelectric actuators. The customer need not access multiple companies to arrange all the set devices. MMI has this sort of differentiation privilege from the other competitive distributors.

P1.1.2 Proprietary Aspects

What are the proprietary aspects of the product, such as patents, trademarks, and trade secrets?

MMI Example

The inventor is Barb Shay, and the patent of the Micro Metal Tube Motor has been filed through SUP. MMI is the patent licensee.

P1.1.3 Innovative Technology

What innovative technology is involved with the product?

MMI Example

Barb Shay is an active developer of the piezoelectric actuators, and her inventions are rather unique. The innovative point is a miniature electric motor made from inexpensive standard parts (piezo-plates and a simple tube).

P1.1.4 Position in Industry

What position does the concept play in the industry, such as manufacturing or distributing?

MMI Example

MMI is an R&D company, which accepts research contracts from federal agencies and companies. Designing and prototyping are MMI's strengths. When the prototype is acceptable, mass production is contracted to Taiwanese partners with lower manufacturing costs.

In parallel, MMI is a sales and distribution company, which distributes materials, components, devices, drive circuits, and design software, relating to the piezoelectric actuators. Customers need not access multiple companies to arrange the complete set of device technologies. MMI has this differentiation (all-in-one) from other distributors.

P1.1.5 Intended Customer

Who is the intended customer?

MMI Example

MMI's products (R&D and imported products) are for their partner companies, manufacturers large and small in the United States, Asia, and Europe. Product distribution customers are mostly U.S. small businesses, R&D contractors, and large corporations in the United States, Europe, and Asia.

P1.1.6 Customer Benefits

What benefits will be delivered to the customer? Explain what problems you are solving for your customer.

MMI Example

MMI provides technology that the customer does not have. It provides necessary piezoelectric devices to customers for developing systems, and accelerates the development of its customer products. Thus, the company logo includes the phrase "Your Development Partner."

P1.1.7 Market Penetration Methods

How will the product be sold to the customer? Explain which distribution channels will be used, such as sales reps, direct sales force, direct mail, telemarketing, and other channels.

MMI Example

Vice President Barb Shay has had a strong connection with Navy and Army laboratories for over 10 years. The acceptance rate for federal SBIR proposals is relatively high for this reason. The R&D division has a close relationship with these R&D contractors.

The sales division created a powerful and attractive company Web site, through which they have averaged 50 hits per day from customers. In addition, a sales representative visits customers (various universities and institutes around College Park) and attends selected exhibitions and trade shows to generate sales. Barb is also advertising MMI to companies that hire her for private consulting.

P1.1.8 Who Will Make or Supply the Product?

Who will make the product? Will it be subcontracted to a manufacturer, or will it be produced in-house?

MMI Example

R&D Division

Products for federal research will be made by MMI engineers. However, for products for commercial companies, MMI sometimes uses an original equipment manufacturer (OEM) partner, specifically, Cheng Kung Corporation in Taiwan, for large quantities.

Sales Division

MMI is a distributor of a consortium of 13 partner manufacturers. However, for software distribution, MMI has three subdistributors in Japan, Korea, and Taiwan.

P1.1.9 Concept Statement

Write your concept statement for your new high-tech company.

MMI Example

MMI specializes in *development* and *commercialization* of piezoelectric actuators, transducers, and their integrated systems, including supporting services, tools, and software. MMI's concept statement is "Your Development Partner," because they provide both the devices (sales) and the technology (R&D) to the customers.

P1.2 Business Mindset

Why are you interested in commercializing your invention in a small venture, rather than in a large firm? Identify the benefits of using a small venture for your commercialization. Relate them to the above discussions, position in industry, innovative technology, and proprietary aspects.

References

1. Data Stream, Japan, 1998. http://www.thomsonreuters.com/products_services/financial/financial_products/investment_management/research_analysis/datastream
2. U.S. Small Business Administration, Small Business Resources, http://www.sba.gov/services/ (accessed March 1, 2008).
3. Schopf, W. Space and Astronomy News, "Is This Life?" http://www.abc.net.au/science/news/space/SpaceRepublish_497964.htm (accessed March 1, 2008).
4. Uchino, K. Comparison of Technology Development Concepts among Japan and the U.S. *Sophia* (Sophia University, Tokyo) 41, no. 2, 1992.
5. Uchino, K., and J. R. Giniewicz. *Micromechatronics*. New York: Dekker/CRC, 2003.
6. Timmons, J. A., and S. Spinelli. *New venture creation*, p. 8. New York: McGraw-Hill/Irwin, 2007.
7. Makino, N. *The Fourth Economic Crisis*, Tokyo: Hiraku, 1997.

Chapter 2

Best-Selling Devices—How to Commercialize Your Invention in the Real World

Have you clearly visualized your product idea in figures or pictures, as I did in the example in Chapter 1? If you cannot presently visualize it, do not worry! This chapter teaches you how to crystallize your vague notion into a solid concept, which will become the first best-selling product for your future company.

2.1 Three Creativities

In the mid-1980s, Akio Morita, former president of Sony Corporation, responded to criticism from a journalist concerning the lack of creativity on the part of Japanese researchers by saying, "Japanese researchers are good at chasing and imitating the original idea for commercialization, but they in general lack creativity" [1]. Mr. Morita suggested that there should be three types of creativity with respect to research and development (R&D) at Sony: "The U.S. people are focusing only on *technological creativity*. But the people must understand there are two more creativities: *product planning creativity* and *marketing creativity*, which are equally important for commercial success."

Matsushita Panasonic's famous color TV technology (black stripe for creating better color resolution) was invented by Philips. Philips could not commercialize it. Matsushita, on the other hand, succeeded after an intensive 3-year development effort. You can decide which company is more "creative" in science and technology. However, it is a fact that only Matsushita profited from this TV development.

Table 2.1 summarizes the three important types of creativity needed when developing an R&D strategy; each will be described in further detail in the following sections. We will consider these creativities in the sequence shown in Table 2.1. A new technology or an idea for a new device

Table 2.1 Three Types of Creativity in Research and Development

Technological creativity	New functions
	High performance
Product planning creativity	Specification (sensitivity, size, power)
	Design
Marketing creativity	Price
	Advertisement

is the first step for a high-tech entrepreneur. Next come the product design and commercialization plans for the first prototype. Finally, marketing promotion is important for making the product a best seller. This development style is called *seed-push*. Note that this sequence is sometimes reversed in a big firm; first market research finds a product need, which is fed back to technology development and finally commercialized. This is called *need-pull*.

2.2 Technological Creativity

There are two different approaches in exercising technological creativity: (1) Find a new functional effect or material, and (2) achieve a high performance or figure of merit (FOM). These are typically called research and development, respectively. A new idea arising from research will create a seed-push market, while development is initiated by a need-pull force.

2.2.1 Discovery of a New Function or Material

Serendipity is often an important factor in discovering a new function in a material or a new phenomenon. Benjamin Franklin, the famous scientist and founding father of the United States of America, discovered that lightning is an electrical phenomenon [2]. Franklin's experiment was done during a thunderstorm. The lightning hit the kite, and he collected an electric charge, as illustrated in Figure 2.1. Two other scientists conducting similar experiments 1 month before and after Franklin were both electrocuted. Franklin was one lucky guy! Franklin became President of Pennsylvania (equivalent to governor today), and the state organization for financially supporting incubator companies in Pennsylvania is named Ben Franklin Technology Partners.

Ivory soap (Proctor & Gamble) was also created by serendipity. William Procter and James Gamble started a candle shop, but because of Thomas Edison's light bulbs, their business declined. In 1879, an employee in their Cincinnati candle factory forgot to turn off a machine when he went to lunch. Upon returning, he found a frothing mass of lather filled with air bubbles. He almost threw the stuff away, but instead decided to make it into soap. Proctor and Gamble sold it as a floating soap with a lot of bubbles. Why was floating soap such a hot item back then? Because clothes were washed in ponds and rivers at that time, a dropped bar of soap would sink and often be unrecoverable. Floating soap had a convenience factor.

Polyvinylidene difluoride (PVDF), a piezoelectric polymer, was discovered accidentally in the early 1970s by Dr. Kawai, and a high-temperature superconducting ceramic was discovered by Doctors Bednortz and Muller in the 1980s. Both are good examples of serendipity.

Figure 2.1 Flying kite experiment in a thunderstorm by Ben Franklin (1752). (From http://www.ushistory.org/franklin/essays/hoffman.htm. With permission.)

A traditional Japanese proverb tells us that every researcher has three lucky chances in his or her life to discover new things. However, most people do not even recognize these chances and lose them. Only people ready to accept these chances can really find new phenomena. A Japanese company executive mentioned that a person who develops one widely commercialized product has the chance to become a general manager; a person who develops two products for the company is guaranteed to be a vice president; and a person who contributes more than three can become president. You can see how difficult it is to develop actual best-selling products. The personality and aptitude of the researcher are, of course, also important factors in recognizing a lucky chance. Example Problem 1.1 assessed your ability to experience serendipity. What was your score?

If you have missed your three chances, what should you do? Quit being a researcher? The following example is dedicated to the unlucky reader who, like myself, missed those lucky chances. We can still research using a more systematic approach, for example, by using our intuition: making use of (1) *secondary effects* and (2) *scientific analogy*. Sections 2.2.1 and 2.2.2 include rather advanced technical, physical, and mathematical items for the front engineer's sake. You may skip this part if you are not considering inventing a high-tech device by yourself.

2.2.1.1 Secondary Effect

Every phenomenon has primary and secondary effects, which are sometimes recognized as linear and quadratic phenomena, respectively. In electrooptic devices, the Pockels and Kerr effects correspond to the primary and secondary effects, where the refractive index is changed in proportion to the applied electric field, or to the square of the applied electric field (this is the basic mechanism for the liquid-crystal display [LCD]). In actuator materials, these correspond to the piezoelectric and electrostrictive effects.

When I started actuator research in the middle of the 1970s, precise displacement transducers (we initially used this terminology) were required in the Space Shuttle program, particularly for deformable mirrors, for controlling the optical pathlengths over several wavelengths (in the order of 0.1 μm). Conventional piezoelectric lead zirconate titanate (PZT) ceramics were plagued by hysteresis and aging effects under large electric fields; this was a serious problem for an optical positioner. Electrostriction, which is the secondary electromechanical coupling observed in centro-symmetric crystals, is not affected by hysteresis or aging (see Figure 2.2) [3]. Piezoelectricity is a primary (linear) effect, where the strain is generated in proportion to the applied electric field, while the electrostriction is a secondary (quadratic) effect, where the strain is in proportion to the square of the electric field (note the parabolic strain curve in Figure 2.2). Their response should be much faster than the time required for domain reorientation in piezoelectrics and ferroelectrics. In addition, electric poling is not required.

However, at that time, most people believed that the secondary effect would be minor, and could not provide a larger contribution than the primary effect. Of course, this may be true in most cases, but my group actually discovered that relaxor ferroelectrics, such as the lead magnesium niobate-based solid solutions, exhibit enormous electrostriction. Remember that the key to new invention is to consider the thing differently from most people, or to hold doubts about normal common sense.

2.2.1.2 Scientific Analogy

Most readers are probably familiar with shape memory alloys, which can revert rather quickly back to their initial shape when subjected to the heat of a cigarette lighter or a hair dryer. The basic principle is a *stress-* or *temperature-induced phase transformation* from the austenite to martensite

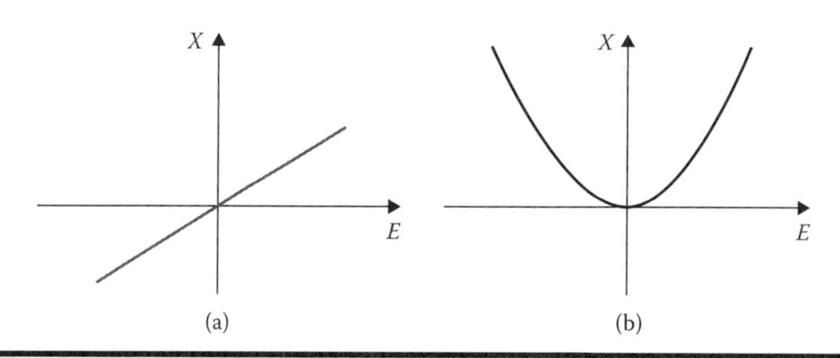

Figure 2.2 (a) Primary effect (piezoelectric effect), and (b) secondary effect (electrostrictive effect).

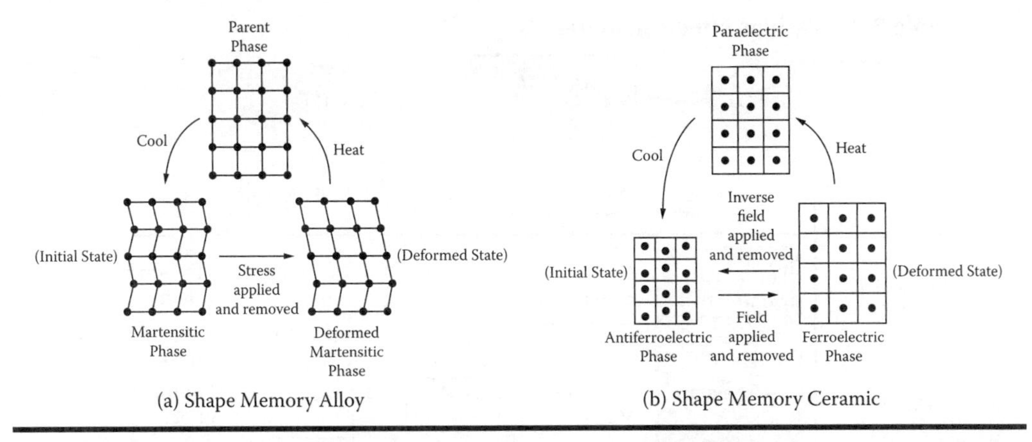

Figure 2.3 Phase transition analogy between (a) shape memory alloy, and (b) shape memory ceramic. (From Uchino, K. *Ferroelectric devices*. New York: Dekker/CRC, 2000. With permission.)

phase. I tried to consider an analogous case among the ferroelectrics (Figure 2.3). Yes, we have an electric field–induced phase transition from an antiferroelectric to ferroelectric phase. This type of phase transition should be much quicker in response and more energy efficient theoretically. After this speculation, we started to investigate lead zirconate–based antiferroelectrics intensively, and discovered the *shape memory effect* in ceramic actuator materials [3]. If you finished your engineering thesis, your knowledge and background helps you in an area different from your original thesis, by using a scientific analogy approach.

2.2.2 Performance Improvement

Starting with material functionality, Table 2.2 lists the various effects relating input (electric field, magnetic field, stress, heat, and light) with output (charge/current, magnetization, strain, temperature, and light). Conducting and elastic materials, which generate current and strain outputs, respectively, for input voltage or stress are well-known phenomena. They are sometimes called *trivial* materials. On the other hand, pyroelectric and piezoelectric materials, which unexpectedly generate an electric field with the input of heat and stress, respectively, are called *smart* materials. These off-diagonal couplings have corresponding converse effects, the electrocaloric and converse-piezoelectric effects. Both sensing and actuating functions can be realized in the same materials. "Intelligent" materials must possess a drive/control or processing function that is adaptive to changes in environmental conditions, in addition to actuator and sensing functions. Ferroelectric materials exhibit most of these effects with the exception of magnetic phenomena. Thus, ferroelectrics are said to be very "smart" materials.

The concept of *composite effects* is very useful, particularly for systematically improving the properties and FOM.

2.2.2.1 Sum Effect

Let us discuss a composite function in a diphasic system to convert an input parameter X to an output parameter Y. Suppose that the volumetric ratio between Phase 1 and Phase 2 is ${}^1v$:${}^2v$ (${}^1v + {}^2v = 1$). Assuming Y_1 and Y_2 are the outputs from Phase 1 and Phase 2, respectively, responding to the input X, the average output Y^* of a composite of Phases 1 and 2 could be an intermediate value

Table 2.2 Various Effects in Materials

INPUT → MATERIAL DEVICE → OUTPUT

INPUT \ OUPUT	CHARGE CURRENT	MAGNET-IZATION	STRAIN	TEMPERATURE	LIGHT
ELEC. FIELD	Permittivity Conductivity	Elect.-mag. effect	Converse piezo-effect	Elec. caloric effect	Elec.-optic effect
MAG. FIELD	Mag.-elect. effect	Permeability	Magneto-striction	Mag. caloric effect	Mag. optic effect
STRESS	Piezoelectric effect	Piezomag. effect	Elastic constant	—	Photoelastic effect
HEAT	Pyroelectric effect	—	Thermal expansion	Specific heat	—
LIGHT	Photovoltaic effect	—	Photostriction	—	Refractive index

Diagonal Coupling ▭ Sensor
Off-Diagonal Coupling = **Smart Material** ▭ Actuator

between Y_1 and Y_2. Y^* may be directly proportional to the volume ratio (linear approximation), or the variation may exhibit a concave or convex shape as a function of a volumetric ratio. In any case, the averaged value Y^* in a composite does not exceed, nor is it less than Y_1 or Y_2. This effect is called a *sum effect*.

An example is a fishing rod, i.e., a lightweight, tough material, where carbon fibers are mixed in a polymer matrix. The density of a composite should be an average value with respect to volume fraction, while a dramatic enhancement in the mechanical strength of the rod is achieved by adding carbon fibers in a special orientation, i.e., along a rod (showing a concave relation).

2.2.2.2 Combination Effect

In certain cases, the averaged value of the output of a composite exceeds both outputs of Phase 1 and Phase 2. Let us consider two different outputs, Y and Z, for two phases (i.e., Y_1, Z_1; Y_2, Z_2). When a FOM for an effect is provided by the fraction (Y/Z), we may expect an extraordinary effect. Suppose that Y and Z follow the concave and convex type sum effects, respectively, as illustrated in Figure 2.4, the combination value Y/Z will exhibit a maximum at an intermediate ratio of phases; that is, the average FOM is higher than either end member FOMs (Y_1/Z_1 or Y_2/Z_2). This is called a *combination effect*.

Certain piezoelectric ceramic/polymer composites exhibit a combination property of g (the piezoelectric voltage constant), which is provided by $d/\varepsilon_0\ \varepsilon$ (d, piezoelectric strain constant; ε, relative permittivity), where d and ε follow the concave and convex type sum effects.

2.2.2.3 Product Effects

When Phase 1 exhibits an output Y with an input X, and Phase 2 exhibits an output Z with an input Y, we can expect a composite that exhibits an output Z with an input X. A completely new function is created for the composite structure, called a *product effect*.

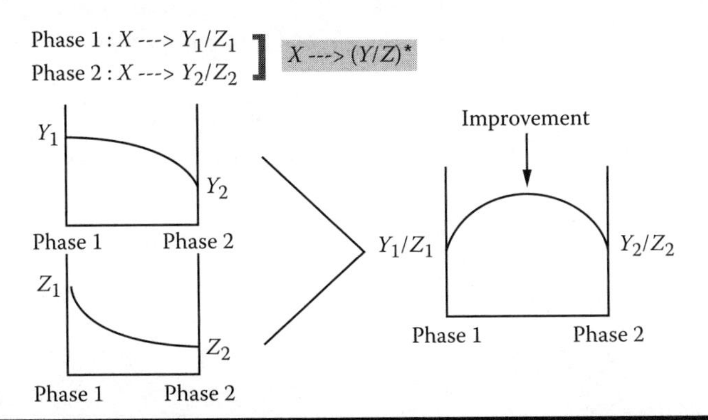

Figure 2.4 Basic concept of the performance improvement in a composite via a combination effect.

I introduce my *functionality matrix* concept here. If one material has a piezomagnetic effect and its converse magnetostrictive effect, the functionality matrix of this material can be expressed by

$$\begin{bmatrix} 0 & 0 & 0 & 0 & 0 \\ 0 & 0 & \text{Magneto-striction} & 0 & 0 \\ 0 & \text{Piezomag. effect} & 0 & 0 & 0 \\ 0 & 0 & 0 & 0 & 0 \\ 0 & 0 & 0 & 0 & 0 \end{bmatrix}$$

On the other hand, a piezoelectric has a functionality matrix of the following form:

$$\begin{bmatrix} 0 & 0 & \text{Converse piezo-effect} & 0 & 0 \\ 0 & 0 & 0 & 0 & 0 \\ \text{Piezoelectric effect} & 0 & 0 & 0 & 0 \\ 0 & 0 & 0 & 0 & 0 \\ 0 & 0 & 0 & 0 & 0 \end{bmatrix}$$

Next we consider a diphasic system of magnetostrictive and piezoelectric materials. When the magnetic field is input first, the expected phenomenon is expressed by the matrix product

$$\begin{bmatrix} 0 & 0 & 0 & 0 & 0 \\ 0 & 0 & \text{Magneto-striction} & 0 & 0 \\ 0 & \text{Piezomag. effect} & 0 & 0 & 0 \\ 0 & 0 & 0 & 0 & 0 \\ 0 & 0 & 0 & 0 & 0 \end{bmatrix} \otimes \begin{bmatrix} 0 & 0 & \text{Converse piezo-effect} & 0 & 0 \\ 0 & 0 & 0 & 0 & 0 \\ \text{Piezoelectric effect} & 0 & 0 & 0 & 0 \\ 0 & 0 & 0 & 0 & 0 \\ 0 & 0 & 0 & 0 & 0 \end{bmatrix} = \begin{bmatrix} 0 & 0 & 0 & 0 & 0 \\ \text{Mag.-elect. effect} & 0 & 0 & 0 & 0 \\ 0 & 0 & 0 & 0 & 0 \\ 0 & 0 & 0 & 0 & 0 \\ 0 & 0 & 0 & 0 & 0 \end{bmatrix}$$

If we start from the electric field input first, the expected phenomenon will be

$$\begin{bmatrix} 0 & 0 & \text{Converse piezo-effect} & 0 & 0 \\ 0 & 0 & 0 & 0 & 0 \\ \text{Piezoelectric effect} & 0 & 0 & 0 & 0 \\ 0 & 0 & 0 & 0 & 0 \\ 0 & 0 & 0 & 0 & 0 \end{bmatrix} \otimes \begin{bmatrix} 0 & 0 & 0 & 0 & 0 \\ 0 & 0 & \text{Magneto-striction} & 0 & 0 \\ 0 & \text{Piezomag. effect} & 0 & 0 & 0 \\ 0 & 0 & 0 & 0 & 0 \\ 0 & 0 & 0 & 0 & 0 \end{bmatrix} = \begin{bmatrix} 0 & \text{Elect.-mag. effect} & 0 & 0 & 0 \\ 0 & 0 & 0 & 0 & 0 \\ 0 & 0 & 0 & 0 & 0 \\ 0 & 0 & 0 & 0 & 0 \\ 0 & 0 & 0 & 0 & 0 \end{bmatrix}$$

Note that the resulting product matrixes include only one component each; magnetoelectric effect or electromagnetic effect component, according to the composite effect sequence. Now is a good time to refresh your memory on linear algebra if you have forgotten this matrix calculation.

Philips developed a *magnetoelectric material* based on this concept [4], which exhibits electric voltage under the magnetic field application, aiming at a magnetic field sensor. This material is composed of magnetostrictive $CoFe_2O_4$ and piezoelectric $BaTiO_3$ mixed and sintered together. Figure 2.5a shows a micrograph of a transverse section of a unidirectionally solidified rod of the materials with an excess of TiO_2. Four finned spinel dendrites $CoFe_2O_4$ are observed in $BaTiO_3$ bulky whitish matrix. Figure 2.5b shows the magnetic field dependence of the magnetoelectric effect in an arbitrary unit measured at room temperature. When a magnetic field is applied on this composite, cobalt ferrite generates magnetostriction, which is transferred to barium titanate as stress, finally leading to the generation of a charge/voltage via the piezoelectric effect in $BaTiO_3$.

My photostrictive materials were also discovered along a similar line of reasoning: functionality matrixes of photovoltatic and piezoelectric effects. The following is an anecdote from the *R&D Innovator* [5].

> I've made a breakthrough that could lead to photophones—devices without electrical connections that convert light energy directly into sound. Perhaps this discovery will help commercialize optical telephone networks. It also could allow robots to respond directly to light; again, without a need for wire connectors. Where did I come up with

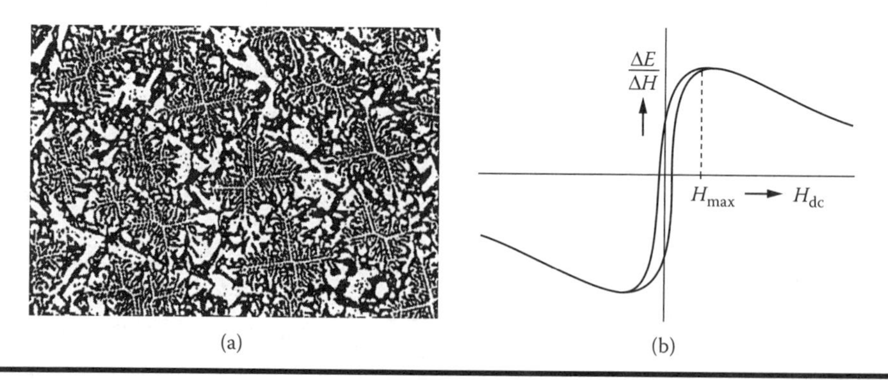

Figure 2.5 (a) Micrograph of a transverse section of a unidirectionally solidified rod of mixture of magnetostrictive $CoFe_2O_4$ and piezoelectric $BaTiO_3$, with an excess of TiO_2. (b) Magnetic field dependence of the magnetoelectric effect in a $CoFe_2O_4$ $BaTiO_3$ composite (at room temperature).

the idea for this light conversion? Not with the sunlight shining through my office window, and not outside feeling the warmth of the sun, but in a dimly lit Karaoke bar.

I've been working on ceramic actuators—a kind of transducer that converts electrical energy to mechanical energy—at the Tokyo Institute of Technology when the trigger for "the light-controlled actuator" was initiated. In 1980, one of my friends, a precision-machine expert, and I were drinking together at a Karaoke bar, where many Japanese go to enjoy drinks and our own singing. We call this activity our "after-5-o'clock meeting," My friend studied micro-mechanisms such as millimeter-size walking robots. He explained that, as electrically controlled walking mechanisms become very small (on the order of a millimeter), they don't walk smoothly because the frictional force drops drastically and the weight of the electric lead becomes more significant.

After a few drinks, it becomes easier to play "what if?" games. That's when he asked, "What if you, an expert on actuators, could produce a remote-controlled actuator? One that would bypass the electrical lead?" To many people, "remote control" equals control by radio waves, light waves, or sound. Light-controlled actuators require that light energy be transduced twice: first from light energy to electrical energy, and second from electrical energy to mechanical energy. These are photovoltaic and piezoelectric effects.

A solar cell is a well-known photovoltaic device, but it doesn't generate sufficient voltage to drive a piezoelectric device. So my friend's actuator needed another way to achieve a photovoltaic effect. Along with the drinking and singing, we enjoyed these intellectual challenges. I must have had a bit too much that night since I promised I'd make such a machine for him. But I had no idea how to do it!

While my work is applied research, I usually come home from scientific meetings about basic research with all kinds of ideas. At one of these meetings, about six months after my promise, a Russian physicist reported that a single crystal of lithium niobate produced a high electromotive force (10 kV/mm) under purple light. His talk got me excited. Could this material make the power supply for the piezoelectric actuator? Could it directly produce a mechanical force under purple light? I returned to the lab and placed a small lithium niobate plate onto a plate of piezoelectric lead zirconate titanate. Then I turned on the purple light and watched for the piezoelectric effect (mechanical deformation). But it was too slow, taking an hour for the voltage to get high enough to make a discernable shape change.

Then the idea hit me: what about making a single material that could be used for the sensor and the actuator? Could I place the photovoltaic and piezoelectric effects in a single asymmetric crystal? After lots of trial and error, I came up with a tungstate-doped material made of lead lanthanum zirconate titanate (PLZT) that responded well to purple light. It has a large piezoelectric effect and has properties that would make it relatively easy to fabricate.

To make a device out of this material, I pasted two PLZT plates back to back, but placed them in opposite polarization, then connected the edges. I shined a purple light to one side, which generated a photovoltaic voltage of 7 kV across the length. This caused the PLZT plate on that side to expand by nearly 0.1% of its length, while the plate on the other (unlit) side contracted due to the piezoelectric effect through the photovoltage. The whole device bent away from the light. For this 20 mm long, 0.4 mm thick bi-plate, the displacement at the edge was 150 μm, and the response speed was 1 second. This fast and significant response was pretty exciting.

Figure 2.6 Photo-driven walking machine.

> Remembering the promise to my friend, I fabricated a simple "light-driven micro walking machine," with two bi-plate legs attached to a plastic board, as shown in Figure 2.6. When light alternately irradiated each leg, the legs bent one at a time, and the machine moved like an inchworm. It moved without electric leads or circuits! That was in 1987, seven years after my promise.
>
> I got busy with my "toy"; but not too busy to attend "after-5-o'clock meetings" in Tokyo's nightclub area. In 1989, at my favorite Karaoke bar, I was talking about my device to another friend who worked for a telephone company. He wanted to know if the material could make a photo-acoustic device—perhaps as a solution to a major barrier in optical-fiber communication. The technology to transmit voice data—a phone call—at the speed of light through lasers and fiber optics has been advancing rapidly. But the end of the line—the ear speaker—limits the technology, since optical phone signals must be converted from light energy to mechanical movement via electrical energy.
>
> I thought my material could convert light flashes directly into sound. I chopped two light beams to make a 180-degree phase difference, and applied each beam to one side of the bi-plate. The resonance point, monitored by the tip displacement, was 75 Hz, just at the edge of the audible range for people! We're now working to fabricate real photo-speakers (I call them "photophones"), and have ideas that may increase the vibration frequency several-fold to reproduce human speech correctly. Photophones could provide a breakthrough in optical communication.
>
> Well, what is my message for you, dear reader? To find a noisy Karaoke bar? Perhaps that's not necessary; but what is necessary is listening to others outside your particular research area: for instance, basic researchers or people with specific, applied objectives.

The above anecdote indicates another important issue: the discovery was motivated by strong customer demand. This is a good example of a need-pull development.

The discovery of *monomorphs* (semiconductive piezoelectric bending actuators) is a similar story [3]. When attending a basic conference of the Physical Society of Japan, I learned about a surface layer generated on a ferroelectric single crystal due to formation of a Schottky barrier. It was not difficult to replace some of the technical terminologies with our words. First, polycrystalline piezoelectric samples were used, with reduction processes to expand the Schottky barrier thickness. We succeeded in developing a monolithic bending actuator. The "rainbow" structure, further developed by Aura Ceramics, is one of the monomorph modifications.

2.3 Product Planning Creativity

2.3.1 Seeds and Needs

I usually suggest product planning divisions reexamine 10-year-old research. Why reexamine "old" technology? There are two key points: (1) It is not really old, it just has not been reexamined since it failed to be developed 10 years ago. (2) The development failure is related to immature supporting technology at that time. Thus, if the *social needs* still exist, there will probably be a good business opportunity because the related patents have probably expired or will soon. More importantly, find the reasons for the lack of success and judge your company's capability to overcome them. I will present two examples here: two-dimensional (2-D) displays and piezoelectric transformers.

In collaboration with Fujitsu General in Japan, we developed a 2-D PLZT display. Electrooptic displays were not new at that time, but there were two major drawbacks for the commercialization: (1) high drive voltage, and (2) expensive manufacturing cost using a sophisticated hot-press furnace. We decided to use the up-to-date nano-powder and tape-casting technologies, in order to overcome the drawbacks; new powder usage provided good transparency to the PLZT devices without using a bulky hot-press furnace, and a tape-casting method realized a mass-production process. We always need to watch carefully for possible supporting technological development, in parallel to 10-year-old research.

Figure 2.7 shows the number of yearly patent disclosures relating to piezoelectric transformers from 1972 until 1999. Two peaks are clear: 1972 and 1998. During the 25 to 30 year gap there was almost no development. Piezoelectric transformers were commercialized for the first time in the beginning

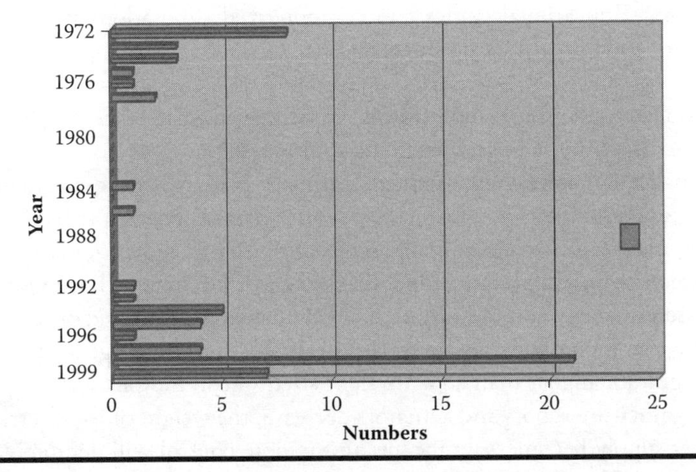

Figure 2.7 Number of yearly patent disclosures related to piezoelectric transformers from 1972 through 1999.

of the 1970s to supply high voltage in color TVs. However, this application disappeared in less than a year, due to ceramic cracking, which destroyed the devices. The second commercialization peak occurred due to three key factors: (1) strong social demand for a laptop computer backlight screen, (2) matured powder technology to provide mechanically strong piezoceramics, and (3) advanced design technology such as finite element method software to simulate electromechanical vibrations.

Tracking future technologies is also important in finding "seeds" for new products. Battelle reports regularly on future technologies. Its 1995 top 10 predictions for 2005 are as follows [6]:

1. *Human genome mapping.* Genetics-based personal identification and diagnostics will lead to preventive treatments of disease and cures for specific cancers.
2. *Super materials.* Computer-based design and manufacturing of new materials at the molecular level will mean new, high-performance materials for use in transportation, computers, energy, and communications.
3. *Compact, long-lasting, highly portable energy sources.* These energy sources, including fuel cells and batteries, will power electronic devices of the future, such as portable personal computers.
4. *Digital, high-definition TV.* A major breakthrough for American television manufacturers—and a major source of revenue—that will lead to better advanced computer modeling and imaging.
5. *Electronics miniaturization for personal use.* Interactive, wireless data centers in a pocket-size unit will provide users with a fax machine, telephone, and computer that contains a hard drive capable of storing all the volumes found in their local library.
6. *Cost-effective "smart" systems.* These systems will integrate power, sensors, and controls, and will eventually control the manufacturing process from beginning to end.
7. *Anti-aging products.* Relying on genetic information to slow the aging process, these will include anti-aging creams that really work.
8. *Medical treatments.* New treatments will use highly accurate sensors to locate problems, and drug-delivery systems that will precisely target parts of the body, such as chemotherapy targeted specifically to cancer cells to reduce the side effects of nausea and hair loss.
9. *Hybrid-fuel vehicles.* Smart vehicles, equipped to operate on a variety of fuels, will be able to select the most appropriate one based on driving conditions.
10. "*Edutainment.*" Educational games and computerized simulations will meet the sophisticated tastes of computer-literate students.

Battelle's prediction hit rate is very high (80%). Thus, you can use Battelle's predictions as a surrogate if you do not have the resources for doing this yourself.

In general, smaller actuators will be required for medical diagnostic applications such as blood test kits and surgical catheters. Silicon microelectromechanical systems (MEMS) are developing rapidly. However, electrostatic forces are generally too weak to move something with sufficient efficiency. Piezoelectric thin films compatible with silicon technology will be much more useful for MEMS. An ultrasonic rotary motor as tiny as 2 mm in diameter, fabricated on a silicon membrane, is a good example (see Figure 2.8) [3]. Even this prototype motor can generate a torque three to four orders of magnitude higher than an equivalent-sized silicon motor.

As the size of miniature robots and actuators decrease, the weight of the electric lead wire connecting the power supply becomes significant, and remote control will definitely be required for submillimeter devices. The photo-driven actuator described in the previous section is a promising candidate for microrobots.

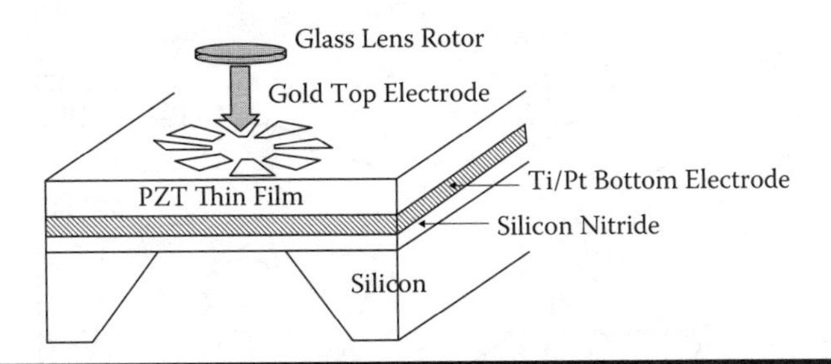

Figure 2.8 Ultrasonic rotary motor as tiny as 2 mm in diameter fabricated on a silicon membrane. (From Uchino, K. *Ferroelectric devices*. New York: Dekker/CRC, 2000. With permission.)

2.3.2 Innovation Obstacles in Technology Management

Innovation obstacles are important issues for technology management. I have observed a pattern: once development has failed, a minimum of 25 to 30 years is required to redevelop the same or similar device. The reason is that once a young researcher fails in development, the experience stops him or her from engaging in similar device development or allowing others in his organization from doing the same after he or she becomes a manager. It takes a generation, 25 years, for the technology to be resurrected, until these "fossil" managers disappear (or retire).

There are two issues to remember:

1. If you are a manager, try not to suppress redevelopment of previously failed devices.
2. If you are a young researcher, reexamine 10-year-old research (half of a generation period). If your boss is a "fossil" type, you should spin off from that institute, and start your own company.

2.3.3 Development Pace

A suitable R&D pace introduces new concepts and products not too early or too late; 3 years for commercialization is a good target for the ferroelectric devices. The Ford Motor Company changed their development pace from 5 to 3 years several years ago, and commercialized the Taurus successfully.

2.3.4 Specifications

Some engineers believe that lowering the drive voltage of a piezoelectric actuator is essential. However, this is not really true for portable equipment if one considers the available battery voltages. Do you know the available battery voltages and voltage supplies in the portable electronic equipment area? The answers are 1.5, 3, 6, 12 (automobile applications,) 24, and 250 V. It is your homework to check the available voltages in your professional field.

For example, when I collaborated with COPAL, a Japanese company, to develop piezoelectric camera shutters using a bimorph structure, we initially used conventional bimorphs driven at around 100 V (see Figure 2.9). But, when we tried to commercialize it, we recognized that we needed an additional 100 V power supply in each camera, which would cost several dollars. Instead we changed the bimorph design, by thickening the piezoelectric ceramic layer, so that it could be driven by 250 V (this voltage is generated in a camera by a cheap power supply conventionally used for a stroboscopic lamp).

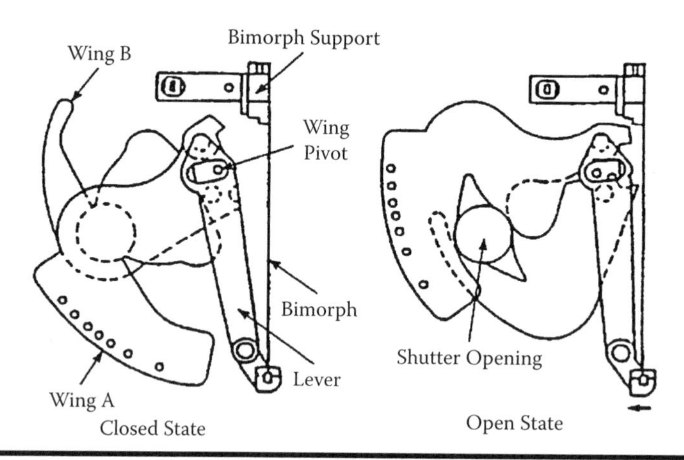

Figure 2.9 Piezoelectric bimorph camera shutter by COPAL, where the PZT layer was intentionally thickened to adapt to a 250 V voltage supply. (From Uchino, K. *Ferroelectric devices*. New York: Dekker/CRC, 2000. With permission.)

Why not use the 100 V bimorph and the 250 V from the battery, and a voltage splitter or resistor? The reason is again additional component costs (10 to 20 cents) in the assembly line.

Product development needs to collect the following necessary information on the specifications:

- Sensitivity
- Size
- Lifetime
- Available power supply

If we extend the above considerations, a device for automobiles should be driven under 12 V (car battery). However, actual piezoelectric actuators for diesel injection valve applications are driven with 160 V for 80 μm PZT layer thickness (see Figure 2.10a). The key restriction not to allow the drive voltage below 12 V per 10 μm PZT layer is not from technological difficulties, but merely from the cost minimization. Figure 2.10b shows prices for both the piezoelectric stack actuator and its drive/control circuit, plotted as a function of drive voltage. Decreasing the drive voltage should decrease the PZT layer thickness dramatically, leading to an increase in the Ag/Pd electrode cost, and manufacturing costs will increase exponentially. On the other hand, the driving circuit cost increases with the required voltage. Accordingly, the total system cost has a minimum around 160 V and a layer thickness around 80 μm. Note that even 10 μm layer actuators are not difficult to manufacture technologically today. The present specification for thickness is merely determined by economics!

2.3.5 Product Design Philosophy

When performance is similar, sales depend strongly on design, color, and so forth. The design must fit social trends as discussed in Section 2.4, Marketing Creativity. In this new era, Y. Hirashima suggests "beautiful," "amusing," "tasteful," and "creative" [7]. A general managerial economics textbook [8] suggests that product demand is determined by six variables: price, income of

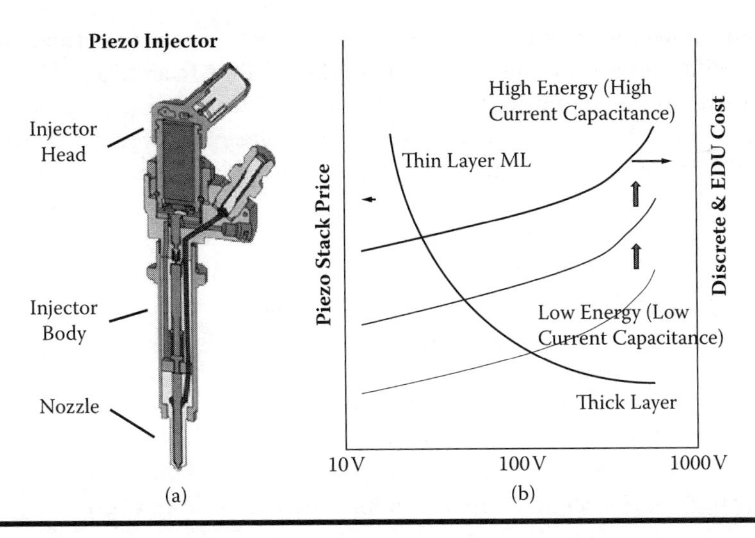

Figure 2.10 (a) Piezoelectric diesel injection valve by Siemens. (b) Cost evaluation for the piezo-stack and drive circuit as a function of drive voltage.

Table 2.3 Difference in Development Concepts between the United States and Japan

United States	*Japan*
Best device	Good device
For military	For civil consumers
With top technology	With improved technology
At expensive cost	At cheap cost
Reliability	Newness, timing

consumers, prices of related products, expected future price of product, number of consumers, and taste patterns of consumers. The last factor has been emphasized in the last 10 years.

Table 2.3 summarizes the differences in development concepts between the United States and Japan. Samsonite aired a TV commercial 10 years ago showcasing its suitcases. In a 3-minute broadcast, a suitcase is thrown from a 10-story building without damage on landing. This ad demonstrates that the suitcase is highly reliable and tough. I appreciate this typical American attitude, but am concerned that they are not worried about the contents of the suitcase, such as glass bottles. A similar concept can be found in a recent TV ad by Maytag. This washing machine ad seems ridiculous. Lots of tennis balls hit the surface of the washing machine, which is meant to show its mechanical toughness. Actually, the value of a washing machine is in how quickly, silently, and efficiently it cleans clothes. The ad did not mention anything about its performance—it merely showed its mechanical toughness.

The promotion of the Nintendo Game Boy is based on very different concepts: newness and timing. The game might not have sold well due simply to misjudgment in timing. Moreover, when Nintendo developed a family computer, the engineers seemed to try to weaken the connector between the software board and the hardware chassis. It seemed to become damaged after about

100 connections, which corresponds to 3 to 6 months of normal usage for a child. When the toy breaks and the child complains to his or her parent, the response (particularly Japanese) may be, "You play computer games too much! It is about time you returned to your studies!" In this scenario, few parents would bother to complain to Nintendo. Of course, the child wants it fixed once he or she realizes how much fun it is. The child may even spend his or her own money to purchase another game machine; if this is true, I admire Nintendo's business strategy. Two comments: (1) Recent game equipment has dramatically improved, so this is no longer true, but it occurred in the older models. (2) If the Game Boy was commercialized first in the United States, it was an easy target by attorneys for product recall with expensive reparations and paybacks. It was lucky for Nintendo that they commercialized it in Japan first.

This basic consumer attitude difference explains why e-mail on a cellular phone system became popular in Japan 10 years ahead of the United States. Camera-phone advertisements have finally been broadcast in the United States in recent years. Americans like Walmart because of its low prices, while Walmart cannot invade Japan or Korea because its products are not the newest versions. They may purchase old products from manufacturers at the lowest cost.

2.3.6 Smart Systems

"Intelligent" or "smart" materials, structures, and systems are often used today. The bottom line of "smartness" is to possess both *sensing* and *actuating* functions. I will now offer my opinion on this issue.

When a new sensing function is required, most researchers try adding another component, leading to a more complex system, which is likely to be more bulky and expensive (I call it "spaghetti syndrome.") Our group is contributing enormously to adding new functions to conventional materials and structures, while reducing the number of components in a system, and aiming for miniaturization and lower cost. The photostrictive actuator is a very good example of an intelligent material. It senses light illumination and generates a voltage/current proportional to the light intensity. Then it produces strains according to this control voltage, leading to the final mechanical actuation.

The design/development concept for smart systems is illustrated in Figure 2.11, using ultrasonic motors. Starting from a traveling wave-type motor with two piezo-actuators and two power supplies, a Philips group moved to a more complex motor with four piezo-actuators in order to achieve better motor performance. However, our group took the opposite approach, simplification, and developed a standing wave-type motor with a single actuator element in order to make it smaller and less expensive. It is your choice to seek merely performance (mostly for military or space applications) or to seek the optimization of performance and cost (for general consumer applications).

2.4 Marketing Creativity

Discipline of Market Leaders [9], authored by Treacy and Wiersema, is an informative guide to understanding marketing creativity. The authors have factored into it the following three basic steps:

1. Choose your customers.
2. Narrow your focus.
3. Dominate your market.

We will consider these steps in detail. General marketing strategy will be discussed in Chapter 8.

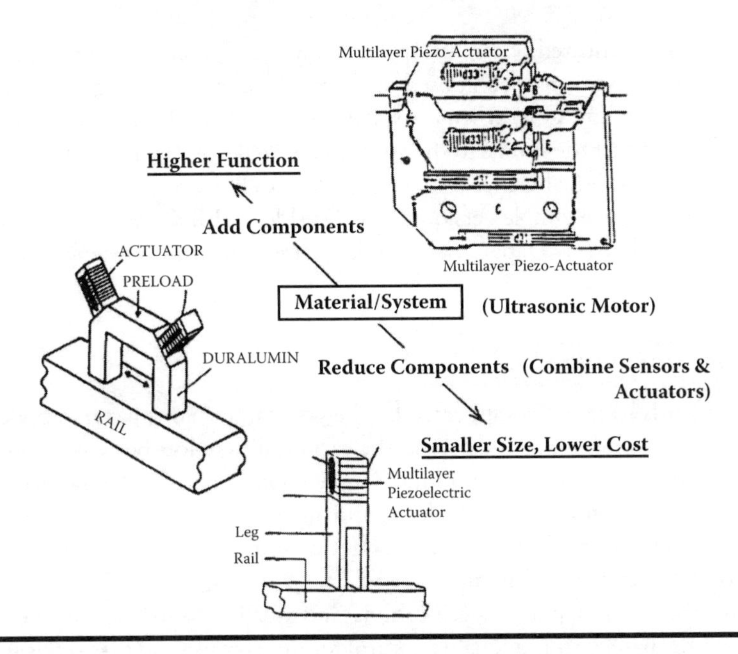

Figure 2.11 Development concept for smart systems, using ultrasonic motors as an example.

2.4.1 *Choose Your Customers*

2.4.1.1 *Domestic or Foreign?*

Let us start by solving the following practical problems which arose during my consulting.

Example Problem 2.1

1. Japanese cars are popular in the United States, but U.S. cars are not popular in Japan. Why?
2. TOTO Washlet (a personal hygiene/cleansing system, for use in the bathroom) is a big hit in Japan, but not in the United States, although TOTO Co. tried to sell it in the United States. Why?

Solutions

1. Lack of development effort. There is a traffic system difference between the two countries: left-hand side in Japan and right-hand side in the United States. While the Japanese auto manufacturers tried to make left-hand-side steering wheel cars, the U.S. manufacturers did not; this presents a serious inconvenience to driving an American car in Japan. Why don't the U.S. auto manufacturers make the right-hand-side steering wheel cars? Is it technologically difficult for them? No, not at all. It is merely the U.S. auto manufacturers' attitude (not customer-oriented, or arrogant to insist) that the customer should use their design!
2. Lack of learning the culture. Japanese toilet facilities don't always have a shower set, thus a special personal hygiene system such as this is convenient. When the restroom possesses bath/shower and toilet facilities together as in American homes, the Washlet may not be necessary.

From the above examples, it becomes clear that, in order to expand business into a foreign country, we need to learn the culture of that country or find a partner in that country.

TOTO Washlet is a sophisticated system using a smart material, a shape memory alloy. The nozzle part of the water jet mechanism is composed of a heater and a nozzle angle control mechanism made of a shape memory alloy. Only when the water is within a suitable temperature range will the shape memory alloy redirect the nozzle to the appropriate angle. When the water is cold, the water jet is angled downward into the flush pot. I recently noticed, however, a relatively large number of American tourists purchase a simple version of the Washlet (with a warming heater function on the sitting plate) when they leave Japan. Middle-class Americans can now appreciate the comfort even if the facility is expensive.

2.4.1.2 Military or Civilian?

Product development is sometimes supported by the government for military applications in the United States. The researcher must understand the various differences between military and civilian commercialization philosophies. For military applications, production quantities are relatively small (several hundred to thousands of pieces), and manual fabrication processes are generally utilized, leading to high prices. The strategy of targeting military customers may be adopted by a small venture company as it is starting up.

The difference between required specifications and quality control is also very interesting. Figure 2.12 shows the basic trends in quality control for military use and mass-consumer products. Due to manual production, the production quality distribution is wider for military products. However, all the products need to be checked for military products. No check is required for mass-consumer products, keeping prices low. To do this, the standard deviation of the production quality must be very small. You should notice that too high a quality of the products is also not good (NG).

Let us consider Toshiba light bulbs. Toshiba is one of the largest light bulb suppliers in Japan, and is where some of my former graduate students are working. Light bulbs typically have an average lifetime of around 2000 h. Their quality control curve has a standard deviation of ±10% (1800–2200 h). If some of the production lots happen to have a slightly longer lifetime of 2400 h, what will happen? A company executive might predict bankruptcy of the division. For this kind of mature industrial field, total sales amount is almost saturated, and this 10% longer lifetime translates directly to a 10% decrease in annual income. Therefore, "too high quality" must be eliminated for mass-consumer products. The researcher needs to understand that simply seeking high quality is not the only goal of the manufacturing company. Of course, Toshiba has the

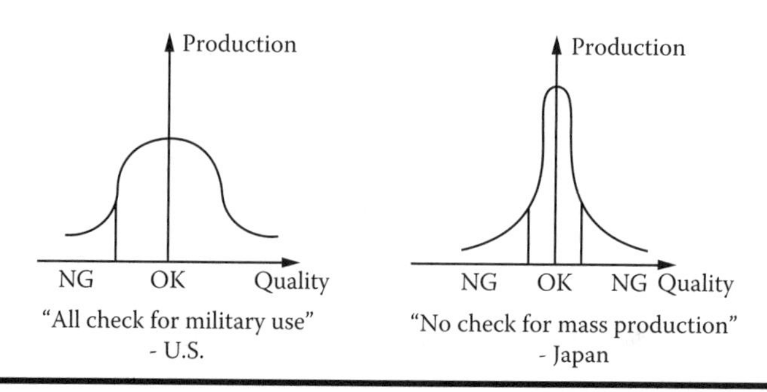

Figure 2.12 Difference between basic trends in quality control for military use and mass-consumer products.

technological capability to extend the lifetime of the bulbs. Toshiba does sell 2400-h-lifetime bulbs. However, the price is exactly 10% higher than the usual 2000-h bulbs.

A final comment: Sometimes, even famous Japanese consumer product companies may contribute to military or governmental applications such as the National Aeronautics and Space Administration (NASA) Space Shuttle program. The main reason is to obtain an aura of high quality for the company's products, leading to very effective advertisements, although the development effort will not bring significant profit directly.

2.4.1.3 Catch the General Social Trends

The market also exhibits trends reflecting cultural characteristics, hence it may gradually or drastically change with time. We consider here changes in Japanese market trends, which must be fully understood before an industry can expand its market globally. A summary is shown in Table 2.4. Japanese people use four Chinese character words to express these trends, as shown at the bottom of Table 2.4 [7].

Table 2.4 Japanese Market Trends over Time

1960s	Heavier	Ship manufacturing
	Thicker	Steel industry
	Longer	Building construction
	Larger	Power plant (dam)
	重厚長大	
1980s	Lighter	Printer, Camera
	Thinner	TV, Computer
	Shorter	Printing time, Communication period
	Smaller	"Walkman," Air conditioner
	軽薄短小	
2000s	Beautiful	Well-known brand apparel
	Amusing	TV game
	Tasteful	Cellular phone (private communication)
	Creative	"Culture" center, Made-to-order shoe
	美遊潤創	

When I was a university student in the 1960s, the most popular departments at my university were metallurgy (for manufacturing steel plates and ships) and electrical engineering (for building power plants), aimed at producing bigger products (heavier, thicker, longer, and larger). However, in the 1980s most Japanese industries became primarily involved in electronics and computer hardware seeking device miniaturization. Piezoelectric actuators, positioners, and ultrasonic motors (my specialty) have been utilized to realize the highest degree of fabrication accuracy. The 1980s keywords were lighter, thinner, shorter, and smaller, all of which are completely opposite to the 1960s keywords.

In the 2000s, the keywords for new products seem to be "beautiful," "amusing," "tasteful," and "creative," according to Hirashima [7]. A good example is the Nintendo Game Boy, a video game system for TVs. It has become popular worldwide among kids (even among Japanese adults under 40). Nintendo used to be a company that supplied Japanese traditional playing cards. At the beginning of the 1970s, when most of the Japanese electronic industries were chasing U.S. technologies in semiconductor devices, a major semiconductor company had a large number of imperfect 8-bit chips (the Japanese technology at that time had such low quality!). Since most of the basic functions of those chips were active, Nintendo decided to purchase them at a very low price, and used them to develop computer-aided toys. The prototype Game Boy did not utilize any advanced technologies, but utilized cheap 8-bit chips with well-known technologies. The key to this big hit was its ability to fit a social trend, amusement, and to firmly attract kids' attention. Because these kids are now adults around 40 years old, they are introducing these toys to their kids. Nintendo is now working for two generations' amusement, using the leading edge information technologies.

Table 2.5 illustrates this new trend with 1998's "Best Hit (Top 10) Products" in Japan. Except for the flat screen TV, which is a technology-oriented (thinner) product, most of the others relate to beautiful, amusing, tasteful, and creative goods. In 1998, Sony was not a computer company. It was technologically a second-rank manufacturer. However, their big hits are mainly due to additional factors such as sophisticated design. Similarly, the iMac is a totally different product

Table 2.5 1998 Best Hit Products in Japan

1.	VAIO Note 505 (Sony) (silver-metallic personal computer)	Thinner, beautiful
2.	iMac (Apple) (inexpensive, sophisticated-designed computer)	Tasteful
3.	Pocket board (NTT Docomo) (mobile telephone designed for ladies)	Amusing
4.	Pocket Pikachu (Tamagochi [computer pet] and walking distance counter)	Amusing
5.	New compact care (further compact size due to the Japanese automobile regulation change)	Smaller
6.	Draft beer (Kirin) (new taste)	Amusing
7.	Fine Pix 700 (Fuji) (better resolution)	Creative
8.	Foreign banks (better services)	
9.	Flat-tube TV (less depth)	Thinner
10.	Viagra (?)	Amusing

concept than the IBM PC. Researchers in the twenty-first century should not only chase the performance of the product, but also the consumer's attitude toward the product.

2.4.2 Narrow Your Focus

After choosing a suitable customer, start narrowing your development focus. The following sections summarize a procedure for narrowing the focus.

2.4.2.1 List All Possible Application Fields

When we invented piezoelectric actuators, initially we considered various application fields such as the following:

- Office equipment (printer, fax machine)
- Cameras
- Automobiles

Can you identify the development "pecking order" among these application areas? I will introduce so-called strategic or managerial decision-making procedures in the simplest way, using this practical example.

2.4.2.2 Start with the Simplest Specifications

Among the possible applications, we tried to find the simplest technological specifications. We considered restrictions by the Japanese Industrial Standards (JIS) first.

2.4.2.2.1 Temperature Range

The standard temperature requirement for office equipment is between –20 and 120°C. For cameras, even though they are used outdoors they are typically held in the hands, thus, the temperature is always maintained between 0 and 40°C. Much above this temperature, the film will be damaged before the camera's failure. Requirements for automobile applications cover a much broader range: between –50 and 150°C.

2.4.2.2.2 Durability

The standard requirement for the lifetime of office equipment such as printers is continuous operation for more than 3 months or 10^{11} cycles. For cameras, it is only 5×10^4 cycles. Imagine how many pictures you take in a year. A 36-exposure roll of film may take months to use! Automobile applications usually require durability of more than 10 years.

In conclusion, the sequence for starting development will be

$$\text{Camera} \rightarrow \text{Office Equipment} \rightarrow \text{Automobile}$$

As we expected, piezo-actuators were first widely commercialized in a camera automatic-focusing mechanism by Canon and in a shutter by Minolta. Then, they were employed in dot-matrix (NEC) and inkjet (Seiko Epson) printers. Since piezo-multilayer actuators have been used by

Siemens in diesel injection valves in the automobile since 2000, we can say the piezoelectric actuator development is in a maturing period.

It is notable that specs for cameras dramatically changed after shifting from film to digital. Digital cameras do not use film or mechanical shutters, leading to much more severe specs. The standard durability is not 5×10^4 cycles anymore, because users started to take 10 times more pictures. The cyclic lifetime specs automatically increased by a factor of 10 with digital cameras.

2.4.2.3 Consider the Cost Performance

We occasionally use a scoring sheet to identify a development target. A sample of how to score is shown in Table 2.6. This table includes various factors that are significant, including financial factors (market and cost) and device performance. We compare the total scores and select the higher priority for development (pecking order).

Example Problem 2.2

Consider the dot-matrix printer which was developed in 1982 by NEC. Adopting a scoring table like the one shown in Table 2.6, compare the cost performances of a bimorph and a multilayer structure.

Table 2.6 Scoring Table for Devices

	Device A			*Device B*		
Costs						
Raw materials cost	0	1	2	0	1	2
Fabrication cost	0	1	2	0	1	2
Labor cost (special skill)	0	1	2	0	1	2
Performance						
Figure of merit	0	1	2	0	1	2
Lifetime	0	1	2	0	1	2
Market						
Design	0	1	2	0	1	2
Production quantity	0	1	2	0	1	2
Maintenance service	0	1	2	0	1	2
Total score (add the scores)						

Figure 2.13 Two typical actuator designs: bimorph and multilayer.

Hint: Since the deformation from the piezoelectric ceramic is small (only 1 μm), we need to amplify the displacement for practical applications. The starting piezo-ceramic has a thin plate design because a reasonable electric voltage can generate a large electric field. By bonding a piezo-ceramic plate onto a metallic plate, small strain is converted to a large bending deformation (*bimorph*). On the other hand, by laminating multiple thin plates, we can increase the displacement in proportion to the number of the laminates (*multilayer*), which is my popular patent. Refer to Figure 2.13 [10].

Solution

Selection criteria for scoring are summarized in Table 2.7: (1) Multilayer structures need large amounts of expensive electrode materials, and the tape-casting requires equipment investment. On the other hand, fabrication is almost automatic. (2) Quick speed, high force, and longer lifetime, which are essential to dot-matrix printers, are possible with multilayer devices. (3) The fabrication process of multilayer actuators (tape-casting) is most suitable for mass production. We adopted the multilayer piezoelectric actuator for this printer development (Figure 2.14).

If the application is for cheaper devices such as inkjet printers or cellular phones, cost should be emphasized. Thus, the *weighted score method* will be applied. Suppose that cost:performance:market weights are 4:1:1; the total scores for bimorph and multilayer are $4 \times 3 + 1 + 3 : 4 \times 2 + 3 + 4 = 16 : 15$. Hence, the bimorph actuator is recommended.

2.4.3 Dominate Your Market

After identifying the target, develop the products according to the following technology and product planning creativity considerations. At the same time, consider a suitable advertising plan and price range.

2.4.3.1 Advertisement (Promotion Strategy)

Naming or selecting a suitable trademark for a device is very important. When I developed co-fired multilayer actuators, they were initially named *displacement transducers*. Of course, this is not an inappropriate name from a physics point of view. However, it was not attractive to customers. The name *positioner* was also used in the mechanics fields.

After discussing this with colleagues at NEC Corporation, the terminology *piezoelectric actuator* was selected, half of which is familiar to electrical engineers (piezoelectric), and the other half of which is familiar to mechanical engineers (actuator). Only people working in this interdisciplinary

Table 2.7 Example of the Scoring Table for Bimorph and Multilayer Actuators (Printer Application)

	Bimorph			*Multilayer*		
Costs						
Raw materials cost	0	√1	2	√0	1	2
Fabrication cost	0	√1	2	√0	1	2
Labor cost (special skill)	0	√1	2	0	1	√2
Performance						
Figure of merit	0	√1	2	0	√1	2
Lifetime	√0	1	2	0	1	√2
Market						
Design	0	√1	2	0	√1	2
Production quantity	0	√1	2	0	1	√2
Maintenance service	0	√1	2	0	√1	2
Total score	7			9		

field can understand the full meaning of this name, making it highly suitable for a device that will be used in an interdisciplinary field.

The details of marketing strategy, the so-called Marketing Four Ps (product, place [distribution channels], price, and promotion) will be discussed in Chapter 8.

2.4.3.2 Determine the Appropriate Price

The profit ratio for a particular sales price depends on the industry category: electronics industries have relatively high profitability—typically 10% in electronic components and 30% in videotapes, as compared with 3–4% for chemical commodities. On the basis of these profit margins, we can estimate the maximum raw materials' cost, labor costs, and so forth. Refer to the rough price calculation presented in Table 2.8. The details will be discussed in Chapter 6, Section 6.1, Accounting Management.

When a company is thinking about starting multilayer actuator production, they need to consider if a tape-casting system really needs to be installed. I usually recommend the installation of a tape-casting system if the production amount exceeds 1 million pieces per year. Otherwise, the conventional cut-and-bond method should be used by hiring several manufacturing technicians, as discussed in Chapter 1. Also, when a company considers purchasing a new robot for automation production, they should consider the price. A typical one-task robot costs $30,000, which can be

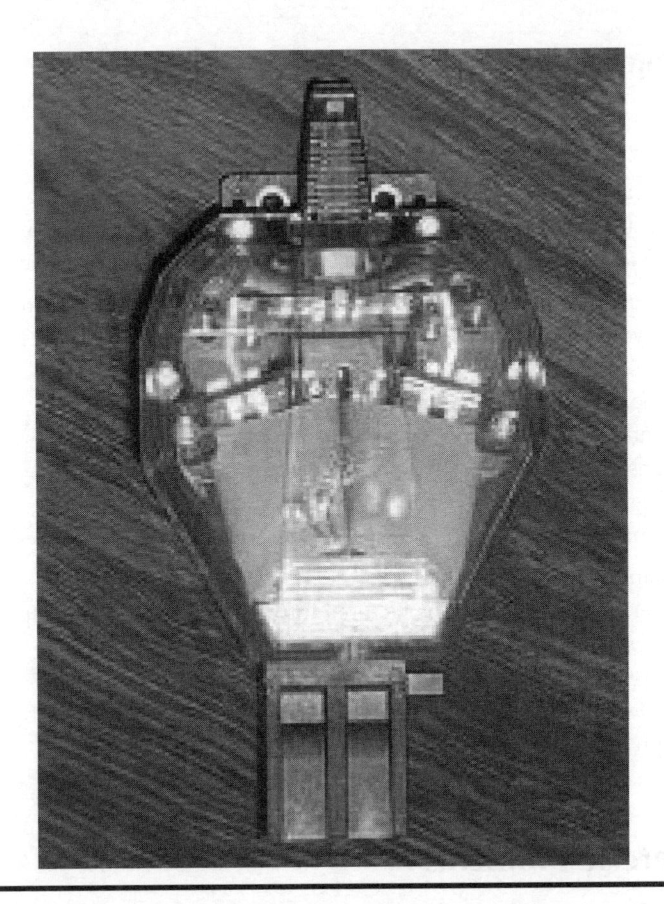

Figure 2.14 Dot-matrix printer head using 24 multilayer piezoelectric actuators (1986, NEC). (From Uchino, K., and J. R. Giniewicz. *Micromechatronics*. New York: Dekker/CRC, 2003. With permission.)

Table 2.8 Price Calculation Sample in the Piezo-Actuator Area

Commercial price (must be comparable to competitive things)	100
Manufacturer's price (varies depending on the distribution channel)	50
Direct materials (raw materials)	10
Direct labor (engineers)	10
Manufacturing overhead	20
Gross profit	10

used for 2 years without a high maintenance fee. On the other hand, an annual salary of $3000 is enough to hire one worker in some countries such as Thailand and Turkey. A manufacturing line having 10 workers corresponds to a robot. Thus, an alternative solution to purchasing a robot is to start a factory in one of these countries. We will discuss detailed financial calculations and production planning in Chapters 6 and 7. The key topics to be learned include break-even analysis, fixed cost, variable cost, and linear programming.

Chapter Summary

2.1 Three essential types of creativity required for best-selling devices are *technological creativity*, *product planning creativity*, and *marketing creativity* (Akio Morita, former Sony President).

2.2 Technology creativity: Discovery of a new function or material depends on serendipity and intuition, among which *secondary effect* and *scientific analogy* approaches are useful. Electrostrictive and shape memory ceramics were invented from these approaches.

2.3 Product planning creativity: This is most essential for a high-tech entrepreneur. This creativity transforms your high-tech idea into a commercial product. The following items need to be considered:
1. Seeds to needs—find the social needs
2. Development pace—prototype 3 years prior to commercialization
3. Specifications—available battery voltage, cost minimization principle
4. Product design philosophy—beautiful, amusing, tasteful, and creative
5. Smart systems—escape from the "spaghetti syndrome"

2.4 Marketing creativity
Effective marketing creativity can be factored into the following three steps:
1. *Choose your customers*—military or civilian, quality control
2. *Narrow your focus*—scoring table for considering cost performance
3. *Dominate your market* [9]—Marketing Four Ps: product, place (distribution channels), price, and promotion (advertisement)

Practical Exercise Problems

P2.1 Product Concept Questions

This exercise will help you write a product brochure for clarifying the product concept. Referring to the example, write your own plan using a scoring table.

P2.1.1 Function of the Product

Discuss the superiority of your product, such as cost, performance, and market, in comparison with the competitive products. Use a scoring table, and discuss it quantitatively.

MMI Example

MMI developed a microultrasonic motor called metal tube type, consisting of a metal hollow cylinder and two PZT rectangular plate actuators. The raw materials cost is lower than the conventional electromagnetic motor's thin copper wire. The metal cylinder motor, 2.4 mm in diameter and 12 mm in length, was driven at 62.1 kHz in both rotation directions. A no-load speed of 1800 rpm and an output torque up to 1.8 mN·m were obtained for rotation in both directions under an applied rms voltage of 80 V. Higher maximum efficiency of about 28% and torque for this small size compared with the electromagnetic type is a noteworthy feature. The key features of MMI micromotors are as follows: (1) they are the world's smallest motors with human finger fighting torque level (much superior to the electromagnetic motors), and (2) their low manufacturing cost is competitive with the lowest electromagnetic motors. Table P2.1 compares the superiority scores for the metal tube and electromagnetic motors for cellular phone camera applications. Miniature

Table P2.1 Scoring Table for Comparing Metal-Tube Ultrasonic and Electromagnetic Motors (Cellular Phone Application)

	Metal Tube			*Electromagnetic*		
Costs						
Raw materials cost	0	1	√2	0	√1	2
Fabrication cost	0	√1	2	0	1	√2
Labor cost (special skill)	0	√1	2	0	√1	2
Performance						
Figure of merit	0	1	√2	√0	1	2
Lifetime	0	1	√2	0	1	√2
Market						
Design (small size)	0	1	√2	√0	1	2
Production quantity	0	√1	2	0	1	√2
Maintenance service	0	√1	2	0	√1	2
Total score		12			9	

size, high torque, and low cost are essential factors for the choice. The metal tube type has advantages in FOM and small size, in comparison with the electromagnetic micromotors.

P2.2 "Don't Read Papers"

Question: My PhD advisor was the late Professor Shoiichiro Nomura at the Tokyo Institute of Technology. He first taught me "Don't read papers" when I joined his laboratory. I sometimes use this when teaching my graduate students. What is the real meaning of this phrase?

Answer: I had top academic grades during my undergraduate period. I read many textbooks and academic journals. Accordingly, whenever Professor Nomura suggested that I study a new research topic, I said things such as "that research was done already by Dr. XYZ, and the result was not promising…." After having a dozen of these sort of negative conversations, partially angrily, partially disappointedly, Professor Nomura ordered, "Hey, Kenji! You are not allowed to read academic papers for a half year. You should concentrate on the following experiment without having any biased knowledge. Having a strong bias, you cannot discover new things. After finishing the experiment and summarizing your results, you are allowed to approach the published papers in order to find whether your result is reasonable, or is explainable by some theory." Initially, I was fearful of getting totally wrong results. However, I finished it. That led to my first discovery: PMN-PT electrostrictive materials.

Remember that knowing too much suppresses innovative work. A real discovery is usually made by a young, less-experienced engineer. Once this engineer becomes an expert professor, unfortunately he or she may lose some creativity.

References

1. Morita, A. Private communication, 1986.
2. Benjamin Franklin paintings by Bernard Hoffman, http://www.ushistory.org/franklin/essays/hoffman.htm (accessed March 2008).
3. Uchino, K. *Ferroelectric devices*. New York: Dekker/CRC, 2000.
4. Uchino, K. How to Develop New Devices. *Solid State Phys* 21: 27–38, 1986.
5. Brill, Winston, J. *R & D Innovator*. 4, no. 3, Winston J. Brill & Associates: Redmond, WA, 1995.
6. Battelle Company Report, 1995.
7. Hirashima, Y. *Product planning in the feeling consumer era*. Tokyo: Jitsumu-Kyoiku, 1996.
8. Thomas, C. R., and S. C. Maurice. *Managerial economics*. New York: McGraw-Hill Irwin, 2005.
9. Treacy, M., and F. Wiersema. *Discipline of market leaders*. Reading, MA: Addison-Wesley, 1996.
10. Uchino, K., and J. R. Giniewicz. *Micromechatronics*. New York: Dekker/CRC, 2003.

Chapter 3

Corporation Start-Up—How to Establish Your Company

This chapter instructs you how to establish your own company once you have targeted a first product.

From my experience, the start-up procedures include the following:

1. Information collection for a start-up: Connect with a mentor (maybe a friend or business tutor) who is or was operating a small business, and ask him or her the following questions:
   a. What was his or her motivation for starting their own venture, instead of taking a job in a larger company? This is important to determine a mindset.
   b. How did he or she evaluate the opportunity and decide on the timing of the company's start?
   c. How much capital did it take to start the company? This is very important.
   d. How did he or she find reliable business advisors, mentors, and partners? From the start, your company cannot be operated in isolation.
   e. How did this friend persuade his or her family that this would be worthwhile, particularly since his or her income may have been reduced by half when the company was launched?

   The purpose of this textbook is to help overcome the following issues.
2. Persuasion of the family and forming a home financial plan
3. Partner and key employee search (covered in Chapter 10)
4. Business plan preparation—important to find investors and lenders (covered in Chapter 4)
5. Investor and loan search (described in Chapter 5)
6. Legal procedure for company start-up

In this chapter, the above items 2, 3, and 6 are detailed.

3.1 The Founder and Team

Many people think that entrepreneurs are their own bosses and completely independent. This is wrong! Entrepreneurs are far from independent, and have to work with many people, including partners, investors, employees, and families.

3.1.1 Persuasion of the Family

The most popular scenario, in my experience, is that one should have a job with a steady income. By starting your own company, you may expect more compensation and mental satisfaction in the future. However, you may experience challenges initially. You may need to work long hours with less compensation during the start-up and initial growth periods. Your lifestyle may change dramatically compared with regular employment, reducing family time and income. An entrepreneur needs to obtain complete understanding and strong support from his or her spouse and family.

Though a corporation is owned by shareholders who have the legal privilege of *limited liability*, the actual situation is usually different for the founder. For example, an emergency short-term loan may be obtained from a bank, with your personal property as collateral, such as a home or other assets. The liability is not completely limited, which may affect your family's security.

3.1.2 The Timmons Model of the Entrepreneurial Process

Figure 3.1 shows the entrepreneurial process model proposed by J. Timmons [1]. Founders need to balance the three driving forces: *opportunity, resources,* and *team*. Since the entrepreneurial process starts with opportunity, the start-up stage exhibits a huge imbalance as illustrated in Figure 3.1a. By putting particularly strong effort into strengthening the team's technological capability and harmony, in addition to financial stability, the entrepreneur can obtain a new balance point as shown in Figure 3.1b.

In addition to support from family, founders need financial sponsors and partners or team members to operate the new company. In the scenario of Micro Motor Inc. (MMI), Barb Shay persuaded the president of Cheng Kung Corporation, Mr. Lenny Chu in Taiwan, to support MMI's initial capital expenditures. Barb had been a consultant to Cheng Kung for a couple of

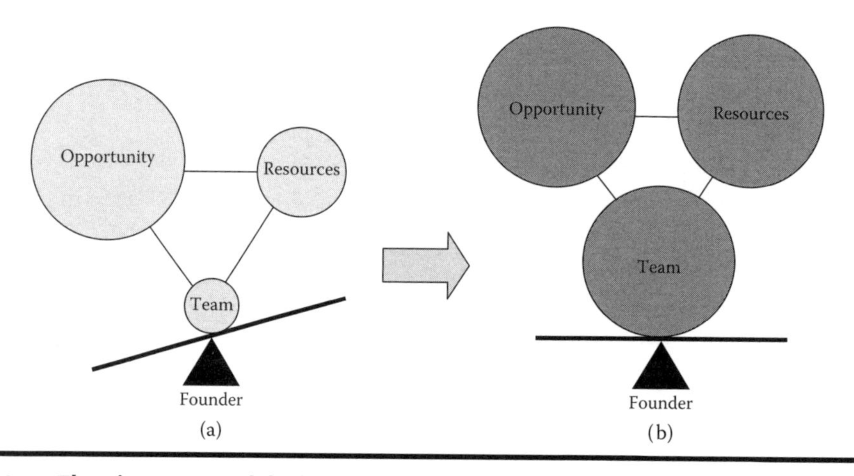

Figure 3.1 The Timmons model of the entrepreneurial process. A huge imbalanced state in the start-up stage (a) should be shifted to a well-balanced state in the initial growing stage (b).

years, and Lenny was interested in operating a U.S. branch of Cheng Kung. Lenny is president and CEO of MMI. He is also a venture tutor for Barb.

Barb Shay recruited Dr. Tom Meyer as director of research and development. He received his Ph.D. at the State University of Pennsylvania (SUP) under Barb's supervision, and worked in a high-tech start-up located in Maryland for 4 years, accumulating small business operational experience.

Chapter 5, "Corporate Capital and Funds," and Chapter 10, "Human Resources" will discuss this in more detail.

3.2 Legal Procedure

3.2.1 Forms of Organization

There are three types of companies: *sole proprietorship*, *partnership*, and *corporation*. The sole proprietorship represents ownership by one person. Its advantages are simplicity of decision making and low cost. On the other hand, the disadvantage is the unlimited liability of the owner. Of all U.S. business firms, 70% are sole proprietorships. However, they represent only 5% of total corporate revenue. Partnerships are similar to sole proprietorships except that there are multiple owners with unlimited liability. Unlimited liability means that the owners are personally liable for debts incurred by the company. If the company must fold, the owners cannot only lose their investment, but are also responsible for repaying creditors.

High-tech entrepreneurs should use a corporation structure. It will enable future expansion more easily and make it easier to issue and own stock. Only 20% of U.S. businesses use this structure. They produce 85% of all sales and more than 60% of the profits. A corporation is owned by shareholders, who have limited liability; legally, they can only lose their investment in the company.

The key for high-tech entrepreneurs is to keep eligibility as a small business in order to seek Small Business Innovation Research (SBIR) and Small Business Technology Transfer Research (STTR) programs run by U.S. government agencies, which provide research funding to small companies.

According to the U.S Small Business Administration [2], a small business is a firm

- With 500 or fewer employees
- With an annual revenue under $5 million

The above regulations can be easily satisfied for the start-up. Moreover, an SBIR awardee must

- Be a company that is at least 51% owned and controlled by one or more individuals who are citizens of the United States, or permanent resident aliens in the United States; or
- Be a company that is at least 51% owned and controlled by another business concern that is itself at least 51% owned and controlled by individuals who are citizens of or permanent resident aliens in the United States.

When you consider foreign investment into your company, you must carefully calculate how much you will ask from outside countries.

You need to determine the following information before contacting attorneys for company setup:

- *Start-up members*—corporate officers
- *Company location*—tentative lease agreement with office space
- *Capital money*—real cash for tentative company operation

3.2.2 Start-Up Members

MMI recruited Lenny Chu as president and CEO; he is a foreigner and the president of Cheng Kung Corporation, another company. If Cheng Kung does not meet the parameters of a small business (i.e., if it has more than 500 employees), this arrangement automatically cancels MMI's eligibility to the SBIR program application. Note also that Lenny or Cheng Kung should not exceed 49% of the stockholders of MMI.

3.2.3 Company Location

High-tech nonsoftware start-ups require laboratory and manufacturing facilities. However, start-ups have difficulty arranging these. Barb decided to choose a company location near SUP (her former employer). Many universities are running company incubators. These usually have conveniently located space along with amenities for starting a company and are a good choice. However, incubator space leases are not actually cheap and are sometimes more expensive than leases for nearby offices. Incubators usually include privileges to the university's facilities with reasonable costs through a research contract.

3.2.4 Capital Money

A company can be started with minimal capital, but the amount differs depending on location and needs. MMI initially set corporate capital at $300,000, in addition to the expected initial SBIR funds of around $300,000. Chapter 5 covers collecting capital and funds in detail.

3.2.5 Legal Process

Ask for quotes from multiple attorneys, and choose a reasonable one. There are ranges in attorney fees, depending on the attorney's level of experience. The key for Barb's MMI case is not to use attorneys who work directly with SUP. Because MMI is a sort of spin-off from the university, MMI may need to license some SUP patents in the future. MMI needs an attorney who is totally independent of SUP and can focus solely on MMI's benefit.

Figure 3.2 shows the actual corporation registration sheets for your reference. You will be required to give public notice of your company's establishment, as shown in Figure 3.3.

In parallel, the corporation needs to establish its corporate bylaws and minutes of the organizational meeting (the first board of directors meeting), which will define the roles of the board members. The following items need to be considered:

- *Board of directors:* chairman and secretary
- *Corporate officers:* president, vice president, and secretary/treasurer
- *Issuance of shares:* usually the corporation issues 1000 shares of common stock initially with a nominal value (par value) of $1.00 per share

Bylaws mandate how a corporation is to be run and operated, and the rights and powers of the shareholders, directors, and officers. Contents typically include the following provisions:

1. The time and place for meetings of officers, directors, and shareholders
2. Number of directors, their tenure, and their qualifications
3. Title and compensation of the corporate officers

2004092-1350
PENNSYLVANIA DEPARTMENT OF STATE
CORPORATION BUREAU

Articles of Incorporation-For Profit
(15 Pa.C.S.)

Entity Number

X Business-stock (§ 1306)
____ Business-nonstock (§ 2102)
____ Business-statutory close (§ 2303)
____ Cooperative (§ 7102)
____ Management (§ 2703)
____ Professional (§ 2903)
____ Insurance (§ 3101)

Name
R

Address

City State Zip Code
Bellefonte, PA 16823

Document will be returned to the name and address you enter to the left.
⇐

Fee: $125

Filed in the Department of State on SEP 29 2004

Secretary of the Commonwealth

In compliance with the requirements of the applicable provisions (relating to corporations and unincorporated associations), the undersigned, desiring to incorporate a corporation for profit, hereby states that:

1. The name of the corporation *(corporate designator required, i.e., "corporation"," incorporated", "limited" "company" or any abbreviation. "Professional corporation" or "P.C")*:

Inc.

2. The (a) address of this corporation's current registered office in this Commonwealth *(post office box, alone, is not acceptable)* or (b) name of its commercial registered office provider and the county of venue is:

(a) Number and Street	City	State	Zip	County
		PA	16803	Centre

(b) Name of Commercial Registered Office Provider — County

c/o:

3. The corporation is incorporated under the provisions of the Business Corporation Law of 1988.

4. The aggregate number of shares authorized: 1,000

Figure 3.2 Corporation registration sheet.

2004092-1351

DSCB:15-1306,2102/2303/2702/2903/3101/7102A-2

5. The name and address, including number and street, if any, of each incorporator *(all incorporators must sign below)*:

Name	Address
Ryan Lee	[redacted]
[redacted]	[redacted]

6. The specified effective date, if any: October 1, 2004
month/day/year hour, if any

7. Additional provisions of the articles, if any, attach an 8½ by 11 sheet.

8. *Statutory close corporation only*: Neither the corporation nor any shareholder shall make an offering of any of its shares of any class that would constitute a "public offering" within the meaning of the Securities Act of 1933 (15 U.S.C. 77a et seq.)

9. *Cooperative corporations only: Complete and strike out inapplicable term:*

The common bond of membership among its members/shareholders is:_______________.

IN TESTIMONY WHEREOF, the incorporator(s) has/have signed these Articles of Incorporation this 28th day of Sept., 2004

Ryan Lee Signature

Kenji Uchino Signature

Figure 3.2 *(Continued)*

3400 East College Avenue

Serving central Pennsylvania and online at www.centredaily.com

Your Life. Your Paper.
CENTRE DAILY TIMES
www.centredaily.com

PROOF OF PUBLICATION

State of Pennsylvania } ss.
County of Centre

______________________ being duly sworn according to law, says that he or she is an agent of the Centre Daily Times, a daily newspaper of general circulation, having its place of business in State College, Centre County, Pennsylvania, and having been established in the year 1898; that the advertisement, a printed copy of which is attached hereto, appeared in said newspaper on the 14th day(s) of October 2004 that affiant is not interested in the subject matter of the notice or advertisement; that all of the allegations contained herein relative to the time, place and character of the publication are true.

Witness Signature

Sworn and Subscribed to before me
this 15th **day of** october
A.D. 2004.

Notary Signature

Notarial Seal
Dustin Musser, Notary Public
Benner Twp., Centre County
My Commission Expires Dec. 3, 2005
Member, Pennsylvania Association Of Notaries

LEGAL NOTICE
NOTICE IS HEREBY GIVEN that Articles of Incorporation have been filed with the Department of State, Commonwealth of Pennsylvania, Harrisburg, Pennsylvania, on September 29, 2004, for the purpose of obtaining a Certificate of Incorporation. The name of the proposed corporation organized under the Pennsylvania Business Corporation Law of 1988, as amended, is:

Inc.
115 E. High St., P.O. Box 179
Bellefonte, PA 16823

KNIGHT RIDDER

Figure 3.3 Corporation proof of publication.

Example

NUMBER 4

SHARES 20

INCORPORATED UNDER THE LAWS OF THE COMMONWEALTH OF PENNSYLVANIA

Micro Motor Inc.

The Corporation is authorized to issue 1,000 Common Shares – Par Value $1.00 each

This Certifies that Barb Shay is the owner of Twenty fully paid and non-assessable Shares of the above Corporation transferable only on the books of the Corporation by the holder hereof in person or by duly authorized Attorney upon surrender of this Certificate properly endorsed.

In Witness Whereof, the said Corporation has caused this Certificate to be signed by its duly authorized officers and to be sealed with the Seal of the Corporation.

Dated

SECRETARY-TREASURER

PRESIDENT

Figure 3.4 Stock shareholder certificate.

4. The fiscal year of the corporation
5. Who is responsible and how the bylaws are to be amended
6. Any rules on the approval of contracts, loans, checks, and stock certificates
7. Inspection of the corporate records book

In the fictitious MMI situation, Lenny Chu is chair, president and CEO, Barb is secretary of the board of directors, and senior vice president and CTO. A total of 100 shares were issued initially to three stockholders (Cheng Kung, Barb, and another American colleague in order to keep the U.S. corporation eligibility with 51% or higher shares by U.S. citizens) for $300,000 capital. Thus, each share had a value of $3000 initially ($300,000 ÷ 100 shares). Figure 3.4 is an example of the stock shareholder certificate.

Chapter Summary

3.1 The company start-up procedure includes the following:
   1. Information collection on the start-up
   2. Persuasion of the family and home financial make-up plan
   3. Partner and key employee search
   4. Business plan preparation—it is important to find investors and loan lenders
   5. Investor and loan search
   6. Legal procedure of company start-up

3.2 Even with a corporation structure, the founder's liability is not completely limited. Understanding and support from the one's family are definitely necessary.

3.3 The Timmons model of the entrepreneurial process: the founder needs to balance three driving forces—*opportunity*, *resources* and *team*.

3.4 Three forms of firm: *sole proprietorship*, *partnership*, and *corporation*. A high-tech entrepreneur should choose a corporation firm, which is owned by shareholders with *limited liability*.

3.5 In order to be eligible for SBIR and STTR programs in the United States, you must start a small business firm that meets the following requirements:
- With 500 or fewer employees
- With annual revenue under $5 million
- At least 51% owned and controlled by one or more individuals who are U.S. citizens or permanent resident aliens

3.6 Required information for starting up a company: (1) Start-up corporate members, (2) company location, and (3) capital money.

Practical Exercise Problems

P3.1 Corporation Structure Questions

Determine the following required information for starting up your company:

1. Start-up corporate members
2. Company location
3. Capital money

MMI Example

- Board of Directors: Chairman Lenny Chu and Secretary Barb Shay
- Corporate Officers: President and CEO Lenny Chu, Vice President, CTO, and Secretary/Treasurer Barb Shay
- Total Capital: $300,000; issuance of shares: MMI issues 1000 shares of common stock, initially 100 issued with an actual value of $3000 each share (par value of $1.00 each)

P3.2 Venture Supporting Organization

Many universities have a venture-supporting organization to advise on and help with how to start a company. Access your university Web site to find this organization, and familiarize yourself with their services. There are two examples below:

- Pennsylvania State University business supporting Web site—http://www.innovationpark.psu.edu/new-business
- Pennsylvania State Organization—financial support, technology and management experience: http://www.cnp.benfranklin.org

References

1. Timmons, J. A., and S. Spinelli. *New venture creation*. New York: McGraw-Hill/Irwin, 2007.
2. U.S. Small Business Administration, Small Business Resource, http://www.sba.gov/services/ (accessed March 1, 2008).

Chapter 4

Business Plan—How to Persuade Investors

You do not need a detailed business plan to persuade family and friends, or to register the company legally. However, one of the most important tasks for the founder is to write the business plan, in order to persuade outside investors, venture capitalists, and banks of the company's viability. There are two major purposes for establishing a clear business plan:

1. The business plan is not for the technology experts, but for investors, who may not be familiar with your technology area. Do not describe it like an academic journal paper! Instead use lay words, excluding professional terminologies. Add simple explanations on professional items. Writing style should be similar to a press release. Remember: the business plan is to persuade outsiders to invest in your idea from a business viewpoint, *not* from a scientific or engineering viewpoint.
2. The founder must visualize his or her business from an outsider's viewpoint. Many high-tech entrepreneurs are too enthusiastic about their inventions, which gets in the way of thinking coherently about business issues, mainly making a profit. The founder can exude confidence in business, but should exclude his or her selfishness in business planning. The business plan should carefully describe the current and projected future financial condition of the company and the development pace from an outsider's viewpoint. Unexpected applications and markets or new business opportunities may be found during brainstorming sessions with outside potential investors.

The example format for the business plan can be found in various places, such as in the textbook *New Venture Creation* [1] and the Web site of the Entrepreneurial Center at Old Dominion University [2]. The following template was used for Micro Motors Inc.'s (MMI) business plan.

Table of Contents

- Executive Summary
- Managent and Organization
  - Management Team
  - Compensation and Ownership
  - Board of Directors
  - Insurance
  - Organization Charts
- Product/Service
  - Purpose of the Product/Service
  - Stage of Development
  - Future Research and Development
  - Trademarks, Patents, Copyrights, Licenses, Royalties
  - Product Liability
  - Production
  - Facilities
  - Suppliers
- Marketing Plan
  - Industry Profile
  - Competition Profile
- Customer Profile
  - Target Market Profile
  - Pricing Profile
  - Gross Margin on Products/Services
  - Break-Even Analysis
  - Market Penetration
  - Advertising and Promotion
  - Future Markets
- Operating and Control Systems
  - Administrative Policies, Procedures and Controls
  - Documents and Paper Flow
  - Planning Chart
  - Risk Analysis
- Growth Plan
  - New Offerings to Market
  - Capital Requirements
  - Personnel Requirements
  - Exit Strategy
- Financial Plan
  - Sales Projections
  - Income Projections
  - Cash Requirements
  - Sources of Financing
- Supporting Documents
  - Financial Worksheets
  - Resumes of Key Personnel
  - Legal Agreements

Marketing Materials
Insurance Documents
Press Releases or Articles

MMI's example business plan in this chapter was prepared by Barb Shay 18 months after the foundation of the company in order to secure additional funds to expand the business. Thus, this scenario mentions the results from years 1 and 2, and is forecasting for the third year and beyond.

4.1 Executive Summary

The executive summary is the most important part of the business plan, and should briefly cover all the contents in the plan. Some investors read only the executive summary in order to screen out inappropriate proposals.

4.1.1 Venture History

MMI Example

MMI was founded by Dr. Barbara Shay on October 1, 200X, as a spin-off company from the State University of Pennsylvania (SUP).

Dr. Barb Shay, founder, vice president, and chief technical officer (CTO) of MMI, is an associate professor at SUP and is well known as an active researcher in piezoelectric actuators. She has been developing various piezoelectric actuators and ultrasonic motors (USMs) for the past 10 years. Recently, her group developed a metal tube-type micromotor consisting of a metal hollow cylinder and two lead zirconate titanate (PZT) rectangular plates, as shown in Figure 4.1a. The assembly is shown in Figure 4.1b. In collaboration with Saito Industries in Japan, her group developed the world's smallest camera module with both optical zooming and auto focusing for cellular phones, as shown in Figure 4.1c. This journalistically sensational product was the trigger for the foundation of MMI. The key features of these micromotors include:

1. World's smallest motors (1.5 mm × 4 mm L) with a torque level (1 mN·m) on a human scale
2. Low manufacturing cost competitive with low-cost conventional electromagnetic motors

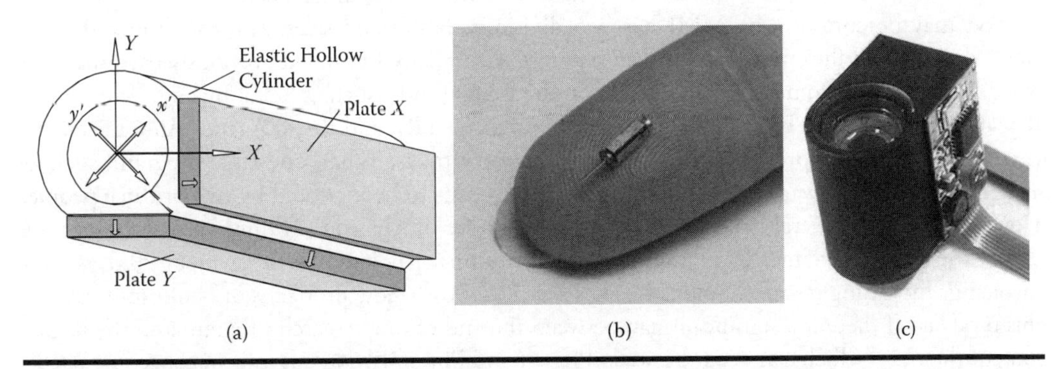

Figure 4.1 Metal tube motor using a metal tube and two rectangular PZT plates. (a) Schematic structure, (b) a world's smallest motor (1.5 mm diameter), and (c) camera zoom mechanism with two metal tube USMs in a Saito cellular phone.

In addition, using Barb's personal connections, MMI established product distribution channels with nine companies worldwide. MMI imports various piezoelectric-related devices, drive circuits, and computer simulation software, and distributes them through North America and Asia.

Even though the company has several products, the first business plan had a single focus—one promising product from a strategic viewpoint, with attractive product pictures on the first page.

4.1.2 Company Description

MMI Example

MMI specializes in the development and commercialization of piezoelectric actuators, transducers, and their integrated systems, including supporting services, tools, and software. In particular, micromotors, piezoelectric actuators, and microrobots are current development targets. The company's mission statement is "Your Development Partner," because MMI supports technology development (R&D division) and provides commercial devices (sales division) to commercial clients and federal agencies.

It has now been 1½ years since the foundation of the company on October 1, 200X. The R&D division has raised $1 million in research funds from federal agencies and private industries. MMI's intellectual property and know-how have gradually increased as the research programs proceed. In the sales division, they have established a product distribution partnership with nine leading industries worldwide. Sales have been increasing (annual sales for 200X are about $200,000). Presently they have eight employees including two corporate officers (president and vice president), four research engineers, one sales engineer, and one office manager.

The R&D division has unique expertise in piezoelectric devices, in particular, micromotors, piezo-actuators, and microrobots. There are no competitive manufacturers at present. MMI possesses several patents (including several pending), and licenses patents from SUP (including the present two patents of the founder, Barb Shay) for manufacturing their products.

The sales division subscribes to an "all-in-one" concept, like e-Bay. Because MMI has two suppliers for each product (e.g., multilayer [ML] actuators from ABC and XYZ, piezo-films from HJK and LMN), customers can compare and choose the best product for their application, through one Web site, MMI.com. This all-in-one concept is different from other competitive distributors.

You should decide your company's mission and describe the mission statement in simple language for the sake of outsiders. MMI's mission statement is "Your Development Partner." A similar company has "Simple Ceramic Motors… Inspiring Smaller Products." GE, Hitachi, and Toyota have "Imagination at Work," "Inspire the Next," and "Moving Forward," respectively.

You may be curious why MMI has two divisions, R&D and sales. You may think that an R&D division is sufficient for a high-tech start-up. A company I founded previously tried that and failed. A pure R&D company relies merely on the R&D funds, such as Small Business Innovation Research (SBIR), Small Business Technology Transfer (STTR), and Broad Agency Announcement (BAA) programs, and private corporations' research contracts. When one large program, such as an SBIR Phase II, is awarded $1 million, manpower needs to be increased by two or three people. However, there are several risks: (1) payment may not be timely, and (2) funding may stop at the end of the program. Some R&D programs are "work first, pay later," with payment 90 days after invoicing. Increasing research funds increases the risk of cash-flow problems or bankruptcy due to this time lag. If there is a significant gap between the end of one research program and the beginning of the next (which can occur between Phase I and Phase II), some layoffs may become necessary. Having another income source is helpful to balance cash flow and reduce the hire–fire cycle. Remember that a "death valley" for start-ups often occurs after the first Phase II program finishes.

4.1.3 Company Organization

MMI Example

MMI is a C-Corporation with three divisions: R&D, sales, and IT/office management. The key management personnel include the following:

- Lenny Chu, MBA, MS—president and CEO of MMI, and chair of the Board of Directors. He is also president of Cheng Kung Corporation, Taiwan, which is the major shareholder and holds a loan to MMI.
- Barbara Shay, PhD—founder, vice president, and CTO of MMI. She is also an associate professor at SUP. She is an active researcher and has worked with piezoelectric actuators and USMs for 10 years.
- Thomas Meyer, PhD—director of the R&D division at MMI. He received his PhD at SUP under Barb's supervision, and worked in a high-tech entrepreneurial start-up located in Maryland for 4 years with experience in small business operations.

Will you be the president of your new company or vice president with a reliable partner as president? Do you have a reliable subordinate engineer?

4.1.4 Marketing Plan

MMI Example

Piezoelectric actuators are directly used in information technologies (microactuators for computers, cellular phones), robotics, and precision machinery (positioners), and provide supporting tools for biomedical equipment, nanotechnologies (nanomanipulation tools), and energy areas (micropumps in fuel cell systems). Thus, the total industry market will significantly increase from the current $10 million per year to $1 billion in 5 to 7 years.

The R&D division targets federal agencies such as DARPA, Navy, Army, National Institutes of Health (NIH), and Telemedicine and Advanced Technology Research Center (TATRC) for research funds (they disperse $3 billion in total). Though MMI does not presently have specific competitors in micromotors or microrobotics, the high growth potential means that competitors will definitely appear in the coming years, which is a risk.

The sales division has an all-in-one, eBay-like concept, which is different from other distributors. MMI has customers for piezoelectric products in federal institutes and universities who are working in the aforementioned IT, biomedical, and energy areas. MMI will advertise in three ways: the MMI Web site, direct e-mail/mail to customers, and customer site visits.

The industry market size should be specific, citing data from reliable sources. What will be your specific tactics for promoting the market and increasing customers?

4.1.5 Company Operations

MMI Example

MMI has spent $400,000 in the first year for eight employees: $300,000 was initial capital and $100,000 was a loan from partner company Cheng Kung Corporation, Taiwan. The current monthly expenses are about $40,000 (second year), which will increase to $50,000 and $65,000 in the third and fourth years, respectively. The revenue target by the end of the third year is $1 million from the R&D programs and $0.5 million from the piezo-product sales.

4.1.6 *Financial Plan*

MMI Example

The start-up capital ($300,000) came from Cheng Kung Corporation, a U.S. citizen private investor, and the founder Barb Shay. In addition, MMI has a loan agreement of $500,000 for the first 3 years from Cheng Kung, which will be paid back in 5 years. There is also a bank loan to cover cash flow problems up to $50,000, which may occur from delays in payment from the government and clients. This short-term loan will be retired as soon as MMI receives research contract payments.

Investors are interested in getting the current and future financial information of the company. Ratio analysis, covered in Chapter 6, Section 6.2, prepares you to answer these questions. The forecast is a plan. Actual performance may deviate from it in the first a couple of years more than expected. However, financial estimates should be as accurate as possible. Otherwise, corporate cash flow problems will occur rapidly.

All three entrepreneurial factors—*opportunity*, *team*, and *resources*—should be satisfied appropriately in order to attract investors.

4.2 Management and Organization

4.2.1 *Management Team*

Please expand upon the information in Section 4.1.3 to provide more detail. Refer also to Chapter 3, Section 3.1.

4.2.2 *Compensation and Ownership*

MMI Examples

Ownership

The MMI stocks are shared by the Cheng Kung Corporation founder (Barb Shay) and a supporting investor (Jackie Wang; see Figure 4.2). The total capital as of the middle of the second year is $330,000. Barb added $30,000 in the second year.

Compensation

- There are two stock option programs: Incentive stock option ($30,000) and compensation stock option ($90,000) for the corporate officer (Barb Shay).
- For the employees: Compensation/payment is a reasonable level, in addition to the employment benefits.

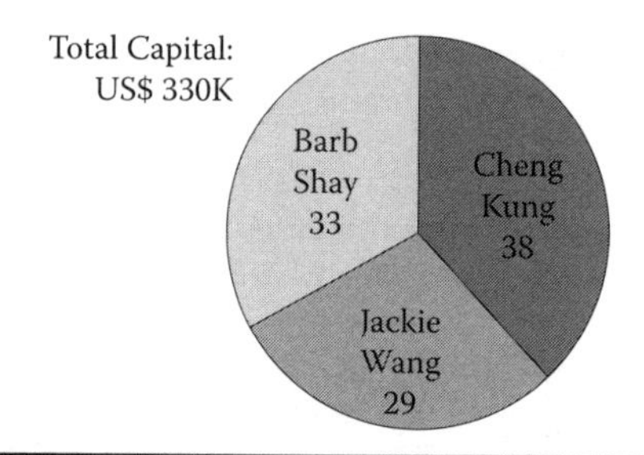

Figure 4.2 Shareholders of MMI (as of the middle of the second year).

- For sales engineers: Base salary plus commission (performance bonus, 2% of the sales amount).
- Employee's benefits: MMI has group medical insurance (except dental) and a pension system (1% coverage). Retirement benefits start in the third year.

Stock options are a popular method to reduce cash expenses by stimulating the corporate officer's performance. There are two types: *compensation stock option* and *incentive stock option*. Employees who receive the former need to pay federal income tax (W-2 income is increased by this amount), while those who receive the latter do not need to pay that tax. However, the maximum percentage of the incentive stock option over the total stock is limited (typically 10%).

MMI Example

In this scenario, Barb received $60,000 for part of her first-year salary through the compensation stock option (20 shares at $3000 per share), leading to additional expenses, i.e., the income tax payment from herself and MMI. She received $30,000 for part of her second year salary through the compensation stock option (10 shares at $3000 per share), and 10 shares at $1.10 per share as the incentive stock option for the second year. She received 40 stocks in total, or 33% of 120 MMI total stocks. The incentive stock option is not considered as compensation, leading to a tax savings. However, there is a regulation: the incentive stock option should not exceed 10% of total stocks. Notice that Barb has not received any cash, but has received stock certificates (corresponding to $90,000) from MMI in these 2 years, even though she paid her own income tax for this nominal additional compensation. In other words, she contributed to accumulate the MMI capital by $90,000 ($300,000 in total). Also note that 10 shares, through the incentive stock option, did not increase MMI capital ($1.10 x 10 = $11). In order to increase her power (number of the stocks) in MMI, Barb decided to take this personal financial risk by negotiating with her husband. Because Barb did not receive any cash, she needed her husband's support to pay her income tax of about $40,000 (corresponding to $90,000) in the first and second years. Also, note that the stock option is a popular tactic to generate corporate "cash flow."

4.2.3 Board of Directors/Advisory Council

Please expand upon the information in Section 4.1.3 to provide more detail.

4.2.4 Infrastructure

MMI Examples

ACCOUNTANTS

- *General accounting service:* Cats and Dogs Co., P.C., which provides advice on corporate tax, tax credit, etc.
- *Payroll accounting service:* HDD Small Business Services, which provides regular payroll service.

LAWYERS

- *Corporate lawyer:* Rob Rendell, who arranges incentive stock options, and other corporate-related issues.
- *Patent attorney:* Matt Smyth, who prepares patent application documents and research contract agreements.

A small start-up company cannot take care of accounting or legal matters by hiring an employee. However, these are essential services. Find accountants and lawyers from university entrepreneur support organizations, or from the Yellow Pages. Compensation for these outside advisors is made on an hourly basis.

4.2.5 Contracts and Agreements

MMI Example

MMI set up product distribution agreements with nine companies worldwide, including the following:

1. Cheng Kung Corporation, Taiwan: PZT actuator components, transducers, and other components
2. XYZ, Japan: ML actuators, piezo-transformers
3. MLN, Japan: USMs
4. Von Boyage, France: Finite element method (FEM) piezo-device simulation software and other products

4.2.6 Insurance

MMI Example

Though MMI leases office space from the Chamber of Business and Industries at College Park, property insurance is mandatory for the tenant. For their products, such as piezoelectric actuators, MMI has product liability (PL) insurance up to $1 million.

Property insurance and *product liability insurance* are essential to manufacturing companies, in addition to the employee health insurance. Note that liability insurance is rather expensive for a start-up company.

4.2.7 Organization Charts

The organization chart of MMI is shown in Figure 4.3.

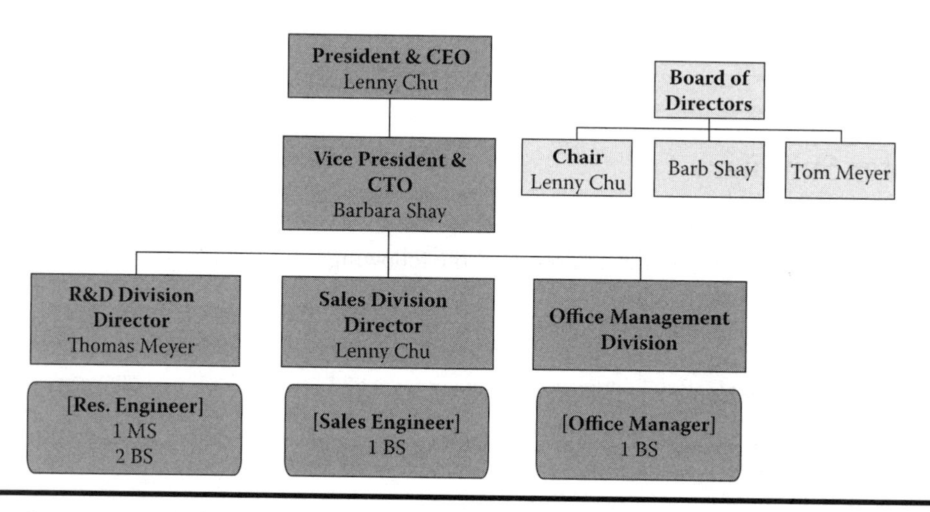

Figure 4.3 MMI organization chart (second year).

4.3 Product/Service

4.3.1 Purpose of the Product/Service

Please expand upon the information in Section 4.1.2 to provide more detail.

MMI Example

MMI's unique and proprietary piezoelectric device technologies can provide improved performance (device design, computer simulation, prototype devices, etc.) to customers in order to improve their devices and equipment production. The R&D fee is set according to a customer's budget, urgency, and required result. When R&D is complete, MMI will arrange mass production using an original equipment manufacturer (OEM), such as Cheng Kung Corporation in Taiwan, in order to reduce the production cost.

The sales division implements the all-in-one concept by setting up two suppliers for each product. Customers can compare and choose the most suitable product for their applications by accessing one site, www.mmi.com. This concept is different from those of other distributors. MMI provides all of the products and services that the piezoelectric actuator customer needs.

Some information has been omitted here; you should expand this description more in your own company's business plan.

4.3.2 Stages of Development

Please expand upon the information in Section 4.1.2 to provide more detail.

4.3.3 Future Research and Development

MMI Example

In addition to the micromotor technology, MMI will expand into high-power piezoelectric transformers, piezoelectric energy harvesting, and microrobots, as illustrated in Figure 4.4.

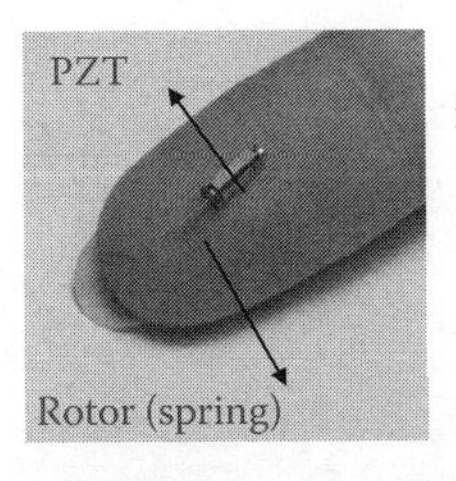

Ultrasonic motors
for medical applications

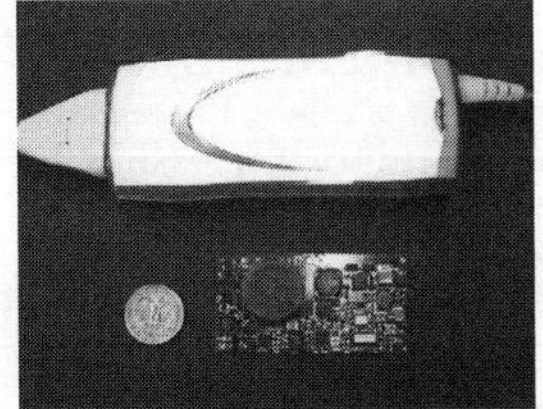

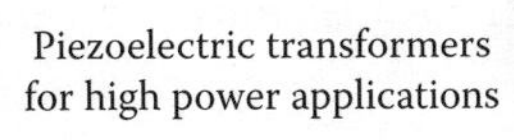

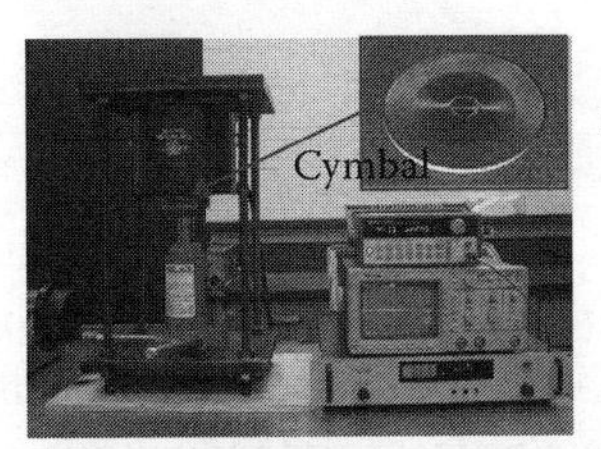

Cymbal transducers
for energy harvesting

Microrobots
for nano factory

Figure 4.4 Future technology areas for MMI.

Investors are interested in the future expansion of the venture in order to synchronize it with their plans.

4.3.4 Trademarks, Patents, Copyrights, Licenses, and Royalties

MMI Example

The MMI company logo is a trademark. A copyright will be pursued for MMI publications such as tutorial CDs (*How to Use FEM Simulation Software*) and company catalogues.

The MMI R&D division continuously seeks patents for protecting the unique technologies developed during research contracts. MMI's R&D agreements with federal agencies and private industries are as follows:

- Intellectual property created belongs to MMI.
- An exclusive license is provided to the client industry, which can then manufacture the products.

MMI also licenses patents filed through SUP.

If you hold a university or national, federal, or state institute-affiliated position, patent assignment to the start-up is limited, depending on employment conditions. Someone who is 100% employed by a university as a professor should not work for the start-up during regular working hours, but rather may work only during nights and weekends. Patent application rights are totally owned by the university (check the employment agreement form). The following example offers an alternative to this situation.

MMI Example

The founder, Barb Shay, is officially employed by SUP for 75% of her time, and by MMI for 25% of her time. Thus, she is eligible to submit a patent from MMI, as long as the content does not conflict with her university's research topics.

This arrangement includes a salary cut from the university, but it is the safest way to keep intellectual property separate.

4.3.5 Government Approvals

One of the most important issues for a high-tech entrepreneur is eligibility to apply for SBIR or STTR programs. Though this was mentioned in Chapter 3, it is worth repeating. According to the Small Business Administration, a small business must meet the following criteria:

- 500 or fewer employees
- Annual revenue under $5 million

More precisely, a small business:

1. Is independently owned and operated, not dominant in the field of operation in which it is proposing, has a place of business in the United States and operates primarily within the United States or makes a significant contribution to the U.S. economy, and is organized for profit.
2. Must be (a) at least 51% owned and controlled by one or more individuals who are citizens of, or permanent resident aliens in the United States or (b) a for-profit business that is at least 51% owned and controlled by another for-profit business that is at least 51% owned

and controlled by one or more individuals who are citizens of, or permanent resident aliens in, the United States.

3. Has, *including its affiliates*, an average number of employees for the preceding 12 months not exceeding 500, and meets the other regulatory requirements found in 13 CFR Part 121. Businesses are generally considered affiliates of one another when, either directly or indirectly: (a) one organization controls or has the power to control the other; or (b) a third party controls or has the power to control both.

One needs to be careful with item 3. Many high-tech small businesses' foundations have research contracts with large companies. If a venture accepts investment from a large company in exchange for stock, the large company is considered an affiliate, so the number of employees (including this affiliate) may be higher than 500. The small business will automatically lose SBIR eligibility.

Another issue is *security clearance* for some of MMI's research engineers for conducting high-security federal defense research programs. If your venture is seeking this direction, at least one engineer in your corporation should obtain a security clearance certificate.

4.3.6 Product/Service Limitations and Liability

A start-up venture has R&D and production capacity limitations, mainly due to the lack of manpower. Even as the number of research contracts increases, there is a risk in increasing the number of researchers because of a 90-day delay between official starting date and payment date, particularly from federal agencies. To compensate for this gap, a start-up needs to have more than $100,000 cash in the bank—a little less than 3 months of operating expenses.

Limitations in the sales division come from a lack of a strategy for sales promotion. There are two reasons: (1) unlike restaurants or retail shops, experienced sales engineers are not readily available, because it is a new field; (2) a customer list for a new venture does not exist. Active promotion strategies, as described in Section 4.4: Marketing Plan, should be used.

Even if company office space is leased, property insurance is usually mandatory for the tenant. For products such as piezoelectric transducers at MMI, the venture should have PL insurance for $1 million (annual cost is around $10,000). PL is the area of law in which manufacturers, distributors, suppliers, retailers, and others who make products available to the public are held responsible for the injuries those products cause. A products liability claim usually originates from the following causes:

- Design defects
- Manufacturing defects
- Failure to warn consumers

Claims may succeed even when products were used incorrectly by the consumer, as long as the incorrect use was foreseeable by the manufacturer. You may remember this lawsuit: A mobile phone company was sued by a customer who swallowed her cellular phone and choked. After that, the mobile phone manual included a ridiculous warning sentence: "Do not swallow the phone."

In order to insulate your company from PL claims, you may request customers of your new prototypes to sign a Waiver Agreement for PL claims: "I, the customer of MMI device, understand that the product is still a prototype for my trial usage only. If the purchased device fails to function, I will not claim for the secondary damages caused by the device failure, but claim merely for the immediate exchange of the device."

4.3.7 Production Facility

Products for military-supported research programs should be manufactured internally by your company in the United States (legal restriction) for small quantities. When production quantities are intermediate, U.S. domestic partners may be used for production. Once the production quantity exceeds 10,000 for consumer applications, an international partner may be chosen, as an OEM manufacturer, to minimize manufacturing costs. Certain countries are available to produce for U.S. government organizations. These can be found in the Commerce Control List on Export Administration Regulations and the related Country Chart at http://www.access.gpo.gov/bis/ear/pdf/738.pdf.

4.3.8 Suppliers

MMI Example

The sales division has nine corporate partners (MMI is an official distributor) including Cheng Kung Corporation in Taiwan (for PZT actuator components, transducers, and other components).

MMI arranges multiple suppliers for each product as backup suppliers.

4.4 Marketing Plan

Refer to Chapter 8 for more details.

4.4.1 Industry Profile

4.4.1.1 Industry Market Research

Information on the start-up's targeted area should be collected before starting your own company. However, the importance of this section is to help the investor to understand the opportunity correctly.

MMI Example

MMI created a table to summarize the developments of piezoelectric actuators in the United States, Japan, and Europe. However, since the details are not relevant to the reader of this textbook, they are omitted. Please also refer to Section 11.1, Strengths-Weaknesses-Opportunities-Threats (SWOT) Matrix Analysis for further information.

The annual sales of ceramic actuator units, camera-related devices, and USMs in 2005 in Japan are estimated to reach \$500 million, \$300 million, and \$150 million, respectively. The total sales may become equivalent to those of the capacitor industry. If these are installed in final actuator-related products, sales are projected to reach \$10 billion. Thus, a bright future is anticipated in many fields of application. Piezoelectric actuators are directly used in information technologies (microactuators for computers, cellular phones), robotics, and precision machinery (positioners), and provide supporting tools for biomedical equipment, nanotechnologies (nanomanipulation tools), and energy areas (micropumps for fuel cell systems). Thus, the market will significantly increase from the current \$10 million annually to \$1 billion in 5 to 7 years. MMI will not be restricted to the domestic market, but will expand worldwide.

I foresee application developments in piezoelectric actuators. In the IT/office equipment application area, reducing production costs is the development target. ML actuators and USMs need to be manufactured for less than \$3 per piece. For robotic applications, nanotechnology is expanding; accordingly, nanopositioning demand will be the key to estimating the future market size. In the biomedical area, price is not the driver, but specs such as very low drive voltage and limited size are critical in design. Lead (Pb) is prohibited when an instrument is inserted into the human body. Prosthetic arms and legs can be designed with more sophistication using USMs. Finally, environmental business is obviously expanding. Accordingly, the demand for actuators and transducers in this area is increasing rapidly. Hazardous waste decomposition, ultrasonic cleaning, and sonochemistry are promising areas.

Sales forecasts can be calculated from the product of the total industry's market size and MMI's targeted share (such as 3% of the worldwide market). If the industry is mature, like semiconductors, the market forecast can be easily obtained from the open literature. However, if the industry is relatively new, like piezoelectric actuators, you will need to collect the data. I occasionally use Google for this purpose. (1) Search "piezoelectric actuators" to find companies manufacturing the devices. (2) Search the financial statements for each company to find the piezo-actuator sales. (3) Add all sales amounts to estimate the industrial market size.

4.4.1.2 Geographic Locations

Where will you start your company? Why? See the example below.

MMI Example

MMI is located in the Science Park area in College Park, Pennsylvania near SUP. The location was originally decided by the founder, Barb Shay, because she is a faculty member at SUP. Additional merits for this location are summarized here:

- MMI can easily access university professors who specialize in piezoelectric devices. At least 20 laboratories at the university will be customers of piezoelectric devices, control devices, and simulation software.
- This area in Pennsylvania is known also as "Piezo Valley" by materials and transducer engineers. There are more than 20 piezo-ceramic industries that support federal agencies such as Navy and Army laboratories. MMI is the North American and Asian distributor for an FEM simulation software used to analyze and design piezo-devices. As a Piezo Valley member, MMI is responsible for training software customers. MMI's sales division is conducting a tutorial seminar on the FEM software, where various piezoelectric products are also promoted.

4.4.1.3 Profit Characteristics

Because your high-tech firm will be an R&D company, typical profit calculations used elsewhere, such as in retail, do not apply. You should develop a sales calculation algorithm for your prototype product. A typical example for a custom-made piezoelectric product is explained in Table 4.1. You will need to consider the following items:

1. Direct labor: Engineering hours (labor rate $25/h + 69.5% indirect cost) include design using computer simulation and manufacturing prototypes.
2. Direct material: Materials cost includes special punching/pressing equipment.
3. Sales hours (labor rate $15/h) and handling charges.
4. General and administrative (G&A) for first R&D product is 17.5%.

Table 4.1 MMI Sales Calculator Software

	B	C	D	E	F	G	H	I	J
1	Labeled NRE				Labled cost of product				PO Total
2		Fill in yellow							
3	Sales Hours	5	$57.40						
4									
5	Engineering Hours	40	$954.00						
6									
7	Labor Total w/ 69.5%		$1,011.40						
8	Materials	2000	$2,000.00						
9	Handling	200	$200.00						
10					hardware quantity	10			
11					hardware cost	80	$800.00		
12									
13	Total Cost		$3,211.40		Total Cost		$800.00		
14	G&A (17.5%)		$3,773.40		G&A (69.5%)		$1,356.00		
15									
16	NRE (with fee 10%)		$4,150.73		hardward (with fee 10%)		$1,491.60		$5,642.33
17									
18					Hardware list price		$149.16		
19					PA sales tax per unit		$8.95		
20					Total PA Sales Tax for Hardware		$89.50		
21									
22					Order Total				$5,731.83

5. Duplication product cost will include G&A 70%.
6. Sales tax (6% in Pennsylvania) is applied only to hardware sales, not R&D efforts.

The total sum of 1 through 4 is billed to the customer (quote). Refer to Chapter 6, Section 6.1: Accounting Management for details.

4.4.1.4 Distribution Channels

MMI Example

MMI structured the distribution channels of the computer simulation software in Asia by setting up subagent agreements with the following three companies:

1. Kimchee, Korea
2. Cheng Kung, Taiwan
3. Hinomaru, Japan

4.4.2 Competition Profile

MMI Example

MMI's R&D division does not have significant competitors in the micromotor and microrobot fields; however, high growth potential will attract multiple competitors in the coming 3 to 4 years.

As mentioned previously, the all-in-one sales division is different from other distributors. To compete, they would need entrepreneur Barb Shay's worldwide human network.

4.4.3 Customer Profile

MMI Example

Customers for piezoelectric products will be found in federal agencies and universities that are working in the high-tech areas, supported by the above federal programs. The sales amount targeted will be $1 million per year.

4.4.4 Target Market Profile

MMI Example

The R&D division targets federal agencies such as DARPA, Navy, Army, NIH, and TATRC to obtain research funds in the range of $30,000 (Phase I) to $800,000 (Phase II). The targeted annual research revenue is $1–2 million.

4.4.5 Pricing Profile

Product sales are usually small for each customer of a R&D-oriented company. Therefore, general calculation sheets for pricing are inappropriate. Refer to Table 4.1 in Section 4.4.1.3, which provides an example for a special piezoelectric actuator estimate.

MMI Example

1. *Research product sales (see Table 4.1)*: The initial product cost includes both R&D costs and hardware costs. For the second purchase order (PO) and beyond, only the hardware costs will be applied.
2. *Partner company's product sales*: Only the hardware cost calculation (right-hand side) is applied.
3. *Federal research program*: Only R&D cost calculation (left-hand side) is applied. There is no need to consider Pennsylvania sales tax.
4. *Private industry research program*: Both R&D and hardware costs are applied. Among entire R&D contract revenue, some is assigned to hardware costs, which will be adopted for future product supply to this client.

4.4.6 Gross Margin on Products and Services

MMI Example

R&D Division

R&D research funds provide 90% of the gross margin (cost is basically for raw materials), most of which should be spent for the R&D research engineers' salaries. The goal is to pay all research engineers' salaries using R&D funds.

Private Company Program

The current number of programs, five (average $50,000 per program), will be sustained, but not expanded, because the gross margin is not high.

Federal Program

Because the gross margin is higher than private company programs, MMI will aggressively seek federal programs, using the following plan:

- *Phase I:* The current number of programs, three (ranges \$80,000 to \$300,000), should be increased to six in 5 years.
- *Phase II:* MMI will target two Phase II programs (average \$800,000) above Phase I programs. This expectation raises the R&D income significantly for fourth and fifth years.

Sales Division

Software (FEM Simulation Software)

FEM software is a lease product (1-year lease). Software profitability is quite high compared to other piezo-devices, with an average price of \$9000 per lease. MMI's target is to increase the number of customers from the current 20 to 60, leading to approximately \$500,000 revenue per year, reflected in the fourth and fifth years.

Hardware (MMI Original Product)

As shown in Table 4.1, this is a small research program with an average price of \$5000 to \$6000 per item. As MMI improves technologies, R&D costs can be reduced, leading to higher gross margins. The strategy is small product lines with reasonable sales, targeting 50 items per year, leading to \$300,000 in revenue.

Hardware (Partner Company Product)

A similar calculation is made for partner company products (average price \$2000 with gross margins typically between 25 and 30%). This category does not require research engineering labor, just sales effort. We target sales around \$400,000 per year.

4.4.7 Market Penetration

MMI Example

Distribution Channels

Sales Representatives

MMI structured the distribution channels of the computer software in Asia by setting three subagent agreements:

1. Kimchee, Korea
2. Cheng Kung Corporation, Taiwan
3. Hinomaru, Japan

Direct Sales Force

MMI's sales division will advertise in three ways: MMI Web site, direct e-mail/mail to the customers, and customer site visit. The customer's inquiry/request is received by phone call, mail/e-mail, or fax.

Direct Mail/Telemarketing

The mailing list for direct mail for the MMI product advertisement is created from the attendance lists of academic conferences and workshops in materials science, nanotechnology, electrical and mechanical engineering, biomedical, and energy engineering.

Advertising and Promotion

R&D division: MMI's R&D research engineer will access the program officer via telephone to discuss MMI proposal possibilities.

Sales division: MMI's sales division will advertise in various ways:

- MMI Web site (update every month)
- Direct e-mail/mail to customers
- Customer site visit (local Pennsylvania, including SUP)
- Trade shows (academic societies, workshops)
- Journal advertisements
- MMI tutorial/promotional seminars (every 3 months)

MMI exhibited its products at seven trade shows and conferences in the first and second year. MMI will continuously seek three to four suitable trade shows every year:

1. Transducer Workshop—College Park, Pennsylvania, May
2. Microrobot Demonstration—College Park, Pennsylvania, March
3. Materials Science and Technology—Cincinnati, Ohio, October

4.5 Operating and Control Systems

4.5.1 Administrative Policies, Procedures, and Controls

MMI Example

Receiving Orders

The sales engineer sends a quote, and receives the order from the customer.

Billing Customers

The sales engineer does this directly.

Paying Suppliers

1. Sales engineer obtains the permission from the sales division director (Lenny Chu).
2. Sales engineer requests that the office manager prepare a check.
3. Vice President Barb Shay signs the check, and the office manager mails the check to the supplier.

Collecting Accounts Receivable

Office manager collects accounts receivable in response to requests by the sales division director for product sales or by the R&D division director (Tom Meyer) for research funds.

Reporting to Management

1. *R&D division:* Research engineer reports to the R&D division director, Tom Meyer, then to the vice president, Barb Shay.
2. *Sales division:* Sales engineer reports to the sales division director, Lenny Chu, then to the vice president, Barb Shay.

Staff Development

The entrepreneur, Barb Shay, regularly provides a weekly 1-hour seminar on piezoelectric actuators in the MMI conference room. Fundamental R&D knowledge is taught to all research engineers.

In order to improve sales, MMI will educate sales engineers by relating fruitful experiences from the president (Lenny Chu), and by sending them to suitable sales promotion seminars. The office manager will learn basic accounting by taking suitable business courses.

Inventory Control

MMI inventory is filed daily on the laptop computer and on the MMI notebook by a research engineer.

Handling Warranties and Returns

According to the customer's request, a sales engineer will handle product returns to the manufacturing partner. MMI's warranty conditions are compatible with the partner company's General Terms of Sale.

Monitoring the Company Budgets

Vice President Barb Shay monitors the company budgets daily, and President Lenny Chu monitors monthly budgets.

4.5.2 Documents and Paper Flow

MMI Example

A transaction is directly prepared by Vice President Barb Shay, with advice from the company's lawyers, if necessary.

4.5.3 Planning Chart

Refer to Chapter 6 for the details.

MMI Example

Product/Service Development

Immediately after an R&D contract is awarded, the R&D division director assigns research engineers to this program, and device design (computer simulation) and prototype manufacturing are conducted in succession. MMI accumulates design and manufacturing know-how by completing research programs.

Financial Requirements

The start-up capital ($300,000) was collected from one company and one private investor (Cheng Kung and Jackie Wang), and the founder Barb Shay. Further, MMI received a loan agreement of $500,000 from the partner, Cheng Kung Corporation. A bank loan was also received for a flexible cash flow of up to $50,000 from a local bank. This will cover 90 days of emergency financial deficit.

Marketing Flow Chart/Market Penetration

R&D Division

MMI's R&D research engineer will access the program officer via telephone to discuss MMI's proposal submission.

Sales Division

MMI's sales division will advertise in various ways, including the following:

- MMI Web site (updated every month)
- Direct e-mail/mail to the customers

- Customer site visit (local Pennsylvania, including SUP)
- Trade shows (academic societies, workshops)
- Journal advertisements
- MMI tutorial/promotional seminars (every 3 months)

MANAGEMENT AND INFRASTRUCTURE

An experienced sales division engineer will be hired in a year, to accelerate sales.

4.5.4 *Risk Analysis*

MMI Example

MMI's R&D division has research capacity limitations, mainly due to lack of manpower. Even though MMI can increase the number of research contracts, it cannot increase the number of research engineers in proportion to the funds due to a 90-day time lag between the official starting date and the payment date from federal agencies. To compensate for this gap, MMI occasionally needs more than $100,000 in cash flow.

Sales increase is limited by the inexperience of the sales engineers in the piezoelectric actuator area. MMI will educate present sales engineers and, in parallel, MMI may need to seek a more experienced senior sales engineer.

4.6 Growth Plan

4.6.1 *New Offerings to the Market*

MMI Example

The R&D division targets federal agencies such as DARPA, Navy, Army, NIH, and TATRC for research funds in the range of $30,000 (Phase I) to $800,000 (Phase II). The targeted annual research revenue is $1–2 million. In contrast, the customers for the piezoelectric products will be found in federal agencies and universities that are working in high-tech areas, supported by the aforementioned federal programs. Targeted sales will be $1 million per year.

Remember that the definition of a small business is (a) 500 or fewer employees, and (b) annual revenue under $5 million. If the total revenue reaches $5 million, you will need to abandon SBIR or STTR programs. Of course, you can apply to a higher level program such as BAA.

4.6.2 *Capital Requirements*

MMI Example

As MMI R&D contracts reach Phase III, MMI will need to set up a manufacturing line or a factory. Capital will be brought in from the partner companies or suitable venture capitalists.

4.6.3 *Personnel Requirements*

MMI Example

MMI will need an experienced sales engineer in a year to accelerate sales.

4.6.4 *Exit Strategy*

MMI Example

The two key indications for an exit strategy are as follows:

1. The break-even point cannot be achieved by the end of the third year.
2. Loans cannot be paid back completely by the end of fifth year.

4.7 Financial Plan

4.7.1 *Sales Projections/Income Projections*

MMI Example

Projections of sales income increase 30% per year (see Table 4.2). The jump in R&D income in the third year is an actual result. MMI can expect a 20% per year increase for R&D income.

4.7.2 *Cash Requirements*

MMI Example

MMI has already spent $810,000 in the first 2 years, for eight employees; $330,000 was the total capital, and $450,000 was a loan from Cheng Kung, Taiwan. The current monthly running cost (expenses) is about $45,000 (second year), which will increase by 10% in successive years. The revenue target by the end of the third year is $500,000 from R&D programs and $200,000 from piezo-product sales which will reach the break-even point.

MMI's R&D division has a cash flow risk. Because of a 90-day time lag between official starting date and payment date from federal agencies, MMI on occasion needs more than $100,000 in cash flow.

4.7.3 *Sources of Financing*

MMI Example

The start-up capital of $330,000 was collected from a partner company, Cheng Kung Corporation in Taiwan, the entrepreneur (Barb Shay), and a private investor (Jackie Wang). Further, MMI received a loan agreement of $450,000 for the first 2 years from Cheng Kung, which should be paid back in 5 years with an annual interest rate of 8%. A bank loan was also received for cash flow flexibility up to $50,000. This short-term loan should be repaid as soon as MMI receives research contract revenue.

Table 4.2 MMI Sales and Income Projection (Unit: US$)

	1st Year	*2nd Year*	*3rd Year*	*4th Year*	*5th Year*
Sales income	166K	150K	195K	253K	330K
R&D income	0	147K	500K	600K	720K

4.7.3.1 *Attached Financial Projections*

- Cash flow for 5 years (Table 4.3)
- Income statement for 5 years (Table 4.4)
- Balance sheet for 5 years (Table 4.5)

The investors are most interested in when break-even occurs, and how much debt is being carried. You will learn how to develop these financial statements in Chapter 6.

Table 4.3 Cash Flow for 5 Years in MMI

	1st Year	*2nd Year*	*3rd Year*	*4th Year*	*5th Year*
Operating activities	($291,000)	($312,000)	$8950	$50,025	$105,194
Net earnings					
Depreciation					
Account receivable					
Inventory					
Accounts payable					
Income taxes	$0	$0	$8950	$50,025	$105,194
Investing Activities					
Marketable securities, properties, plant, equipment	$0				
	1st Year	*2nd Year*	*3rd Year*	*4th Year*	*5th Year*
Financing Activities					
Borrowed funds	$113,000	$336,000			
Loan payments	$0	$0	$8950	$50,025	$105,194
Share capital	$240,000				
Net cash flow	$62,000	$24,000	$0	$0	$0
Opening cash	$0	$62,000	$86,000	$86,000	$86,000
Closing cash	$62,000	$86,000	$86,000	$86,000	$86,000
Nonoperating income (Cheng Kung support)	$240,000 (Capital)	$336,000	0	0	0
Accumulated loan to Cheng Kung (interest rate 8%)	$113,000	$449,000	$475,970	$464,023	$395,951

Table 4.4 Income Statement for 5 Years in MMI

	1st Year	*2nd Year*	*3rd Year*	*4th Year*	*5th Year*	
Sales income	$166,000	$150,000	$195,000	$253,500	$329,550	*30% increase per year
R&D income	$0	$147,000	$500,000	$600,000	$720,000	*20% increase per year (real number for 1,2,3 years)
Cost of goods sold	$39,000	$72,000	$86,400	$103,680	$124,416	*20% increase per year
Gross margin	$127,000	$225,000	$608,600	$749,820	$925,134	
Operating expenses (salaries, fringes, and office rent, etc.)	$418,000	$537,000	$590,700	$649,770	$714,747	*10% increase per year
Operating profit	($291,000)	($312,000)	$17,900	$100,050	$210,387	
Profit before tax	($291,000)	($312,000)	$17,900	$100,050	$210,387	
Income taxes	$0	$0	$8950	$50,025	$105,194	*50% of profit before tax
Net income	($291,000)	($312,000)	$8950	$50,025	$105,194	*To be spent for returning loans
Nonoperating income (Cheng Kung support)	$240,000 (Capital)	$336,000	$0	$0	$0	
Accumulated loan (to Cheng Kung)	$113,000	$449,000	$475,970	$464,023	$395,951	*Interest rate = 8%

Table 4.5 Balance Sheet for 5 Years in MMI

	1st Year	*2nd Year*	*3rd Year*	*4th Year*	*5th Year*	
Assets						
Current Assets						
Cash (loan and capital)	$353,000	$336,000	$0	$0	$0	
Accounts receivable (sales)	$166,000	$150,000	$195,000	$253,000	$329,550	*30% increase per year
Accounts receivable (R&D)	$0	$147,000	$500,000	$600,000	$720,000	*20% per year (real number for 1, 2, 3 years)
Inventory	$0	$0	$0	$0	$0	

(Continued)

Table 4.5 Balance Sheet for 5 Years in MMI (*Continued*)

Fixed Assets						
Land	$0					
Building	$0					
Equipment	$0					
Vehicles	$0					
Total assets	$519,000	$633,000	$695,000	$853,500	$1,049,550	
Liabilities and Equities						
Current Liabilities						
Accounts payable	$418,000	$537,000	$590,700	$649,770	$714,747	*10% increase per year
Current portion (long-term debt)	$0	$0	$8950	$50,025	$105,194	
Income taxes	$0	$0	$8950	$50,025	$105,194	*50% of profit before tax
Dividends payable						
Long-Term Liabilities						
Long-term loans	$113,000	$336,000	$0	$0	$0	
Owners' equity						
Share capital						
Retained earnings	($12,000)	($240,000)	$86,400	$103,680	$124,415	
Total Liabilities and Equities	$519,000	$633,000	$695,000	$853,500	$1,049,550	

Chapter Summary

4.1 The business plan is important for persuading investors, venture capitalists and banks.

4.2 The business plan is typically composed of the following:

1. Executive Summary
    a. Management and organization
    b. Product or service
    c. Marketing plan
    d. Operating and control systems
    e. Growth plan
    f. Financial plan
2. Supporting documents

4.3 R&D-based product price includes the direct labor cost of the researchers and indirect G&A cost (manufacturing overhead), in addition to direct materials cost.

4.4 Growth/Financial Plan: The growth rate of your firm should be realistic (not too optimistic or pessimistic), which may be justified by the industry market growth rate.

4.5 Essential Three Financial Projections:
1. Cash flow for 5 years
2. Income statement for 5 years
3. Balance sheet for 5 years

Practical Exercise Problems

P4.1 Business Plan

Begin to write a business plan for your start-up by referring to the sample business plan of MMI. Use a realistic growth rate for your business plan, taking into account the growth rate in your firm's industry category. In order to complete this business plan, you may need to wait until you read Chapters 6, 7, and 8.

P4.2 Reviewers of the Business Plan

Who should prepare the business plan? Of course, you should, in collaboration with your corporate partners (corporate officers, president, vice presidents, division directors, etc.). Who do you think are the best people to review your business plan? Internal review by your corporation partners will be the first step.

If you have a close personal friend who is a corporate director of a company that may be a potential competitor, would you ask him to review your business plan? Discuss the merits and demerits of this issue.

P4.2.1 Merits

1. Outsiders usually point out items which insiders often forget from too much focus or narrow brainstorming, such as unexpected marketing areas and product lines.
2. In particular, a potential competitor may already know "what not to do," which is important feedback for an entrepreneur.

P4.2.2 Demerits

1. You will be giving away your strategy to a possible competitor.
2. Depending on your friend's personality, he or she may mislead you intentionally.

P4.2.3 Comments from the Author

1. Disclosing your business plan to a third party is illegal from a confidentiality clause in your employment agreement. Even though you are the president or vice president, you should follow the nondisclosure agreement (NDA). When you find disclosure is necessary, you need to ask the permission of the board of directors.

2. You should sign an NDA between your firm and your friend. The information should not be disseminated to the competitor's company or wider, but limited to your friend.
3. I advise you to do this only when your industry is expanding greatly. I usually persuade a director of a partner company with the suggestion, "Let us make 1 + 1 = 2.5, not 1.5." Company mergers are occasionally effected to restructure or downsize to reduce the cost. When the industry is already mature and the total market is limited, one company's share increase means another competitive company's loss. However, if the industry is growing remarkably, two competitive firms' collaboration can actually increase the total sales, as long as the collaboration is going well. Under this assumption, if your friend's company may be a future collaborator in business, asking him or her to review the business plan may not be a bad idea. You may find a future sponsor or partner by disclosing your company's unique technology, which your friend's company may not have.

References

1. Timmons, J. A., and S. Spinelli. *New venture creation*. New York: McGraw-Hill/Irwin, 2007.
2. Business Plan Template, Old Dominion University, http://www.damiani.net/bplan/bplan.htm (accessed March 11, 2008).

Chapter 5

Corporate Capital and Funds—How to Find Financial Resources

As an entrepreneur, you will need money to start up a company. You can collect investment money for the initial company stocks from your partners or get loans from banks, in addition to your own money. We will learn how to collect start-up money in this chapter.

In parallel, it is time to write a research and development (R&D) proposal for Small Business Innovation Research (SBIR), Small Business Technology Transfer (STTR), and other funding sources. Successful proposal writing and presentation will be explained in the latter part of this chapter.

5.1 Debt and Equity—Financial Resources at the Start-Up Stage

5.1.1 Stock or Loan

How does the firm determine the corporate capital? Some suggest stock be sold, while some may advise to borrow money. What is the difference between stock and loan from the corporate financial viewpoint?

- *Stock*: You do not need to pay back the initial investment amount, but you need to pay the *stock dividend* with the effective annual return rate around 10–12%, higher than the interest rate of loan. However, the dividend distribution may not start in the first 5–7 years after the start up.
- *Loan*: You must pay back the total amount of the principal and the interest by the end of the loan period. The interest rate is usually lower than the stock return rate, typically 5–10%. A short-term loan (less than 1 year) has a lower interest rate than a long-term loan (3–5 years).

Though an initial money collection financed by selling stocks of the firm might appear acceptable at first glance, when the majority of the stock is shared by outside investors, you lose the control of the firm. On the other hand, the use of short-term debt might increase the overall risk of the firm, because bankruptcy follows once the loan cannot be returned. A typical suggestion is a 50:50 mixture of equity (stock) and debt (loan).

5.1.2 Partnership

During the development of a new technology, the entrepreneur must have collaborated or been acquainted with some companies in a similar industry. These firms are good candidates for setting up a partnership or a joint venture to collect the start-up money.

MMI Example

Micro Motor Inc. (MMI) established a partnership with Cheng Kung Corporation, Taiwan at the start-up stage. Barb Shay had been consulting for them for a couple of years; her persuasion story was as follows:

1. Global sales expansion of Cheng Kung requires a U.S. branch–MMI will take the role.
2. MMI will be an R&D center of Cheng Kung, so that Cheng Kung will obtain steady consultation and updated products for the next generation.

The initial capital, $300,000, was shared by Cheng Kung (45%), Jackie Wang, a U.S. private investor (35%), and Barb Shay (20%). Barb's originally contributed $60,000 via her compensation stock option. The share by Cheng Kung, an international firm, should be less than 49%. Then, the loan agreement up to $500,000 was set for the initial 3 years with Cheng Kung. Barb added $30,000 capital in the second year as her compensation stock option ($90,000 in total), in order to solidify her control power (40 stocks, 33% of total 120 shares).

5.1.3 Venture Capital/Angel Money

When an entrepreneur is searching for a venture capital investor, a good source is *Pratt's Guide to Venture Capital Sources*, published by Venture Economics, as well as the VentureOne Web site at http://www.ventureone.com, in the U.S. case. Entrepreneurs can also seek referrals from accountants, lawyers, investment and commercial bankers, and business people who are knowledgeable about professional investors. Nowadays, many venture capitalists approach the university's intellectual property office or research contract office, as well as each research center directly, to find an embryonic technology and a start-up entrepreneur. Thus, it is not difficult to find venture capitalists.

However, you should be aware that there are some aggressive venture capitalists who will consider your firm as one product, i.e., buying it cheaper and selling it higher. They may not consider helping you to grow the firm. Thus, start-up entrepreneurs need to seek investors who

1. Are considering new financing proposals and can provide the required level of capital.
2. Are interested in companies at the particular stage of growth.
3. Understand and have a preference for investments in the particular industry (i.e., some are interested only in medical applications using MMI's piezoelectric devices).
4. Can provide good business advice and contacts in the business and financial community.
5. Are reputable, fair, and ethical and with whom the entrepreneur gets along.
6. Have successful track records of 10 years or more advising and building smaller companies [1].

5.1.4 Bank Loans

A bank loan is helpful for covering only a short-term financial deficit, because the loan allowance for a start-up company is very limited (typically $50,000 immediate allowance from a main bank with an annual interest rate of around 8% = prime rate + 1–2%).

5.2 Research Funds—How to Write a Successful Proposal

Unlike the entrepreneur who is starting up a restaurant or a retail shop, the high-tech entrepreneur has the privilege to apply for various R&D funds from federal and state governments.

5.2.1 Small Business Innovation Research (SBIR) Programs

One of the most important issues for a high-tech entrepreneur is eligibility approval by the Small Business Administration to apply for SBIR or STTR programs. (Review Chapter 4, Section 4.3.5: Government Approval.) There are strict detailed definitions for a small business, in addition to two simple regulations: (1) 500 or fewer employees, and (2) annual revenue under $5 million.

You need to be careful with item 3 in Section 4.3.5. The motivation for the high-tech small business foundation comes occasionally from a research contract from a large company. If your venture accepts some investment from a big company by sharing the company stocks (such as 10% of the capital stocks), the big company is considered an affiliate, leading to a number of employees (including the affiliate) higher than 500. This means that your small business will automatically lose SBIR eligibility. Because Cheng Kung has 400 employees, MMI introduced their investment. However, if their employee number increases to more than 500, Cheng Kung's investment to MMI should be reconsidered. One possible solution is to transfer Cheng Kung's stocks to another small business corporation or a venture capitalist. However, care must be taken not to change the corporate control power according to the stock transfer.

SBIR and STTR programs are similar, but the details are different. The SBIR program does not require inclusion of the academic institute in their collaboration, but the STTR program does. At least 30% of the research money should go to the academic institute, because of the program's aim (technology transfer from the university to the industry). Almost all federal agencies (Department of Defense, Department of Energy, Defense Advanced Research Projects Agency, Navy, Army, Air Force, National Institutes of Health, National Aeronautics and Space Administration (NASA), Homeland Security, and so forth) provide SBIR and STTR programs.

5.2.2 Successful Proposal Writing

5.2.2.1 Finding a Suitable Solicitation

The most popular Web site for searching SBIR/STTR programs is http://www.sba.gov/SBIR/indexsbir-sttr.html. The home page is shown in Figure 5.1. Program solicitations of major private industries can be found at http://www.ninesigma.com/solution-providers/innovation-alliance-program. Identify the most suitable program for your company's technology.

5.2.2.2 Writing a Successful Proposal

A proposal consists of the following 10 sections:

1. Cover page, including the submitter's information
2. Executive summary, including the most attractive picture or figure of your proposed product
3. Chapter 1: Background, including market trends, and how much your company can expect for the revenue based on this investment
4. Chapter 2: Literature survey, subsectioned into several of the previous studies, with a final summary table in terms of merit/demerit of each previous design

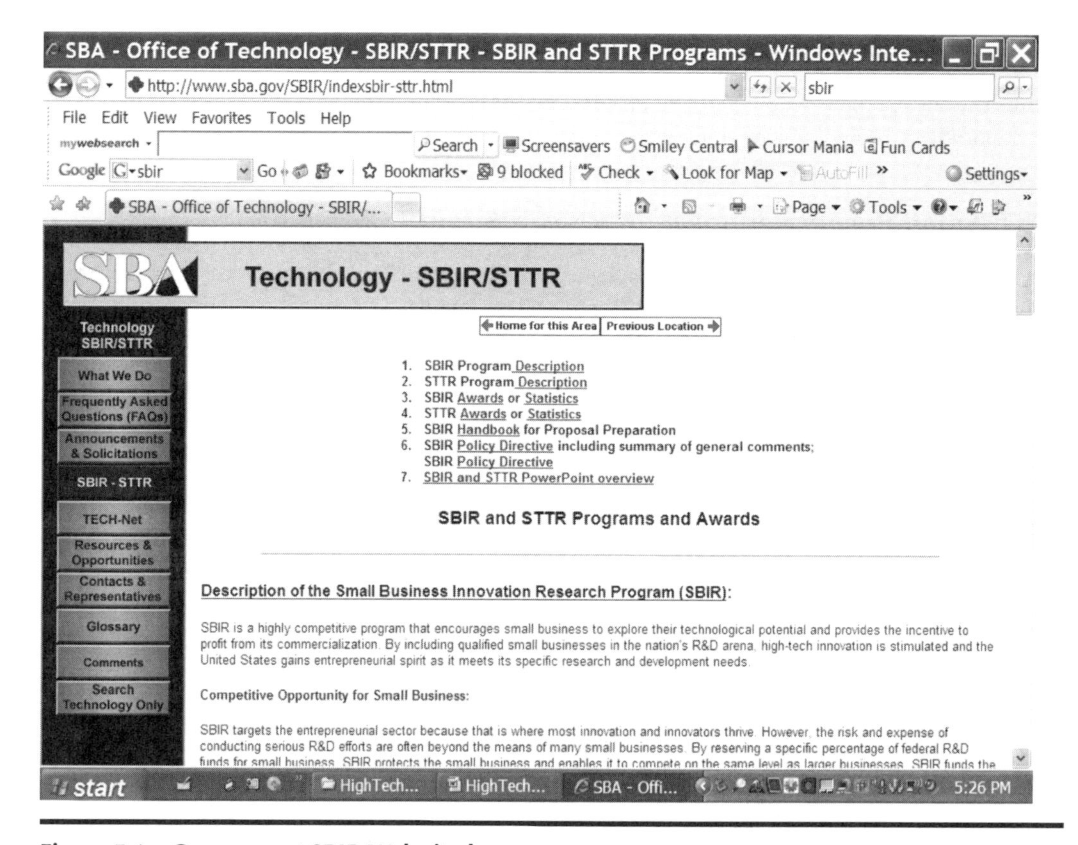

Figure 5.1 Government SBIR Web site home page.

5. Chapter 3: Your proposed design, explicitly mentioning your targeted specs, and how your design overcomes the previous problems
6. Chapter 4: Approach and milestones, consisting of a detailed discussion, including the time and money (related with the budget calculation)
7. Chapter 5: Others, which include patent application and future manufacturing plan
8. Chapter 6: References, principal investigator's capability (bio), and facility introduction
9. Chapter 7: Budget summary
10. Chapter 8: Appendix, including the support letter from the partners

5.2.2.2.1 Cover Page

MMI Example

This portion may be on a separate cover page:

Agent RFP #10667-1	Submission Date: October 1, 20XX
Title: Ultra-Miniature Actuator Development	
Agent Point of Contact:	Tom Raynolds, Eight Sigma Inc., raynolds@eightsigma.com

GENERAL INFORMATION

Title: Ultra-Miniature Actuator with Piezoelectrics Date: September 28, 20XX
Name of Primary Proposer: Dr. Barbara Shay
Company/Institution: Micro Motor, Inc.
Telephone: 814-XXX-XXXX, E-mail: barbs@mmi.com
Address: Street Address, College Park, PA 19837

5.2.2.2.2 Executive Summary

The executive summary should be included in the actual proposal, but in order to reduce redundancy in this chapter, this part is omitted here.

5.2.2.2.3 Background

This portion can be replaced by an introduction and market research discussion. In any case, describe the specific target here! If the proposal page allowance is long, include an attractive executive summary with the most attractive figure or picture of your proposed design.

Target Specifications. Eight Sigma is seeking a proposal for design and development of an ultra-miniature actuator. The target specifications include the following (refer to Figure 5.2):

- Size: within a cylindrical shape, OD 5 mm, ID 3 mm, and height 0.5 mm
- Displacement: 300 μm with suspension function at four arbitrary points
- Generative force: higher than 0.5 mN
- Power supply: max ±100 V and 2 W

5.2.2.2.4 Literature Survey

This portion is the literature survey, including patents.

Background—Previous Studies. Actuators made from piezoelectric ceramics are suitable for miniature types smaller than 10 mm due to their excellent electromechanical energy conversion rate and efficiency. Piezoelectric motors have demonstrated mechanical power density of 10 times higher than electromagnetic motors when the size is reduced, i.e., for a motor level less than 35 W. Efficiency on small piezoelectric motors is also higher, because the electromagnetic motors exhibit significant Joule loss due to the reduced thickness of the coil wire with reduced motor size.

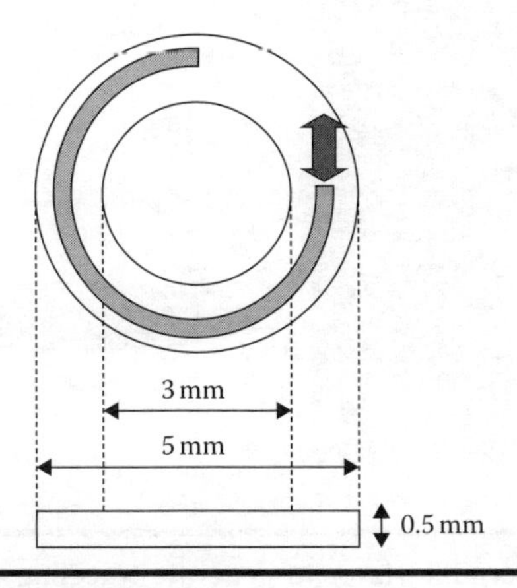

Figure 5.2 Targeted size of the ultra-miniature actuator.

The current proposal addresses two designs in order to satisfy the targeted specifications: a curvilinear ultrasonic motor (USM) and an impact drive mechanism.

5.2.2.2.4.1 Curvilinear Piezoelectric USMs — In collaboration with State University of Pennsylvania (SUP), MMI developed a curvilinear piezoelectric USM for rotating a ball. The principle is to generate a traveling wave on the surface of a curved piezoelectric ceramic strip, and to drive a cylinder or a ball put on this curved surface. The lead zirconate titanate (PZT) ceramic achieved the desired radius of curvature by replicating a prescribed form without further machining of the PZT ceramic. The dimensions for this segment are indicated in Figure 5.3a.

Standard USMs are activated in the d_{33} mode. This means that the poling and applied electric field directions are the same, so only one set of electrodes are necessary. However, our curvilinear actuators are activated in a d_{15} mode because of the use of a high electromechanical coupling factor k_{15} in perovskite piezoceramics. Figure 5.3b depicts the finite element method (FEM) simulation result, showing surface wave generation inside the curved PZT strip. The consequence of d_{15} mode of activation is that the ceramic must be poled along its arc length (electrodes placed on the ends). Once poling is completed, the end electrodes are removed, and new electrodes are placed at the top and bottom surfaces. This means that the electric field has to be applied through the thickness (top to bottom) so as to activate the d_{15} shear mode.

The actuators were successfully demonstrated, operated in shear mode to induce a controllable elliptical motion on the track ball.

5.2.2.2.4.2 Impact Drive Mechanism — A competitive technology is an impulse motor. Konica-Minolta developed a smooth impact drive mechanism (SIDM) for cellular phone camera modules using a multilayer piezo-element [2]. The idea comes from the "stick-and-slick" condition of the ring object attached on a drive rod in Figure 5.4a. By applying a saw-shaped voltage (40 to 50 V) to a multilayer (ML) actuator, alternating slow expansion and quick shrinkage are excited on a drive friction rod (see Figure 5.4b). A ring slider placed on the drive friction rod will stick on the rod due to friction during a slow expansion period, while it will slide during a quick shrinkage period, so that the slider moves from the bottom to the top. In order to obtain the opposite motion, the voltage saw shape is reversed. Compared to the USMs, the impulse motor is simpler in manufacturing, but the one-tenth smaller holding force may be a problem.

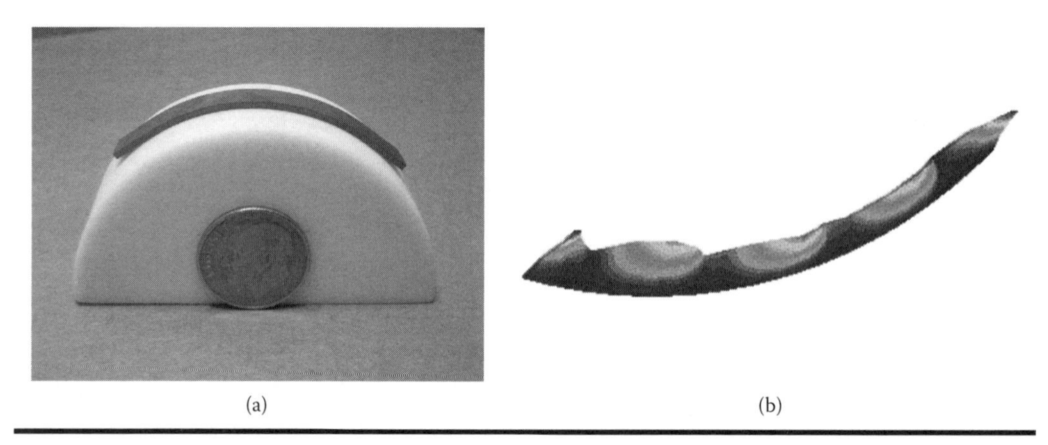

(a) (b)

Figure 5.3 (a) Curved PZT ceramic stator to generate traveling wave on the surface. (b) FEM simulation for vibration of the stator.

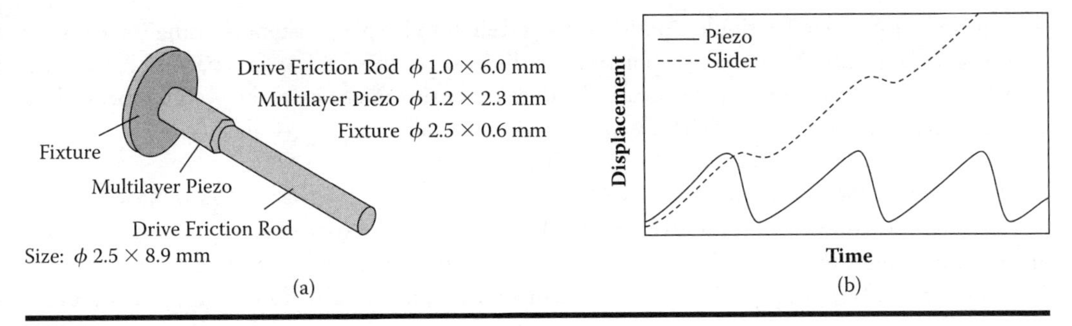

Figure 5.4 (a) Illustration of the SIDM structure composed of an ML actuator and a friction rod. (b) Displacement of the piezo-element and the slider. Notice its "stick-and-slick" motion.

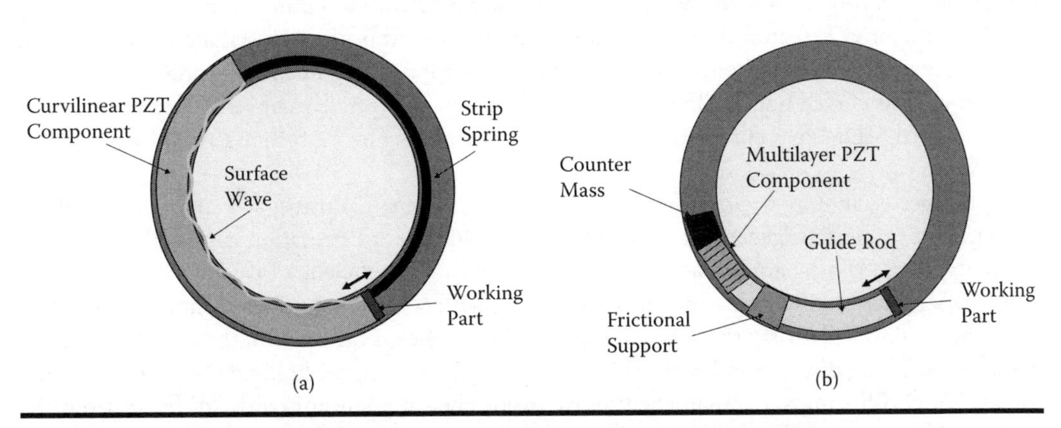

Figure 5.5 (a) Miniature curvilinear piezoelectric USM with an outer diameter of 5 mm. A curved PZT ceramic with the polarization direction parallel to the strip is used to excite the k_{15} mode for generating a traveling wave on the inner surface. (b) Miniature impact drive motor with a micromultilayer piezoelectric actuator. Balance between a countermass and a frictional support is key to this design. We will also adopt a guide rod to obtain a smooth movement in a curved hollow.

The previous discussion introduces only two literature citations, including MMI's contribution. If the page limit allows, I recommend you collect more than seven literature sources. At the end of this section, compare these seven groups' work in a table, and discuss the merits and demerits for deciding your design.

5.2.2.2.5 Your Proposed Design

Make your detailed discussion on your unique design proposal.

Technical Approaches. MMI will miniaturize the aforementioned two actuation mechanisms into the required spec dimensions, as schematically illustrated in Figure 5.5 a and b. The outer and inner diameters of this hollow cylinder are 5 and 3 mm with a height of 0.5 mm.

5.2.2.2.5.1 Curvilinear Piezoelectric USMs — We will fabricate curved PZT strips with an inner radius of 1.5 mm, a height of 0.5 mm, and a curvature angle from 90° to 180°. We will use a spring strip with a frictionless plastic such as Teflon to keep a constant pressing force on the traveling wave surface. Note that the polarization direction of this PZT strip is parallel to the strip length in order to use the k_{15} mode effectively. Because of this microscale, the poling process does not have an arcing problem, which we experienced in the previous large-scale motors. Figure 5.5a shows a miniature curvilinear piezoelectric USM.

The expected motor specifications are (1) speed = 10 mm/s; (2) thrust = 100 mN, which will be the same for suspension/generative force; (3) stroke is infinite (one rotation is possible), with any stopping point; (4) mechanical output power = 1 mW, with an efficiency of 10–15%; and (5) drive voltage = 20–30 V_{rms} at 200–400 kHz (resonance frequency) with consuming power around 10–20 mW.

5.2.2.2.5.2 Impact Drive Mechanism

—We will shrink the total size to roughly half of the Konica-Minolta original design. We utilize the ML actuator with 0.5 mm × 1 mm × 1.5 mm long as an impulse drive actuator. A counter mass is put at one end of the ML actuator, and a friction slider is put at the other end of the actuator. We will adopt a long guide rod in order to smoothly move inside this curved hollow. The challenge will be how to effectively convert a linear movement of the original SIDM to a curved movement in our design. Figure 5.5b shows a miniature impact drive motor.

The expected motor specifications are (1) speed = 2 mm/s; (2) thrust = 1 mN, which will be the same for suspension/generative force; (3) stroke is infinite (one rotation is possible), with any stopping point; (4) mechanical output power = 0.01 mW, with efficiency 1%; and (5) drive voltage = 20–30 V_{rms} at 100–200 kHz (resonance frequency) with consuming power around 10–20 mW.

Either design will satisfy the targeted specifications. Like an earthworm, the curvilinear USM will move any long distance inside the tube in Figure 5.2, keeping the holding thrust much higher than 0.5 mN. Our impact actuator will move inside the tube intermittently (different from the USM), still keeping sufficient holding thrust, which is lower than the camera module spec. The merits and demerits of these two designs are summarized in Table 5.1.

5.2.2.2.6 Approach

The required development tasks will include the following:

Brainstorming: *Technical discussion with the final customer and specific design selection* to confirm the motor specifications, applications, and, most importantly, the four arbitrary stop points required, which determines MMI's specific approach.

Table 5.1 Comparison among a Curvilinear Ultrasonic Motor and an Impulse Drive Motor

	Curvilinear Ultrasonic Motor	*Impulse Drive Motor*
Merits	High thrust	Simple drive circuit (pulse)
	No audible noise	Simple design, light weight
Demerits	Complex high-frequency drive circuit	Low thrust
		Intermittent drive noise (audible)

Task 1: *FEM computer simulation* for optimizing the piezo-ceramic and PZT curved strip sizes for the specific design selected after the brainstorming.
Task 2: *Motor prototypes manufacturing*, after finalizing the structure of the stator. We will measure the magnitude of the displacement on the stator surface and design the rotor so that it will touch at an optimum point on the stator surface. We will also construct the pulse drive mechanism using the ML actuator. Packaging the stator in a thin hollow cylinder will be challenging.
Task 3: *Motor experimental characterization* is a challenge because the size of the prototype is small, and characterizing such a small motor without disturbing the rotation is difficult. We will adopt the transient response method to obtain speed vs. load characteristics of the motor.

The proposed development is limited to the prototype of the piezoelectric ultra-miniature actuator and does not include the design of a specific electronic driving circuit. MMI will use lab amplifiers to operate the motor in resonance mode for a curvilinear motor and in a pulse mode for an impact drive motor, during the characterization. Future driving circuit integration is open to further discussions with the final customer.

5.2.2.2.7 Milestone

Milestone and Cost Estimation. The 1-year period R&D milestone is illustrated below. MMI will request $100,000 for this 1-year development. MMI will use a full set of piezo-motor characterization facilities equipped at SUP.

Work Plan/Milestones Overview	*Timeline Schedule (Est.)*					
Month	*1/2*	*3/4*	*5/6*	*7/8*	*9/10*	*11/12*
Continuous brainstorming on motor specifications (MMI/client)						
TASK 1. FEM computer simulation						
TASK 2. Prototype manufacturing						
TASK 3. Motor experimental characterization						

5.2.2.2.8 Others

If this proposal is for SBIR/STTR, you should choose a development partner from a university or national institute. Even for other programs, you may choose a partner who can cover the development, which your company cannot do by itself. In this scenario, you should consider the budget distribution between your company and a partner company.

Intellectual Properties and Business Plan. A patent (U.S. number omitted) of the shear-type linear USM (in the present proposal, curvilinear motors are covered) was issued in 20XX, under the submitter's name through SUP. Thus, MMI can license from SUP. On the contrary, the impact drive type is patented by Konica-Minolta. We need to negotiate the cross-license with Konica-Minolta, by submitting a new patent on our curved, inner tube moving-type mechanism.

Prototypes will be prepared at MMI partially using SUP's facilities, and tested by the client. Once passed for the preliminary test, small-scale production (100–1000 pieces) will begin at the

MMI facility. Finally, large quantity production (more than 1 million pieces) will be shifted to the Cheng Kung factory in Taiwan.

5.2.2.2.9 References

MMI Example

Technical and Commercial Capabilities

Lead Dr. Barbara Shay, active researcher in piezoelectric actuators, is vice president and CTO of MMI, College Park, Pennsylvania. She is also an associate professor of electrical engineering (EE) at SUP. After being awarded her PhD degree from SUP, she became a research associate in the EE Department at the university. Her research interests are in development of piezoelectric actuators and USMs. She has authored 80 papers, 2 books, and 3 patents in the ceramic actuator area. She is a U.S. citizen.

Selected Publications (from a total of 80; 3 patents, 1 pending)

Shay, B. 2007. Ultrasonic motors. *Journal of Precision Engineering* XX(3):00–000.

Shay, B. 2007. Ceramic actuators: Principles and applications. *Materials Research Society Bulletin* XX(4):00–000.

Shay, B. 2008. *Piezoelectric actuators and ultrasonic motors.* New York: Academic Press.

MMI (College Park, Pennsylvania), an outgrowth of SUP founded in 2004, is now considered one of the world's foremost piezo-motor design centers. The MMI staff (including B. Shay) has been directly responsible for many of the innovations in piezoelectric USMs over the past decade. One of their products, the world's smallest metal tube motor, was awarded the Smart Product Implementation Award from SPIE in 20XX. MMI is also a main vendor of piezo-motor technology and products in the United States imported from worldwide manufacturers. MMI has a Taiwanese industrial partner, Cheng Kung Corporation, which helps in the mass production of piezo-motors at the cheapest cost.

If the proposal includes a research partner (subcontract), attach a letter of support from that institute.

Dear Dr. Shay,

Electrical engineering at State University of Pennsylvania is the world leader for development and testing of piezoelectric devices and motors. We are pleased to collaborate in the proposed Phase I DARPA STTR program "Ultra-Miniature Actuator Development." I have read your proposal and believe the proposed work is technically sound and commercially useful. I certify that we will be available to work on this program during the Phase I time frame.

Based on our discussions, I estimate that the proposed work will require $30,000 for the 9-month program.

Thank you for recognizing our superior developing and characterization capabilities of piezoelectric devices and motors. We hope the proposal is successful, and the collaboration will continue on this interesting topic.

Sincerely,

David Robinson, PhD
Professor, Electrical Engineering

5.2.3 Successful Proposal Presentation

The keys to a successful presentation are introduced in this section.

5.2.3.1 Structure Presentation Visuals

1. There are two types of visuals: linking and supporting. Linking slides make the trunk, and supporting slides make the branches. Figure 5.6 presents an example structure.
2. Spend an average of 15 seconds for one linking slide, and 2–3 minutes for one supporting (main content) slide. Thus, a 10-minute presentation = 7 slides, a 15-minute presentation = 10–12 slides, a 30-minute presentation = 25–30 slides.
3. Separate completely your proposed design from the other persons' studies. Once your proposal starts, do not discuss the previous study's designs.

5.2.3.2 Words–Visual Suggestions

1. Use high contrast between the background and characters: dark characters on a light-color background or light-color characters on a dark background.
2. Use the bold version of the font with a minimum font size of 20 points, preferably 24 points. The title should be 30–40 points.
3. Do not use long sentences; use a bullet-point writing style.

5.2.3.3 Figures–Visual Suggestions

1. Ensure that the resolution of the pictures is high enough, and the type size used for the axis labels is large enough.

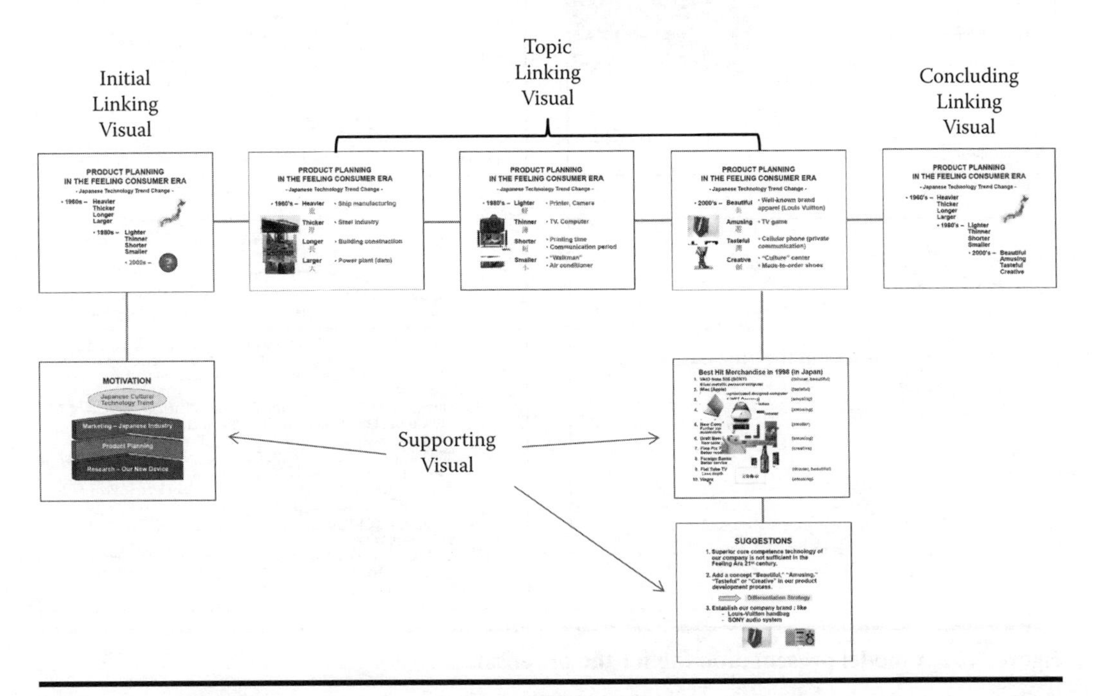

Figure 5.6 An example structure of a visual presentation.

2. Cited pictures and tables require citation references. No reference citation indicates that the data is your own material.

Figure 5.7 shows an example presentation file with key comments to remember.

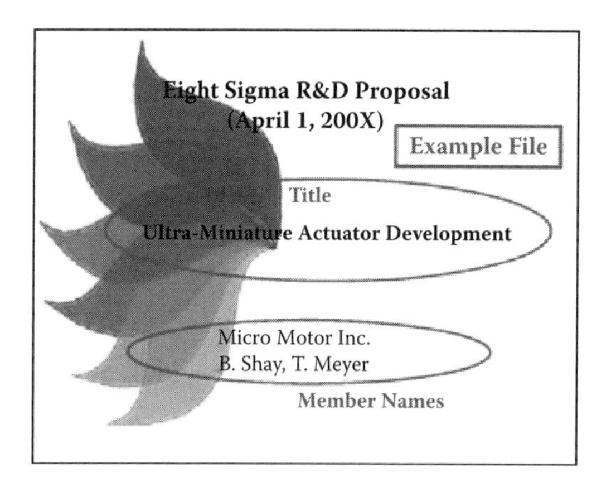

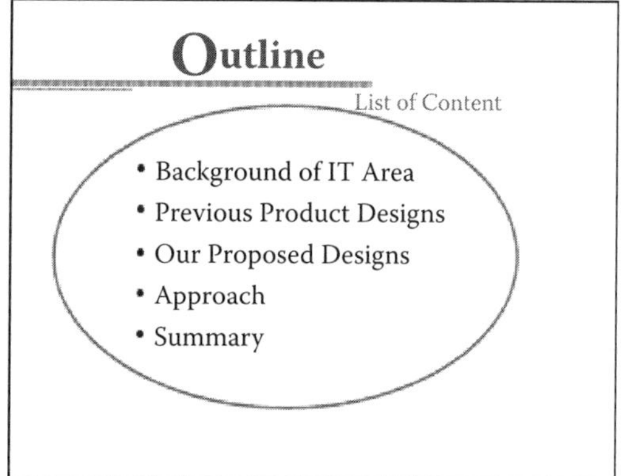

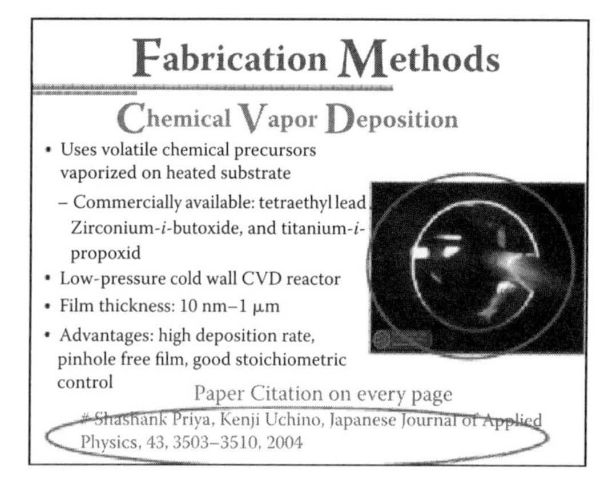

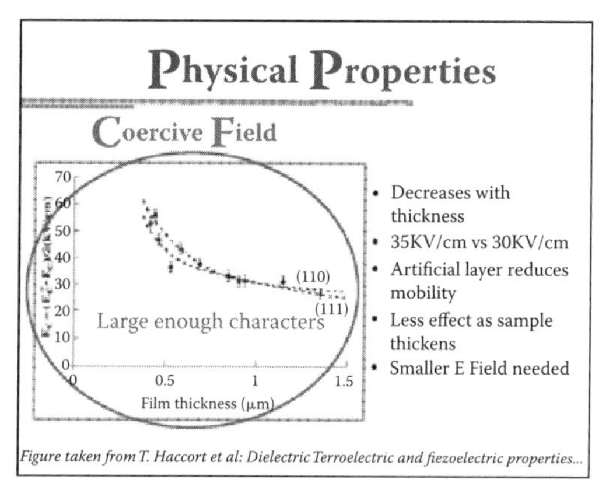

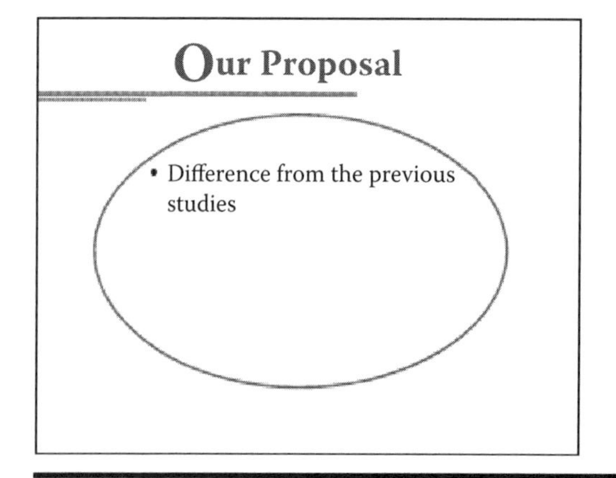

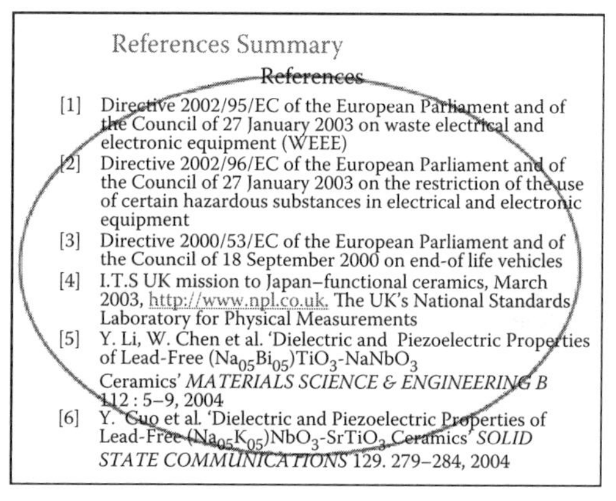

Figure 5.7 A model presentation file for the proposal.

Chapter Summary

5.1 Stock or loan

- *Stock:* You do not need to pay back the initial investment amount, but you need to pay the stock dividend with the effective annual return rate around 10–12%, higher than the interest rate of the loan. However, the dividend distribution may not start in the first 5–7 years after start up.
- Loan: You must pay back the total amount of the principal and the interest by the end of the loan period. The interest rate is usually lower than the stock return rate, typically 5–10%. A short-term loan (less than 1 year) has a lower interest rate than a long-term loan (3–5 years).

5.2 SBIR/STTR program application eligibility requires

- 500 or fewer employees
- Annual revenue under $5 million
- At least 51% owned and controlled by one or more individuals who are citizens or permanent resident aliens in the United States

5.3 Proposal standard format:

1. Cover page
2. Executive summary
3. Chapter 1: Background
4. Chapter 2: Literature Survey
5. Chapter 3: Proposed Design
6. Chapter 4: Approach and Milestones
7. Chapter 5: Others
8. Chapter 6: References, Bio, and Facility Introduction
9. Chapter 7: Budget Summary
10. Chapter 8: Appendix

Practical Exercise Problems

P5.1 Proposal Writing

Write a proposal for your targeted product or invention for a particular application. The proposal should not exceed 10 pages in total, including a one-page budget sheet (last page).

P5.2 Project Report Modification Practice

A researcher wrote a quarterly summary report of an R&D project to the fund agency:

> We theoretically explored the loss difference between the resonance and antiresonance modes in piezoelectric materials. Analytical solution proved the deviation of the resonance quality factor Q_A from the antiresonance quality factor Q_B in the k_{31} vibration mode. At the same time, we developed a simple method to determine the piezoelectric loss factor tanθ' through admittance/impedance spectrum analysis.

There seems to be nothing wrong in this summary to describe the research content. This is almost perfect for an abstract for an academic journal manuscript. However, you need to understand the difference between the academic journal paper and the R&D proposal or report for the sponsor. Readers of an academic journal are usually experts in that field with a strong knowledge of the subject matter. They may need only the final new discovery/conclusion and the detailed analytical process or experimental procedure. On the contrary, the readers of your R&D report are typically outside your technological area, such as a general manager in your company and a program officer of a federal agency. The important issue is expressing how your result will impact the sponsor's purpose. Taking this into account, please modify your previous report first, before seeing the modification example below.

Modification Example

We theoretically explored the loss difference between the resonance and antiresonance modes in piezoelectric materials. The analytical solution proved the deviation of the resonance quality factor Q_A from the antiresonance quality factor Q_B in the k_{31} vibration mode, which was an excellent support for our previous suggestion to use the antiresonance drive as an alternative for the piezoelectric motor, transducer, transformer, etc. This is also complementary for IEEE Standard, in which Q_A is assumed equal to Q_B. The most suggestive conclusion is the importance of "piezoelectric loss," which has been neglected by most of the previous researchers.

At the same time, we developed a simple, easy and user-friendly method to determine the piezoelectric loss factor $\tan\theta'$ through admittance/impedance spectrum analysis. Our method is based on the importance of "piezoelectric loss" and will change the FEM computer simulation algorithm drastically in the Navy community. This program is supported by Office of Naval Research.

Notice that the above modifications are primarily made to provide further information on how our result is important and what merit will be delivered to the sponsor, the Office of Naval Research.

P5.3 Presentation File Preparation

Prepare a PowerPoint presentation file for the proposal explanation (Section 5.1).

References

1. Sapienza, H. A., and J. A. Timmons. 1989. Launching and building entrepreneurial companies. *Proceedings of the Bobson Entrepreneurship Research Conference*, May 1989, Babson Park, MA.
2. Okamoto, Y., R. Yoshida, and H. Sueyoshi. 2004. Konica Minolta Tech. Report, 1, p. 23.

Chapter 6

Corporate Operation—Survival Skills in Accounting and Financial Management

Even though the firm size is small for the high-tech start-up, accounting and financial management are important from both tax return and company-survival viewpoints. Annual tax return forms must legally be submitted to the Internal Revenue Service (IRS). This chapter summarizes what an engineering corporate executive needs to learn in order to operate a corporation.

The high-tech firm sells the products to generate revenue, purchases and pays for necessary goods from other companies, pays employees' salaries, and finally creates profit for future growth. The most important equation

$$\text{Profit} = \text{Sales revenue} - \text{Cost} \tag{6.1}$$

suggests that the company should maximize product sales, and/or minimize the cost, in order to increase the firm's profit. Taking into account

$$\text{Sales revenue} = \text{Unit price} \times \text{Number of products sold} \tag{6.2}$$

the company should increase the number of sales (through marketing and promotion) or set the unit sales price to obtain the necessary profit. The concept *break-even point* comes from the zero profit in Equation 6.1. In the fiscal year end, as a summary of daily accounting items, the company needs to prepare three basic types of financial statements: *income statement*, *balance sheet*, and *cash-flow statement*. Because this is a legal requirement, the company executive needs to learn these as a survival skill. You will also need to learn how to reduce corporate tax, and how to improve cash flow.

Once the company gradually expands, financial decisions must be made in various situations: Shall we invest to set up a new manufacturing line? Shall we issue new stocks or acquire a bank loan to collect necessary money for the company growth? We will consider these financial issues in the latter part of this chapter.

From an engineer's viewpoint, accounting and finance are rather similar—both work with money—but their differences found in the mathematical formula seem to lie in the concept "time is money"; that is, the *time value of money* (present value of future $1 and future value of present $1). Accounting treats money in the short term, while financial management treats money in the long term.

6.1 Accounting Management—Sales and Payroll

Most small firms use a professional accounting office for official accounting documents such as a corporate tax return form. However, it is too costly to enlist the services of a professional accounting office for daily bookkeeping, and thus it is recommended that this task be conducted in your firm, by you or an office manager. There are many computer software programs for small business accounting: Quickbooks, Microsoft Money Plus, BusinessVision32, and so forth, most of which are based on Microsoft Excel functions. Figure 6.1 shows an introductory page of Quickbooks. Be sure to use the same accounting software as the one your accounting office uses, or follow the recommendation by the accounting office. We will start with the bookkeeping basics.

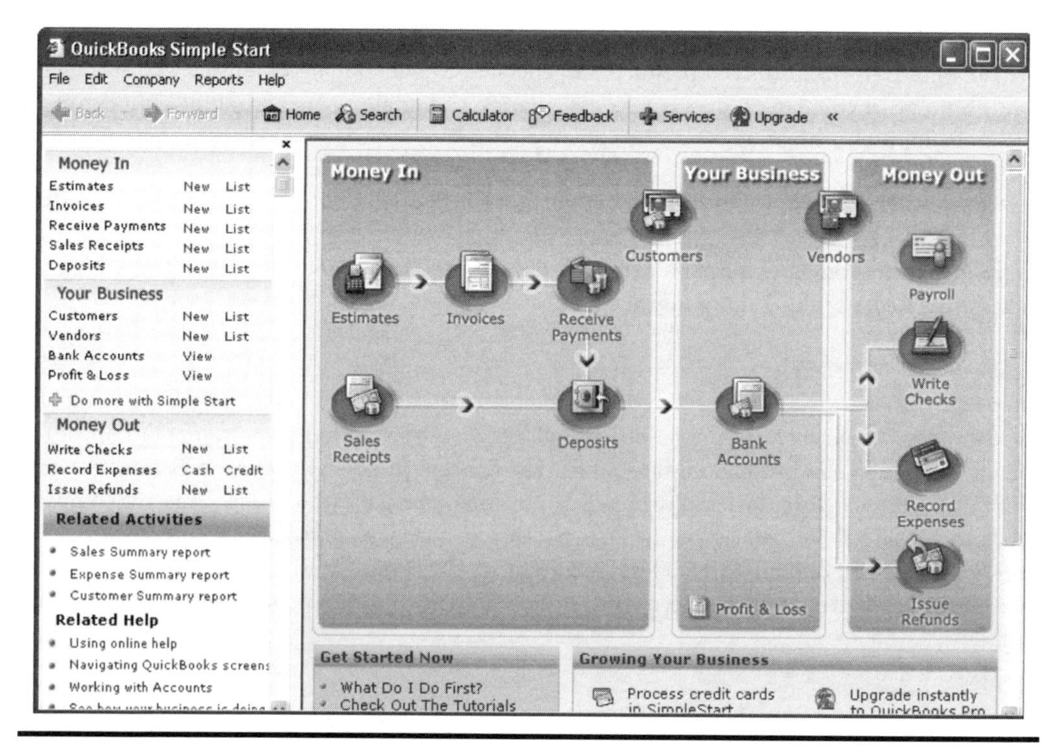

Figure 6.1 Introductory page of Quickbooks.

6.1.1 Daily Accounting

6.1.1.1 Product Costs and Period Costs

Both product and period costs should be considered "costs" in accounting [1]. A *product cost* (or inventoriable cost) is a cost assigned to goods that were either purchased or manufactured for resale. The product cost is used to value the inventory of manufactured goods or merchandise until the goods are sold. In the period of the sale, the product costs are recognized as an expense called *cost of goods sold*. The cost of goods sold is equal to the product cost (identified with the goods "sold").

The product cost of manufactured inventory includes all of the costs incurred during its manufacture. The product cost for a retailer (e.g., the sales division of Micro Motor Inc. [MMI]) is equal to the cost of purchase plus the cost of transportation. The product cost for a manufacturer (e.g., the research and development [R&D] department of MMI) is equal to the *manufacturing cost*.

Manufacturing cost includes the following:

1. *Direct material:* Raw material that is consumed during the manufacturing process
2. *Direct labor:* The cost of salaries, wages, and fringe benefits for personnel who work directly on the manufactured products
3. *Manufacturing overhead:* All other costs of manufacturing; indirect material, indirect labor (salaries of production department supervisors, etc.), and other manufacturing costs. Other manufacturing costs include:
   a. Depreciation of plant/equipment
   b. Property taxes, insurance
   c. Utilities (electricity, water, etc.)
   d. Service/support departments (equipment maintenance, etc.)

All costs that are not product costs are called *period costs*. Period costs are recognized as expenses during the time period in which they are incurred. General R&D, selling (salaries, commissions, and travel costs of sales personnel, and the costs of advertising and promotion), and administrative costs (salaries of corporate executives, costs of the accounting, legal, and public relation activities) are treated as period costs.

6.1.1.2 Recording Journal Entries in the Ledger

Now let us consider the journal entry process into the ledger using an example case from MMI. MMI received a purchase order of 20 pieces of ultrasonic metal tube motors (refer to Figure P1.1), and planned to manufacture 25 pieces (5 extra pieces are made in case there are deficits; these may be sold in the future), starting from the raw materials (lead zirconate titanate [PZT] ceramic plates and metal tubes). The quote was issued for $14,000 in total (20 pieces at $700 per unit). The starting date was June 1, 20XX, and the due date was June 30 (i.e., 30 days). Ledgers typically contain three successive steps in product manufacturing:

- Work-in-process inventory
- Finished-goods inventory
- Cost of goods sold

We will consider *debit* and *credit* relations for each inventory/process (this is called *double-entry bookkeeping*).

6.1.1.2.1 Purchase of Material

MMI purchased raw materials (PZT ceramic plates and metal tubes) from an outside supplier on June 1. The purchase is recorded with the following journal entry: raw-material inventory has PZT and metal materials (debit), and MMI owes $2500 to the supplier (credit).

Journal Entry 1	*Debit*	*Credit*
Raw-material inventory	2500	
Accounts payable		2500

6.1.1.2.2 Use of Direct Material

On June 1, 20XX, the material requisitions were submitted, and the motor assembly started.

Journal Entry 2	*Debit*	*Credit*
Work-in-process inventory	2000	
Raw-material inventory		2000

Note that direct materials cost is proportional to the number of products manufactured.

$$\begin{aligned}\text{Direct material cost} &= \text{Unit materials cost} \times \text{Number of products manufactured}\\ &= \$80 \times 25 \text{ pieces} = \$2000 \qquad (6.3)\end{aligned}$$

Extra materials were purchased in case of manufacturing loss or waste. If the materials can be used in the next manufacturing lot, the raw materials can remain in the raw-material inventory.

6.1.1.2.3 Use of Direct and Indirect Labor

At the end of June, the cost-accounting department uses the labor time records filed during the month to determine the direct-labor costs. MMI spent 50% of the time of one engineer ($25/h × 100 h/month at monthly salary of $5000) for the motor manufacturing. The direct labor should also include fringe benefits such as health insurance and pension contributions from MMI, which are about 20% of the salary.

$$\text{Direct labor} = \$5000 \times 50\% \times (1 + 0.20) = \$3000$$

Note that direct labor cost is proportional to the amount of working time.

$$\begin{aligned}\text{Direct labor cost} &= \text{Unit labor cost and fringe (per hour)} \times \text{Number of working time (hours)}\\ &= \$25 \times 1.20 \times 100 \text{ hours} = \$3000 \qquad (6.4)\end{aligned}$$

The production supervisor's salary (the R&D division director in MMI), which is not charged to any particular job, is *indirect labor*. Other examples would include custodial employees and security guards, but these will not be relevant to a small enterprise. Since the director takes care of five projects, he or she might spend 20% of their time for this task (monthly salary of $10,000 with 20% of fringe benefit).

$$\text{Indirect labor} = \$10{,}000 \times 20\% \times (1 + 0.20) = \$2400$$

Journal Entry 3	*Debit*	*Credit*
Work-in-process inventory	3000	
Manufacturing overhead	2400	
Wages payable		5400

6.1.1.2.4 Application of Manufacturing Overhead

Manufacturing overhead includes the following general and administrative (G&A) fixed costs:

- Rent on factory building—$1500 per month
- Depreciation on equipment—$1000 per month
- Utilities (electricity, gas, water, etc.)—$300 per month
- Property taxes—$200 per month
- Insurance—$300 per month (= $3600/12 months)
  - Total—$3300

Since MMI is running five projects, manufacturing overhead for this project can be considered $3300/5 = $660. The manufacturing overhead costs are not traceable to any particular job.

Journal Entry 4	*Debit*	*Credit*
Manufacturing overhead	660	
Prepaid rent		300
Accumulated depreciation—equipment		200
Prepaid utilities		60
Prepaid property taxes		40
Prepaid insurance		60

Various manufacturing overhead costs were incurred during June, and these costs were accumulated by debiting the manufacturing overhead account. However, in order to issue a *quote* to the customer before starting the motor manufacture, the above procedure is not very convenient. Because we do not know how much the actual manufacturing overhead will be, MMI uses a so-called *predetermined overhead rate*. Costs incurred for the above example are as follows:

- Direct materials and labor costs = $2000 + $3000 = $5000
- Manufacturing overhead (indirect labor and G&A) costs = $2400 + $660 = $3060
- Overhead costs/direct costs = $3060/$5000 = 61.2%

MMI uses the predetermined overhead rate of 69.5% averaged over a couple of years, which is slightly larger than the above particular example. Refer to MMI's Sales Calculator Software introduced in Chapter 4, Section 4.4.5.

6.1.1.2.5 Selling and Administrative Costs

During June, MMI incurred selling and administrative costs for this particular project as follows:

- Salaries of sales personnel: 5 h at \$12/h = \$60
- Salaries of management: 2 h at \$60/h = \$120
- Promotion and advertising: \$100
- Office supplies used: \$20

Note that selling and G&A costs are *period costs*, not product costs. Thus, the period costs are not included in cost of goods manufactured or sold. Also note that these costs seem to be fixed costs (regardless of the production amount, but just for one project) in MMI's case.

Journal Entry 5	*Debit*	*Credit*
Selling and G&A expenses	300	
Wages payable		180
Accounts payable (advertisement)		100
Office supplies inventory		20

6.1.1.2.6 Completion of Production

The motor manufacture (25 pieces) was completed in June. The following journal entry records the transfer of these job costs from work-in-process inventory to finished-goods inventory. Note that we use the predetermined overhead rate (rather than the actual manufacturing overhead) in this stage (direct cost \$5000 × (1 + 0.695) = \$8475). Also, the selling and administrative costs are not included in these job costs.

Journal Entry 6	*Debit*	*Credit*
Finished goods inventory	8475	
Work-in-process inventory		8475

Therefore, the actual manufacturing cost per unit should be calculated, with the actual manufacturing overhead, as (\$5000 + \$3060)/25 pieces = (\$8475 − \$415)/25 pieces = \$322.40. This number should be used for cost of goods sold in the next section.

6.1.1.2.7 Sale of Goods

Twenty motors were sold for \$700 each at the end of June. The cost of each unit sold was \$322.40, as calculated above. The following journal entries were made.

Journal Entry 7	*Debit*	*Credit*
Accounts receivable	14,000	
Sales revenue (for 20 motors)		14,000

Journal Entry 8	*Debit*	*Credit*
Cost of goods sold (for 20 motors)	6448	
Finished-goods inventory		6448

The difference of the amount between finished-goods inventory in journal entry 6, \$8475, and in journal entry 8, \$6448, is the remainder (\$1612 = \$322.4 unit cost × 5 motors) in the finished-goods inventory on June 30, and the *overapplied overhead* is \$415. The cost of goods sold here has already been adjusted by the overapplied overhead.

6.1.1.3 Accounting Schedules and Income Statements

Now, we summarize the accounting process in three Excel spreadsheets. The *schedule of cost of goods manufactured* lists the manufacturing costs applied to works in process (Table 6.1). A total of \$500 remains in raw material inventory on June 30. Note that the overhead applied (predetermined overhead rate with 69.5% in MMI) is used for works in process, rather than the actual

Table 6.1 Schedule of Cost of Goods Manufactured

MMI-MotorManufactur... M

Home | Insert | Page L | Formul | Data | Review | View | Add-In

H23 | fx =SUM(H6:H21)

	A–F	G	H
1	Micro Motor Inc.		
2	**Schedule of Cost of Goods Manufactured (Piezo Ultrasonic Motors)**		
3	For the Month of June, 200X		
4			
5	**Direct Material:**		
6	Raw-material inventory, June 1	\$ -	
7	Add: June purchases of raw material	\$ 2,500	
8	Raw material available for use	\$ 2,500	
9	Deduct: Raw-material inventory, June 30	\$ 500	
10	Raw material used		\$ 2,000
11			
12	**Direct Labor**		\$ 3,000
13			
14	**Manufacturing Overhead**		
15	Indirect labor	\$ 2,400	
16	Rent on factory building	\$ 400	
17	Depreciation on equipment	\$ 200	
18	Insurance	\$ 60	
19	Total actual manufacturing overhead	\$ 3,060	
20	Add: Overapplied overhead*	\$ 415	
21	Overhead applied to work-in-process*		\$ 3,475
22			
23	Cost of Goods Manufactured		\$ 8,475
24			
25	* Applied manufacturing overhead = Direct material and labor costs x 69.5%		
26	= \$ 3,475		

COG-Manuf | COG-Sold | IncomeSt

Ready | 100%

Table 6.2 Schedule of Cost of Goods Sold

MMI-MotorManufacturing [C... M–

Home | Insert | Page Layout | Formulas | Data | Review | View | Add-Ins

G11 | fx =SUM(G9,-G10)

	A	B	C	D	E	F	G	H
1	**Micro Motor Inc.**							
2	**Schedule of Cost of Goods Sold (Piezo Ultrasonic Motors)**							
3	**For the Month of June, 200X**							
4								
5		Finished-Goods Inventory, June 1					0	
6	Add:	Cost of Goods Manufactured*					$ 8,475	
7		Cost of Goods Available for Sale					$ 8,475	
8	Deduct:	Finished-Goods Inventory, June 30					$ 1,612	
9		Cost of Goods Sold					$ 6,863	
10	Deduct:	Overapplied Overhead**					$ 415	
11	**Cost of Goods Sold (adjusted for overapplied overhead)**						$ 6,448	
12								
13								
14	* From the Schedule of Cost of Goods Manufactured in Table 6.1							
15								
16	**The company closes underapplied or overapplied overhead into Cost of Goods Sold.							
17	Hence, the $415 balance in overapplied overhead is deducted from the Cost of Goods							
18	Sold for the month.							
19								

COG-Manuf | **COG-Sold** | IncomeStateme

Ready | 120%

overhead. Therefore, we added the overapplied overhead on the cost of goods manufactured, as shown in the last line of the schedule, totaling $8475. This is the amount transferred from work-in-process inventory to finished-goods inventory during June, as recorded in journal entry 6.

A schedule of cost of goods sold is displayed in Table 6.2. This schedule shows the June cost of goods sold and details the changes in finished-goods inventory during the month. The final number in finished-goods inventory is based on the actual manufacturing overhead. Because the cost of goods manufactured is estimated on the basis of the predetermined overhead rate, the overapplied overhead should be subtracted to obtain the *adjusted cost of goods sold*. The Excel spreadsheet in Table 6.3 displays the company's June income statement. As the income statement exhibits, income before taxes is $7252, from which income tax expenses of $2901 (at a typical income tax rate of 40%) is subtracted, yielding *after-tax net income* of $4351.

Example Problem 6.1

If MMI made a quote for the selling price of each motor as follows, what would the net profit be?

a. 20 motors by $600 each
b. 20 motors by $400 each
c. 20 motors by $300 each

Table 6.3 June Income Statement (MMI, Piezo-Motors)

MMI-MotorManufact... M

Home | Insert | Page L | Formu | Data | Review | View | Add-In

G11 | f_x =SUM(G9,-G10)

	A	B	C	D	E	F	G
1	**Micro Motor Inc.**						
2	**Income Statement (Piezo Ultrasonic Motors)**						
3	**For the Month of June, 200X**						
4							
5		Sales Revenue (20 units @ $700)					$ 14,000
6	Less:	Cost of Goods Sold*					$ 6,448
7		Gross Margin					$ 7,552
8	Less:	Selling and Administrative Expenses					$ 300
9		**Income Before Taxes**					$ 7,252
10	Deduct:	Income Tax Expenses (@ 40%)					$ 2,901
11		**After-Tax Net Income**					$ 4,351
12							
13	* From the Schedule of Cost of Goods Sold in Table 6.2						
14							

IncomeStatement | BE Analysis

Ready 120%

Solution

Using the Excel Spreadsheet template in Table 6.3, by changing box G5 to 12,000, 8000, and 6000, respectively, we obtain the net profit of (a) $3151, (b) $751, and (c) $748 (income before tax is already –$748). By equating the Sales revenue = Cost of goods sold + Selling and administrative expenses = $6448 + $300 = $6748, we can obtain the minimum motor price (for zero profit) of $337.40. If MMI sells the remaining motors in the finished goods inventory, the company can still expect additional profit later.

6.1.2 *Financial Statements*

A firm needs to generate their annual financial report by law, which includes the *income statement, balance sheet,* and *cash-flow statement.* These financial statements should be prepared according to *generally accepted accounting principles* (GAAP). MMI usually hires an accounting professional such as a Certified Public Accountant (CPA) for this preparation. Because MMI is a small firm without common stocks, we use here their partner company, Saito Industries, as an example to consider the more general aspects in financial reports.

6.1.2.1 *Income Statements*

The income statement represents the profitability of a business over a period of time. This is an extension of Table 6.3 (income statement for one project) to a 1-year period. Table 6.4 shows an

Table 6.4 Income Statement for Saito Industries, Year Ending December 31, 20XX

	(In Millions)
Sales	\$100.0
Less: Cost of goods sold	\$59.0
Gross margin	\$41.0
Less: Selling and administrative expenses	\$15.0
Less: Depreciation (building etc.)	\$12.0
Earnings before interests and taxes (EBIT)	\$14.0
Less: Interest charges	\$0.5
Earnings before taxes (EBT)	\$13.5
Less: Income tax expenses (at 40%)	\$5.4
After-tax net income (EAT)	\$8.1
Preferred stock dividends	\$1.1
Earnings available to common stockholders	\$7.0
Shares outstanding	5 million
Earnings per share	\$1.4

income statement for Saito Industries in the year 20XX. Remember the necessary terminologies and the formula:

$$\text{Gross margin (Gross income)} = \text{Sales} - \text{Cost of goods sold} \tag{6.5}$$

$$\begin{array}{c}\text{Earnings before interests and taxes (EBIT)} \\ \text{(Operating income)}\end{array} = \begin{array}{c}\text{Gross margin} - \text{(Selling and administrative} \\ \text{expenses + Depreciation)}\end{array} \tag{6.6}$$

$$\text{Earnings before taxes (EBT)} = \text{EBIT} - \text{Interests} \tag{6.7}$$

$$\text{Earnings after taxes (EAT) (Net income)} = \text{EBT} - \text{Taxes} \tag{6.8}$$

Depreciation is indicative of the potential for tax reduction arising from general wear and tear or the obsolescence of equipment or property (building, etc.). This situation will be further investigated in Section 6.1.4. *Net income* (or net loss) represents the net profitability of the firm. This is commonly referred to as its *bottom line*. Note also that EAT is what really counts (net income) rather than what is earned (operating income = EBIT).

6.1.2.2 Balance Sheets

The balance sheet is a snapshot of a business at a particular point in time, as shown in Table 6.5. It reveals financial resources the company owns (*assets*), debts it owes to the others (*liabilities*),

Table 6.5 Balance Sheet for Saito Industries as of December 31, 20XX

	(In Millions)
Assets	
Current Assets	
Cash	$15.5
Marketable securities	$3.0
Accounts and notes receivable	$4.0
Inventories	$16.5
Total current assets	$39.0
Property, Plant, and Equipment	
Buildings, machines, and equipment	$190.0
Less: accumulated depreciation	$(19.5)
Land	$5.5
Total property, plant, and equipment	$176.0
Other Assets	
Receivables due after 1 year	$8.5
Other	$1.5
Total assets	$225.0
Liabilities and Shareholders' Equity	
Current Liabilities	
Accounts payable	$25.0
Accrued liabilities	$6.5
Current maturity of long-term debt	$2.0
Federal income and other taxes	$12.5
Dividends payable	$1.0
Total current liabilities	$47.0
Other Liabilities	$6.0
Long-Term Debt	$27.0
Total liabilities	$80.0
Shareholders' Equity	
Preferred stock	$15.0
Common stock	$45.0
Additional paid-in capital	$20.0
Retained earnings	$65.0
Total shareholders' equity	$145.0
Total liabilities and shareholders' equity	$225.0

and its net worth (*shareholders' equity*). Assets are listed in order of liquidity, while liabilities are listed in order of claim. The balance sheet is prepared directly corresponding to the double-entry bookkeeping introduced in Section 6.1.1, Daily Accounting. Items on the debit side belong to assets, while items on the credit side belong to liabilities. On the debit side, accounts receivable, raw materials, and finished-goods inventories are shown in assets. On the credit side, accounts payable, and accrued liabilities (including wages payable) are shown in liabilities. Because of the original double-entry principle, the balance sheet needs to maintain the equation:

$$\text{Assets} = \text{Liabilities} + \text{Shareholders' equity} \tag{6.9}$$

Actually, the shareholders' equity is adjusted to keep Equation 6.9; more precisely, the retained earnings (companies' profits) are usually adjusted.

6.1.2.3 Cash-Flow Statements

The cash-flow statement exhibits sources and uses of cash over a given period of time. Refer to Table 6.6 for Saito Industries in 20XX. The focus is on generating income and honoring obligations (e.g., loans and other debts). It is notable that the concept of profitability is different from cash flow. A firm's balance sheet may reveal assets that substantially outweigh liabilities (i.e., profitability). However, if these assets are not collectible (e.g., bad debts or accounts receivable in

Table 6.6 Cash-Flow Statement for Saito Industries, Year Ending December 31, 20XX

Cash Flow	*In Millions*
Sources: Operating Activities	
Net earnings	$8.5
Accounts and notes receivable	$(2.5)
Inventories	$(5.0)
Depreciation	$12.0
Accounts and notes payable	—
Federal income and other taxes	$(1.0)
Total	$12.0
Sources: Investing Activities	
Marketable securities	$0.5
Property, plant, and equipment	$(16.0)
Total	$(15.5)
Sources: Financing Activities	
Preferred stock (sale)	$3.0
Common stock (sale)	$2.5
Total	$5.5
Cash, net change	$2.0

Table 6.6 Cash-Flow Statement for Saito Industries, Year Ending December 31, 20XX *(Continued)*

Cash Flow	*In Millions*
Cash, on January 1	$13.5
Cash, on December 31	$15.5

arrears) or liquid (e.g., inventory) within a given time frame, then it is more than just theoretically possible that the firm might be unable to meet its obligations and be forced to file for bankruptcy even though financial statements indicate profitability. As the saying goes, "Positive cash flow, not profit (on the books), pays for lunch."

6.1.3 Demand, Supply, and Market Equilibrium

6.1.3.1 Market Equilibrium

The equation

$$\begin{aligned}\text{Profit} &= \text{Sales revenue} - \text{Cost}\\ &= \text{Unit price} \times \text{Number of product sold} - \text{Cost}\end{aligned}$$

is again our starting point. We will learn the first term (product of price and quantity) in this section, in addition to how the product price and production quantity are set from a viewpoint of the basic microeconomics theories.

Demand (of consumers) and supply (of producers) of the product depend on various factors. We will adopt here the simplest functions with only the product price P as a variable. As shown in Figure 6.2, we assume that the demand for the product from the consumer should decrease with increasing the product price, while the supply of the product from the manufacturer should increase with increasing the price. Let us denote the demand (quantity: Q_d) and supply functions (quantity: Q_s) as $f(P)$ and $g(P)$, respectively, in terms of one variable of price P. We also assume that market equilibrium occurs when the demand and supply quantities are the same:

$$Q_d = f(P)$$
$$Q_s = g(P)$$

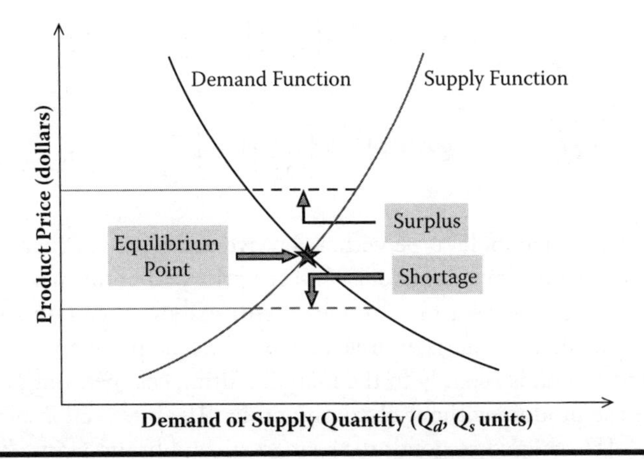

Figure 6.2 Demand and supply curves and market equilibrium.

$$\text{Equilibrium condition: } Q_d = Q_s = Q_E$$

In order to approximate the general functions into a linear relationship with price P around the market equilibrium point, we will adopt Taylor's expansion series around the equilibrium price, P_E:

$$Q_d = f(P) = f(P_E) + \left(\frac{\partial f}{\partial P}\right)_{P=P_E} (P - P_E) \tag{6.10}$$

$$= a + bP$$

$$Q_s = g(P) = g(P_E) + \left(\frac{\partial g}{\partial P}\right)_{P=P_E} (P - P_E) \tag{6.11}$$

$$= h + kP$$

Note that b and k should be negative and positive, from the initial assumption. By completing Equations 6.10 and 6.11, we obtain the equilibrium price P_E and demand/supply quantity Q_E as

$$P_E = \frac{(a - h)}{(k - b)} \tag{6.12}$$

$$Q_E = \frac{(ak - bh)}{(k - b)} \tag{6.13}$$

When the price is a little higher (by ΔP) than P_E (i.e., $P_E + \Delta P$), $Q_s = h + k(P_E + \Delta P) = Q_E + k\Delta P > Q_d = a + b(P_E + \Delta P) = Q_E + b\Delta P$, there will be a *surplus* (the quantity supplied exceeds the quantity demanded). Note that b and k are negative and positive, respectively. To the contrary, when the price is a little lower (by $-\Delta P$) than P_E (i.e., $P_E - \Delta P$), $Q_d = a + b(P_E - \Delta P) = Q_E - b\Delta P > Q_s = h + k(P_E - \Delta P) = Q_E - k\Delta P$, there will be a *shortage* (the quantity demanded exceeds the quantity supplied).

These linear relationships are popular because of their mathematical simplicity. Let us consider further the consumer's expenditure $P \cdot Q_d$ and the producer's revenue $P \cdot Q_s$, which are provided by the product of price and quantity (see Figure 6.3):

$$P \cdot Q_d = P(a + bP) = b\left[P + \left(\frac{a}{2b}\right)\right]^2 - \left(\frac{a^2}{4b}\right) \; [b < 0] \tag{6.14}$$

$$P \cdot Q_s = P(h + kP) = k\left[P + \left(\frac{h}{2k}\right)\right]^2 - \left(\frac{h^2}{4k}\right) \; [k > 0] \tag{6.15}$$

Note that both curves are a parabolic type with a concave and a convex shape, respectively. The expenditure exhibits the maximum amount of $|a^2/4b|$ at a price half of the maximum price ($|a/2b|$), while the producer's revenue shows a monotonous increase above a certain threshold $P = -h/k$ (since h seems to be negative, this value is positive), below which no production will be taken by the producer. This price corresponds roughly to the manufacturing cost per unit product. No manufacturer will produce the product without obtaining profit. The intersection between two curves (Equations 6.14 and 6.15) again corresponds to the market equilibrium point, P_E and Q_E.

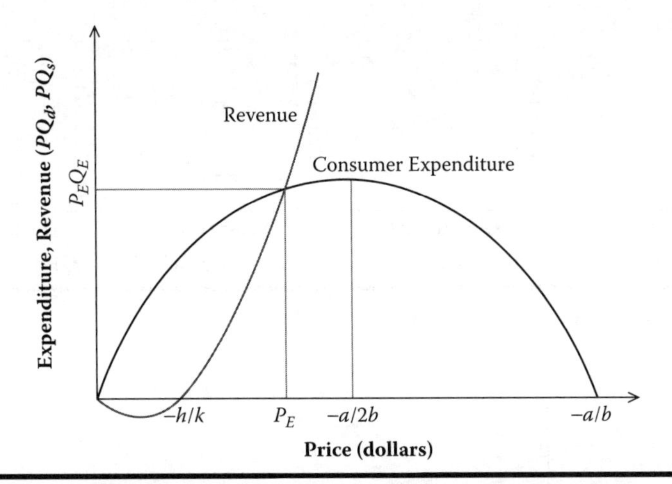

Figure 6.3 Consumer expenditure and producer revenue curves plotted as functions of product price.

6.1.3.2 Demand Elasticity

Demand elasticity is defined as the responsiveness of consumers to changes in price, which is more precisely referred to as *price elasticity of demand.*

Economists often use an index of the ratio (% change in Y)/(% change in X), which is mathematically formulated by $d(\log(y))/d(\log(x)) = (x/y)(dy/dx)$, as learned in freshmen differential calculus. Price elasticity of demand, E, can be defined as

$$E = \frac{d(\log Q_d)}{d(\log P)} = \left(\frac{dQ_d}{dP}\right)\left(\frac{P}{Q_d}\right) \tag{6.16}$$

$$= \frac{\text{(Percent change in quantity of demand)}}{\text{(Percent change in price)}}$$

As you may have already noticed, E is always negative, because the change in quantity demanded is accompanied by the change in price in the opposite direction. For simplicity, the negative sign is often neglected (i.e., the absolute value is considered).

When $|E| > 1$, we call it *elastic*; when $|E| = 1$, it is *unitary*; and when $|E| < 1$, it is *inelastic.* An Apple iPod or new computer game software may be a necessary "cool" good for students, and they may not care about the price (sometimes they pay a premium to obtain it); that is inelastic. The same good may not be a "cool" good for senior guys like me. They may not purchase it until the price decreases to a reasonably cheap level; that is elastic. Figure 6.4 shows various demand curves corresponding to elastic and inelastic products, and Table 6.7 summarizes the determinants of elasticity of market demand. Regarding MMI's metal tube motors, for example, when they are used for medical endoscope or catheter applications, they are the necessary key components. Thus, the price is inelastic. A partner medical company will not accord MMI a significant price reduction. On the other hand, when the motor is used for cellular phone camera applications, a new function phone may not be a "must." Thus, the price is really elastic. Without procuring the cheapest price, a partner electronic company will not purchase the motor.

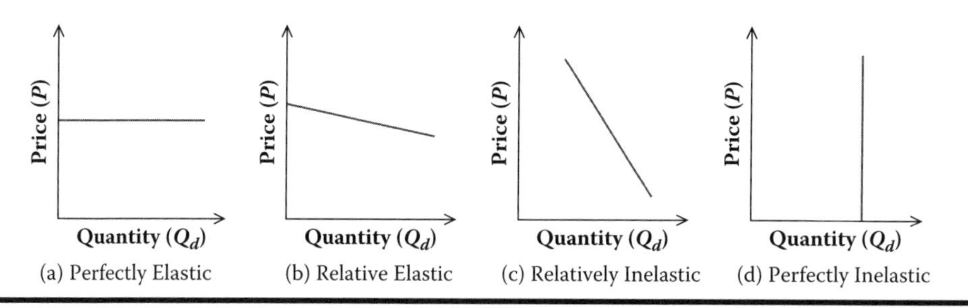

Figure 6.4 Price elasticity of demand curves.

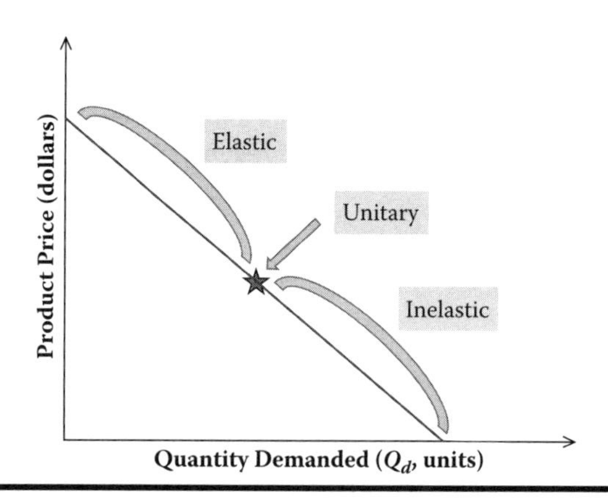

Figure 6.5 The elasticity of a linear demand curve.

Table 6.7 Determinants of Elasticity of Market Demand

Elastic	*Inelastic*
Luxury	Necessity
Many substitutes available	Few substitutes available
Price is a large part of income	Price is a small part of income
Long-run time period	Short-run time period

Let us obtain the demand elasticity for the case of the linear relationship in Equation 6.10:

$$E = \frac{bP}{a + bP} \tag{6.17}$$

For $P \to 0$, $E \to 0$; for $P \to -a/b$, $E \to -\infty$; and for $P = -a/2b$, $E = -1$. The demand elasticity of a linear demand curve is illustrated in Figure 6.5, which shows that the upper half of the demand line is elastic, while the lower half is inelastic. At the halfway point, the demand is unitary elastic ($E = -1$).

6.1.4 Break-Even Analysis

Now, let us consider the second term of the following equation:

$$\begin{aligned} \text{Profit} &= \text{Sales revenue} - \text{Cost} \\ &= \text{Unit price} \times \text{Number of product sold} - \text{Cost} \end{aligned}$$

You learned the classification of manufacturing costs and period costs in Section 6.1.1. Another classification of costs is introduced here: *variable costs, fixed costs*, and *semivariable costs*. A variable cost changes in direct proportion to a change in the production level, such as the number of products and working hours of services. Direct materials, direct labor, and sales commissions belong to this category. On the other hand, a fixed cost remains unchanged, regardless of the level of production. Lease or rent fees of facilities and equipment, depreciation, property taxes, insurance, and the salary of corporate executives belong to this category. Semivariable costs are exemplified by utilities (electricity, water, etc.) and repairs/maintenance, which are charged by the sum of the base fee and the amount of usage (such as wattage).

6.1.4.1 Break-Even Analysis Method

We use the multilayer (ML) manufacturing business of MMI as an example (recall Chapter 1, Section 1.3.1: Background of the Case Study). There are two techniques for making multilayered devices: the *cut-and-bond method* and the *tape-casting method*. As you already learned, the cut-and-bond method does not require expensive facilities, but requires a lot of manpower and labor, typically a multiple-worker fabrication line (i.e., direct labor-oriented method; six workers make one team). To the contrary, the tape-casting method requires an expensive fabrication facility, but almost no labor, which is suitable for mass production. The manufacturing cycle is typically biweekly (i.e., one lot comes out every 2 weeks).

MMI received a purchase order from a Japanese company, Saito Industries, to supply ML ultrasonic motors (USMs) for a zoom/focus mechanism of the next generation cellular phone camera modules. The terms were a total quantity of 200,000 units at a price of $5, each within a period of 90 days.

MMI must decide whether they will purchase equipment in the production process or use only manpower. By installing mass-production equipment, MMI can virtually eliminate labor in the production of inventory. At high volume, MMI will do well, as most of the production costs are fixed. At low volume, however, MMI could face difficulty in making the payment for the equipment. So, how much will changes in volume affect cost and profit? At what point does the firm break even (zero deficit)?

We need to consider in general the following constraints:

- *Price constraint:* If the motor price is unreasonably low, MMI should decline this purchase order. This should be higher than a sum of the manufacturing cost and the period cost.
- *Production capability:* Both equipment and manpower have some production capability.
- *Space constraint:* MMI factory space can accommodate a maximum of two sets of mass-production equipment and/or 10 production lines (or 60 workers).
- *Production period:* One lot routine takes 2 weeks, and a typical delivery due date is in 90 days (or 6 biweekly).

Let us summarize here the necessary accounting data:

1. Purchase order: 200,000 pieces of ML USM at $5 per piece (total sales = $1 million)
2. Tape-casting facility and maintenance: $300,000 per set
   Production capability: 42,000 pieces biweekly
   Maximum number of equipment: 2 sets
3. Human manufacturing line (6 workers): $6250 biweekly per manufacturing line
   Production capability per line: 2100 pieces biweekly
   Maximum number of workers: 10 lines/60 workers
4. Raw material costs (PZT ceramic powder, Ag/Pd electrode, etc.) = $0.4 per unit ML
5. Manufacturing overhead = $0.6 per unit ML

First, is there a possibility to use the cut-and-bond method in the United States? Because direct labor requires $3.0 per unit (= $6250/2100 pieces), the total variable cost is $4.0 (= $3.0 + raw materials $1.0) per unit. Because the sales price is $5 per unit, we can get positive gross income. However, in this scenario, in order to satisfy the purchase order production, we need 96 production lines (= 200,000/2100). Even if we spend 3 months (6 biweekly), we need to hire 6 workers 16 lines = 96 workers. It is beyond MMI's space availability. Even if we can find the temporary space, it may be difficult in practice to hire 96 workers for only 3 months for this project in the United States.

Second, the tape-casting method is considered. The total production cost (TC) for producing Q pieces of motors is represented in terms of the fixed cost (FC) and the total variable cost (TVC; variable cost per unit [direct material cost + manufacturing overhead in this case]: VC) as

$$\begin{aligned} \mathrm{TC} &= \mathrm{FC} + \mathrm{TVC} = \mathrm{FC} + Q \times \mathrm{VC} \\ &= 300{,}000 + 1.0Q \end{aligned} \tag{6.18}$$

Note that Q is accumulated every 42,000 pieces biweekly per equipment, leading to five successive biweekly periods (i.e., less than the 90 days due). The total revenue is given with the sales price per unit (P) as

$$\begin{aligned} \mathrm{TR} &= Q \times P \\ &= 5.0Q \end{aligned} \tag{6.19}$$

A *break-even chart* is shown in Figure 6.6. Note that the total cost linear line starts from a biased position (FC = $300,000) and the total revenue line starts from the origin (zero). By equating Equations 6.18 and 6.19, we obtain a famous equation for the break-even point BE (number N of products)[3]:

$$\begin{aligned} \mathrm{BE} &= \frac{\mathrm{FC}}{(P - \mathrm{VC})} \\ &= \frac{\text{Fixed cost}}{(\text{Price} - \text{Variable cost per unit})} \\ &= \frac{\$300{,}000}{(\$5 - \$1)} \\ &= 75{,}000 \text{ pieces} \end{aligned} \tag{6.20}$$

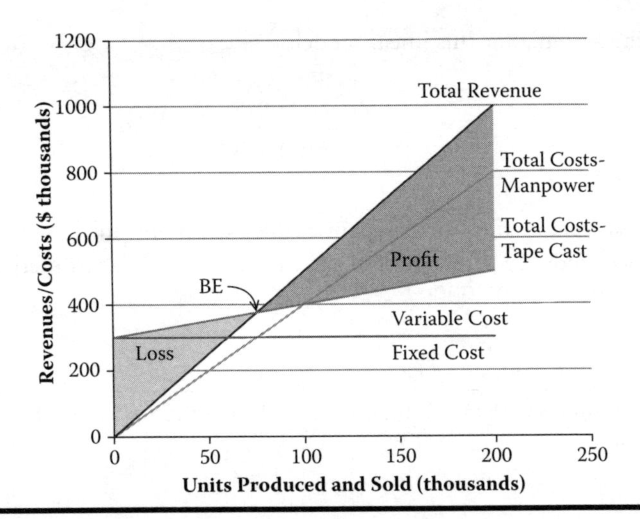

Figure 6.6 Break-even chart for the ML production. The chart is also indicative to the selection between the mass-production equipment and workers' line.

By introducing the *contribution margin* (CM), defined as price minus variable cost per unit, Equation 6.20 can be written as

$$\mathrm{BE} = \frac{\mathrm{FC}}{\mathrm{CM}}, (\mathrm{CM} = P - \mathrm{VC}) \tag{6.21}$$
$$= \frac{\text{Fixed cost}}{\text{Contribution margin}}$$

The CM is the amount of revenue that is available to contribute to covering fixed expenses after all variable expenses have been covered. Taking into account CM $4 (= $5 – $1) in MMI's case, as the fixed cost of equipment is quite high ($300,000), and minimum sales of 75,000 units are required to compensate all the fixed cost.

By inputting $Q = 200{,}000$, we obtain TC = 500,000. Since the total sales amount is $1 million, we can expect the operating income (or EBIT) = $500,000, and which is compared with EBIT of $200,000 for the cut-and-bond method.

6.1.4.2 Degree of Operating Leverage

Degree of operating leverage (DOL) is defined as

$$\mathrm{DOL} = \frac{d(\log \mathrm{EBIT})}{d(\log Q)} = \left[\frac{d(\mathrm{EBIT})}{dQ}\right]\left[\frac{Q}{\mathrm{EBIT}}\right] \tag{6.22}$$
$$= \frac{\text{Percent change in operating income}}{\text{Percent change in unit volume}}$$

DOL is a measure of effectiveness of income increase over the production increase. DOL should be computed only over a profitable range of operations. Taking into account

$$\mathrm{EBIT} = \mathrm{TR} - \mathrm{TC} = Q(P - \mathrm{VC}) - \mathrm{FC} \tag{6.23}$$

we obtain the following formula for this linear model:

$$\text{DOL} = \frac{Q(P - \text{VC})}{Q(P - \text{VC}) - \text{FC}} \tag{6.24}$$

Thus, the closer DOL is computed to the company break-even point, the higher the number will be due to the divergence tendency of Equation 6.24. In the above tape-cast method case, DOL is calculated for $Q = 100{,}000$ and 200,000:

$$\text{DOL} = \frac{100{,}000(\$5 - \$1)}{100{,}000(\$5 - \$1) - 300{,}000} = 4.0 \quad \text{or} \quad \frac{200{,}000(\$5 - \$1)}{200{,}000(\$5 - \$1) - 300{,}000} = 1.6$$

In other words, if the purchase order number is only 100,000 pieces (rather close to the BE point), a slight increase in the production reflects a dramatic increase in operation income, which seems to be an attractive option. However, a slight reduction in the sales causes a dramatic decrease in income, and the equipment introduction seems to be rather risky, leaving a large debt or loan. This is analogous to a physics *lever rule*; the force is amplified inversely proportional to the distance between the fulcrum and the force point. The break-even point and the quantity of products sold correspond to the fulcrum and the force point, respectively.

Finally, let us consider the intersect between two total cost lines for the tape-cast and man-power methods in Figure 6.5, which indicates the point of new equipment introduction. Using the following equation similar to BE, we obtain the minimum number of units, below which the workers line seems to be less costly.

$$\frac{\text{Fixed cost}}{\text{Variable cost per unit for cut-and-bond} - \text{Variable cost per unit for tape-cast}}$$

$$= \frac{\$300{,}000}{\$4 - \$1} = 100{,}000 \text{ pieces}$$

6.1.4.3 Marginal Analysis

There are some limitations to the aforementioned break-even analysis. First, the above analysis selects only one price for a particular product, regardless of the quantity, and then proceeds to determine how much a firm has to sell at this price to break even. In order to consider the possibility of different quantities demanded by consumers at different prices (i.e., the price elasticity of demand in Section 6.1.3), a complete schedule of prices and break-even points needs to be constructed. Consumers expect a price decrease with an increase in the order quantity, leading to revenue saturation for the producer. Second, the above analysis assumes that a firm's average variable cost per unit product is constant, regardless of the quantity. Under certain circumstances, it is quite possible for the firm's average costs to either decrease or increase as more of a good is produced. In order to improve the above linear model, we will introduce a *nonlinear model*. Because the unit price or cost is not constant, but varies in the nonlinear model, we will use marginal values, which correspond to the derivative of the parameter in terms of production activity.

- Marginal revenue (MR) – *dR/dQ*. Increment of the revenue per unit increment of product quantity
- Marginal cost (MC) – *dC/dQ*. Increment of the total cost (or variable cost) per unit increment of product quantity

We will take into account the following two laws:

1. *Law of Diminishing Marginal Utility:* The level of demand or satisfaction derived from a product diminishes with each additional unit consumed. This creates saturation of the revenue and price reduction with an increase in the quantity.
2. *Law of Diminishing Returns:* Though additional units of labor may contribute to increased productivity in absolute numbers, each such additional unit contributes relatively less than the preceding unit to the productivity. Let me remind you of MMI's ML production with the manual cut-and-bond method. In that scenario, to satisfy the production of the purchase order, we would need 96 production lines. Even if we spent 3 months (6 biweekly), we would need to hire 6 workers × 16 lines = 96 workers. This is beyond MMI's space availability. Even if we resort to a temporary space, the productivity per unit labor (or unit line) will be drastically decreased, because of fewer machines, tools, space, or other inputs per worker. Also in such a case, the additional cost (temporary space rent) needs to be considered.

From the above discussion, we can expect the revenue and manufacturing cost curves as shown in Figure 6.7a. The revenue curve is concave-parabolic (simple *diminishing marginal utility law*), while the cost curve is reverse-S like. The productivity increases with increasing quantity initially due to the learning curve of the workers, leading to the cost saturation tendency. However, over a certain production quantity, the manufacturing cost starts to increase again.

Assume that the firm is in the short run, so some costs are fixed. Let the inverse demand for a monopoly (no other company's competitive factors are considered) be $P = P(Q)$. Thus, total revenue is $R(Q) = P(Q) \cdot Q$. The monopoly profit function is

$$\Pi = R(Q) - \text{TVC}(Q) - \text{TFC} \tag{6.25}$$

where TVC(Q) is total variable cost and TFC is total fixed cost.

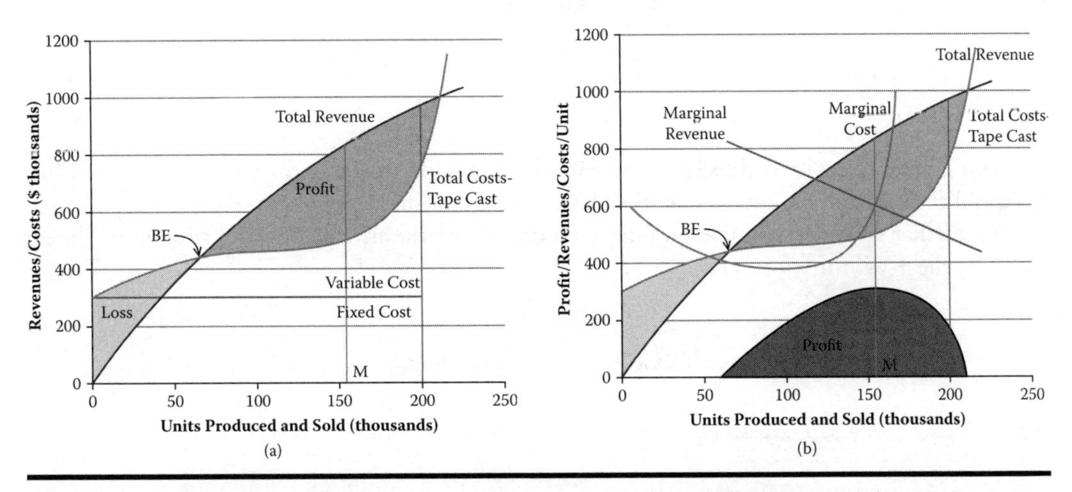

Figure 6.7 (a) Nonlinear revenue and manufacturing cost curves. (b) The maximum profit is obtained at the quantity of the intersection between the marginal revenue and marginal cost curves.

The first-order condition for profit maximization requires

$$\frac{d\Pi}{dQ} = \frac{dR}{dQ} - \frac{d\text{TVC}}{dQ} = 0 \tag{6.26}$$

The second-order condition for a maximum is that at the equilibrium quantity

$$\frac{d^2\Pi}{dQ^2} = \frac{d^2R}{dQ^2} - \frac{d^2\text{TVC}}{dQ^2} < 0 \tag{6.27}$$

Equation 6.26 dR/dQ is marginal revenue and $d\text{TVC}/dQ$ (or $d\text{TC}/dQ$) is marginal cost, so choosing the quantity of output that maximizes profit requires that marginal revenue equal marginal cost: MR = MC. Solve Equation 6.26 for the equilibrium output Q_E so the equilibrium price is $P_E = P(Q_E)$.

If $\Pi = P(Q_E)Q_E - \text{TVC}(Q_E) - \text{TFC} > 0$, the firm makes a profit, but if $\Pi = P(Q_E)Q_E - \text{TVC}(Q_E) - \text{TFC} < 0$, the firm experiences a loss. Even in this loss case, the firm should produce Q_E rather than shutting down, as long as $P(Q_E)Q_E - \text{TVC}(Q_E) > 0$. If the firm shuts down, TFC remains as a loss, which is larger than $|P(Q_E)Q_E - \text{TVC}(Q_E) - \text{TFC}|$. Thus, the operation condition of the firm is given by

$$P(Q_E) \geq \frac{\text{TVC}(Q_E)}{Q_E} \tag{6.28}$$

where the right-hand-side term $\text{TVC}(Q_E)/Q_E$ is the average variable cost.

Next, we demonstrate that the profit-maximization price and quantity must lie on the elastic portion of demand. Because

$$\text{MR} = \frac{dR}{dQ} = P(Q) + Q\left(\frac{dP(Q)}{dQ}\right) = P\left[1 + \left(\frac{Q}{P}\right)\left(\frac{dP}{dQ}\right)\right] = P\left[1 + \frac{1}{E}\right]$$

where E is the *price elasticity of demand*, in equilibrium

$$\text{MR} = \text{MC} = P\left[1 + \frac{1}{E}\right] \tag{6.29}$$

MC and P must be positive, thus $[1 + 1/E] > 0$, which requires that $|E| > 1$ (remember $E < 0$). Thus, in equilibrium P_E and Q_E must lie on the elastic portion of the demand curve.

We further consider a simple nonlinear model, where the inverse demand function is linear and only the TVC function is nonlinear:

$$P(Q) = A - BQ$$
$$\text{TVC}(Q) = DQ - EQ^2 + FQ^3$$

where A, B, D, E, and F are all positive constants. The profit function is therefore

$$\Pi = R(Q) - \text{TVC}(Q) - \text{TFC} = AQ - BQ^2 - DQ + EQ^2 - FQ^3 - \text{TFC}$$

For profit maximization, Equation 6.26 becomes

$$\frac{d\Pi}{dQ} = A - 2BQ - D + 2EQ - 3FQ^2 = 0$$

So, from the equation $(A - D) + 2(E - B)Q - 3FQ^2 = 0$, we obtain the two solutions for Q_E

$$Q_E = \frac{2(E-B) \pm \sqrt{4(E-B)^2 + 12F(A-D)}}{6F}$$

one of which (+) satisfies the second-order condition Equation 6.27:

$$\frac{d^2\Pi}{dQ^2} = 2(E-B) - 6FQ = -\sqrt{4(E-B)^2 + 12F(A-D)} < 0$$

Finally, using this Q_E, we will obtain the equilibrium solution, under the condition: $P(Q_E) \geq \text{TVC}(Q_E)/Q_E$,

$$P_E = A - BQ_E$$

$$\Pi = AQ_E - BQ_E^2 - DQ_E + EQ_E^2 - FQ_E^3 - \text{TFC}$$

6.1.5 *Tax Reduction Considerations*

Referring to the income statement (Table 6.4), it is clear that a firm's net income or "bottom line" can be enhanced by a reduction in tax obligations, as it is by increased revenues and operating income. The firm typically accomplishes this on a number of different fronts.

6.1.5.1 Timing of Purchases and Bad-Debt "Write-Offs"

Most accounting is made on an *accrual basis*. Revenues are recognized on transactions during the tax period in which they actually occur. Deferring the transaction even a single day, from December 31 to January 1, causes revenues to be recognized in the more recent year. If income is high in a given year, the firm may benefit from being able to write off deductible expenditures against such income in the same year, effectively reducing its taxable income. In practice, the invoices for the sales or for the R&D contract funds can be sent after January 1. If, however, income was very low or the firm has already written off a great deal for this year, it may elect to defer additional purchases until the following year when it can use the deductions. Suppose the earnings before taxes (EBT) in 2007 are \$100,000, and that in 2008 are expected to be −\$20,000; by postponing the invoicing of \$20,000 by only one day, we can save the tax by \$8000 (20,000 × tax rate 40%) for 2007. The corporate tax in 2008 will be zero regardless of this treatment.

6.1.5.2 Depreciation

The government allows the firm to reduce its tax liability by acknowledging the decrease in value of equipment and property relating to wear and tear or obsolescence. Depreciation policies can change to reflect the attempts of government to stimulate growth of particular industries or of the economy, in general.

Through recent tax legislation, assets are now classified according to nine categories that determine the allowable rate of depreciation write-off. Each class is referred to as a *modified accelerated cost recovery system* (MACRS)[3]. Some references are also made to asset depreciation range (ADR), or the expected physical life of the asset. Most assets can be written off more rapidly than the midpoint of their ADR. Table 6.7 shows the various categories for depreciation, linking the depreciation write-off period to the midpoint of the ADR and the depreciation percentages for each category. The rates shown in Table 6.8 are developed with the use of the half-year convention, which treats all property as if it were put in service in midyear.

The depreciation schedule for MMI's tape-casting equipment is explicated here, and the result is shown in Table 6.9. Since the manufacturing equipment is categorized as a 7-year MACRS, we adopt 8 years for depreciation period. If we start from earnings before depreciation and taxes (EBDT), EBT is given by (EBDT – depreciation). Taxes are calculated with a tax rate T of 40% as 0.4 (EBDT – depreciation); then earnings after taxes (EAT) will be $(1 - T)$ (EBDT – depreciation) = 0.6 (EBDT – depreciation). Because the depreciation is pure cash, the final cash flow can be calculated as

$$(1-T)(\text{EBDT} - \text{Depreciation}) + \text{Depreciation} = (1-T)\text{EBDT} + \text{Depreciation} \times \text{Tax rate } T \quad (6.30)$$

Table 6.8 Categories for Depreciation Write-Off and Depreciation Percentage in MACRS

Class	*Items*	*Depreciation Year*							
		1	*2*	*3*	*4*	*5*	*6*	*7*	*8*
3-year MACRS	All property with ADR midpoints of 4 years or less	0.333	0.445	0.148	0.074				
5-year MACRS	Property with ADR midpoints of 4–10 years; automobiles, light trucks, technology equipment (computers and research-related properties)	0.200	0.320	0.192	0.115	0.115	0.058		
7-year MACRS	Property with ADR midpoints of 10–16 years; most types of manufacturing equipment, office furniture and fixtures	0.143	0.245	0.175	0.125	0.089	0.089	0.089	0.045

ADR = asset depreciation range.

Table 6.9 Annual Depreciation and Cash Flow Benefit Related to the Purchase of Equipment

Items	*Year*							
	1	*2*	*3*	*4*	*5*	*6*	*7*	*8*
Depreciation base (tape-casting equipment)	$300,000	$300,000	$300,000	$300,000	$300,000	$300,000	$300,000	$300,000
Depreciation rate	0.143	0.245	0.175	0.125	0.089	0.089	0.089	0.045
Annual depreciation	$42,900	$73,500	$52,500	$37,500	$26,700	$26,700	$26,700	$13,500
Tax rate (at 40%)	0.40	0.40	0.40	0.40	0.40	0.40	0.40	0.40
Additional cash flow, or tax saving amount	$17,160	$29,400	$21,000	$15,000	$10,680	$10,680	$10,680	$5400

The first term denotes the profit created from the actual production business. Additional cash flow is the *tax savings amount*, related to the equipment depreciation deduction, and is provided by

$$\text{Additional cash flow} = \text{Depreciation base (Equipment price)} \times \text{Depreciation rate} \times \text{Tax rate } T \quad (6.31)$$

Because the total of 8-year depreciation should be equal to the initial equipment price ($300,000), the total tax saving should also be provided by the product of equipment price and tax rate ($120,000). In other words, the initial equipment price ($300,000) is overestimated and requires readjustment from an investment viewpoint.

6.1.5.3 *Tax Credit*

According to Wikipedia [4], the term *tax credit* describes two different concepts:

1. A recognition of partial payment already made toward taxes due
2. A state benefit paid to employees through the tax system, which has the effect of increasing (rather than reducing) net income

Tax credits are characterized as either *refundable* or *nonrefundable*, or, equivalently, *nonwastable* or *wastable*. Refundable or nonwastable tax credits can reduce the tax owed below zero, and result in a net payment to the taxpayer beyond their own payments into the tax system, appearing to be a moderate form of *negative income tax*. Examples of refundable tax credits include the earned income tax credit and the additional child tax credit in the United States. A nonrefundable or wastable tax credit cannot reduce the tax owed below zero, and hence cannot cause a taxpayer to receive a refund in excess of their payments into the tax system.

There is another modified category of refundable tax credit for high-tech entrepreneurs. Each state in the United States promotes encouraging programs for high-tech start-ups involving tax savings. For instance, Pennsylvania offers a program called the Keystone Innovation Zone (KIZ) Tax Credit Program. Further information can be found here: http://www.newpa.com/programDetail.aspx?id=56.

The KIZ Tax Credit Program was established in 2004 to create designated geographic zones to foster innovation and create entrepreneurial opportunities, aligning the combined resources of educational institutions, private businesses, business support organizations, and venture capital networks. These "knowledge neighborhoods" provide companies that have been in operation for less than 8 years and are located in a KIZ, with a maximum of $100,000 in tax credits annually. With a total pool of up to $25 million available to KIZ companies per year, the Tax Credit Program contributes to the ability of young KIZ companies to transition through the stages of growth. This program also provides a tradability component, which is crucial to young companies that do not yet have a tax liability. A KIZ company may claim a tax credit equal to 50% of the increase in that KIZ company's gross revenues in the immediately preceding taxable year. A tax credit under this program for a KIZ company shall not exceed $100,000 annually. For the purposes of the KIZ Tax Credit, the term "gross revenues" may include grants received by the KIZ company from any source.

MMI Example

Operates in a KIZ zone within the targeted industry segment for that KIZ (now its fourth year). In year 1, MMI's gross revenues resulting from activities within the KIZ (excluding their activities in the partner company, Cheng Kung Corporation) amounted to $100,000. In year 2, MMI business grew and their gross revenues amounted to $260,000. In year 3, MMI could apply for a KIZ tax credit of $80,000 (50% of the $160,000 increase in gross revenue from year 1 to year 2). This tax credit was traded or purchased in practice by a local bank, and MMI received $72,000 in cash (10% trading charge). Fortunately, MMI's gross revenues in year 3 grew further to $700,000. MMI was eligible to apply for a maximum KIZ tax credit line of $100,000 in year 4. 50% of their gross revenue increase ($440,000) exceeds the maximum allowable tax credit line.

This is an actual cash refund, which provides a significant benefit to a small high-tech venture. I recommend entrepreneurs find out if there is a similar tax credit program in their state.

6.1.6 Cash Flow Analysis

Here I will remind you of the saying, "Positive cash flow, not profit (on the books), pays for lunch." Even if a Small Business Innovation Research (SBIR) program is awarded to your firm, the actual cash usually arrives 90 days after the awarded date (this is in the usual government accounting payment agreement), due to the size of the federal organization's system. However, the R&D task should be started immediately after the awarded date in order to meet the development milestones in the proposal (usually the program is for 6 to 9 months). Because the firm needs to pay wages for the researchers, rent, utilities, and other payments for goods such as raw materials, this 2- to 3-month delay period is critical, and sometimes fatal, for a small firm. The financial deficit may force your firm to take a risky loan (i.e. a high interest rate).

MMI's sales division reached a business agreement with a Japanese company, Tanaka Corporation, on USM distribution in the United States: MMI receives 30% off the retail price (in other words, 30% gross margin of the total sales at MMI) under the condition that MMI pays up front to receive the products from Tanaka. MMI's business with the customers is based on the credit sales, and their experience has shown that 30% of sales receipts are collected in the month of sale and 60% in the following month, and the remaining 10%, 2 months later. MMI sold $50,000 of the motors in September (actual sales), and has forecast credit sales for the fourth quarter of the year (based on the federal research progressing pace) as follows:

August (actual)	$40,000
September (actual)	$50,000
Fourth Quarter	
October	$100,000
November	$70,000
December	$80,000

Let us prepare a schedule of cash payments and receipts for MMI covering the fourth quarter (October through December) and provide information on how much they will need to borrow in a short run.

As you can see from Table 6.10, due to the time lag between cash payments and collections, a sales increase in October will cause a net cash flow deficit. MMI may borrow this amount ($6000) from the main bank in a short run, or more likely MMI will keep the daily base cash flow, typically 10% of the average monthly sales ($7000 to 10,000), in their main bank.

Table 6.10 Schedule of Cash Payments and Receipts for MMI Covering the Fourth Quarter

	August	*September*	*October*	*November*	*December*
Sales	$40,000	$50,000	$100,000	$70,000	$80,000
1. Collections: 30% of current sales			$30,000	$21,000	$24,000
2. Collections: 60% of previous month's sales			$30,000	$60,000	$42,000
3. Collections: 10% of the sales 2 months ago			$4000	$5000	$10,000
Total cash receipts			$64,000	$86,000	$76,000
Cash payments to Tanaka			$70,000	$49,000	$56,000
Net cash flow (in a month)			$(6000)	$37,000	$20,000

6.2 Financial Management—Fundamentals of Finance

6.2.1 *Key Financial Ratios*

Financial ratios are used to weigh and evaluate the operating performance of the firm, in order to compare it with that of other companies in a similar industry category[3]. While an absolute value such as earnings of $50,000 may appear satisfactory for a small firm, it may be considered small for a large firm. For this reason, investors and financial managers emphasize ratio analysis.

6.2.1.1 *Price-Earnings Ratio*

The most popular ratio is the *price-earnings* (P/E) *ratio.* This refers to the multiplier applied to earnings per share to determine the current value of the common stock:

$$\text{Earnings per share} = \frac{\text{Earnings available to common stockholders}}{\text{Number of shares outstanding}} \tag{6.32}$$

$$\frac{\text{P}}{\text{E}} = \frac{\text{Current price of the common stock}}{\text{Earnings per share}} \tag{6.33}$$

In the case of Saito Industries, refer to the income statement in Table 6.4. From the EAT, the preferred stockholders receive $1.1 million and the remaining amount of $7 million is the earnings that are available to common stockholders. Because the number of shares outstanding is 5 million, earnings per share can be calculated as $1.40 (= $7 million/5 million shares). If the current price of the stock (market value) is $35 per share, P/E becomes 25 (= $35/$1.4).

The P/E ratio is influenced by the earnings and the sales growth of the firm, the risk (or volatility in performance), the debt-equity structure of the firm, the dividend payment policy, the equity management, and various other factors. The P/E ratio indicates expectations about the future of a company. Firms expected to provide returns greater than those for the market in general with equal or less risk often have P/E ratios higher than the market P/E ratio.

However, the P/E ratio can be confusing. When a firm's earnings drop suddenly (even approaching zero), its stock price, though also declining, may not follow the magnitude of the fall-off in earnings. Though the company would still be exhibiting an increasing P/E ratio, this does not mean a high expectation for this firm.

6.2.1.2 *Financial Analysis*

There are 13 significant ratios in four primary categories, which are summarized in Table 6.11.

Example Problem 6.2

Using the income statement and balance sheet provided in Tables 6.4 and 6.5, compute the profitability, asset utilization, liquidity, and debt utilization ratios for Saito Industries.

Solution

Profitability Ratios (M = millions):

1. Profit margin = net income/sales = 8.1M/100M = 8.1%
2. Return on assets = net income/total assets = 8.1M/225M = 3.6%
3. Return on equity = net income/stockholders' equity = 8.1M/145M = 5.6%

Table 6.11 Thirteen Significant Financial Ratios in Four Primary Categories

Profitability Ratios	1. Profit margin	$\frac{\text{Net income}}{\text{Sales}}$	
	2. Return on assets (investment)	$\frac{\text{Net income}}{\text{Total assets}}$	Profit margin × Asset turnover
	3. Return on equity	$\frac{\text{Net income}}{\text{Stockholders' equity}}$	$\frac{\text{Return on assets}}{\left(1-\frac{\text{Dept}}{\text{Assets}}\right)}$
Asset Utilization Ratios	4. Receivable turnover	$\frac{\text{Sales (Credit)}}{\text{Receivables}}$	Ratios relate the balance sheets (assets) to the income statement (sales)
	5. Average collection period	$\frac{\text{Accounts receivable}}{\text{Average daily credit sales}}$	
	6. Inventory turnover	$\frac{\text{Sales}}{\text{Inventory}}$	
	7. Fixed asset turnover	$\frac{\text{Sales}}{\text{Fixed assets}}$	
	8. Total asset turnover	$\frac{\text{Sales}}{\text{Total assets}}$	
Liquidity Ratios	9. Current ratio	$\frac{\text{Current assets}}{\text{Current liabilities}}$	Usually > 1
	10. Quick ratio	$\frac{\text{Current assets} - \text{Inventory}}{\text{Current liabilities}}$	Usually > 1
Debt Utilization Ratios	11. Debt to total assets	$\frac{\text{Total dept}}{\text{Total assets}}$	
	12. Times interest earned	$\frac{\text{Income before interest and taxes}}{\text{Interest}}$	
	13. Fixed charge coverage	$\frac{\text{Income before fixed charges and taxes}}{\text{Fixed charges}}$	

Asset Utilization Ratios:

4. Receivable turnover = sales (credit)/accounts receivable = 100M/4M = 25x
5. Average collection period = accounts receivable/avg. daily credit sales = 4M/(100M/360) = 14.4 days
   Average daily credit sales = annual sales/360 days
6. Inventory turnover = sales/inventory = 100M/16.5M = 6.1×
7. Fixed asset turnover = sales/fixed assets = 100M/176M = 0.57×
   Fixed assets = property, plant, and equipment
8. Total asset turnover = sales/total assets = 100M/225M = 0.44×

Liquidity Ratios:

9. Current ratio = current assets/current liabilities = 39M/47M = 0.83×
10. Quick ratio = (current assets – inventory)/current liabilities = (39 – 16.5)M/47M = 0.48×

Debt Utilization Ratios:

11. Debt to total assets = total debt/total assets = 80M/225M = 36%
    Total debt = total liabilities
12. Times interest earned = IBIT/interest = 14M/0.5M = 28×
13. Fixed charge coverage = income before fixed charges and taxes/fixed charges = 15M/1.5M = 10×
    Fixed charges = lease payments + interest. We cannot compute this number because the income statement does not have the information on lease payments. Because lease payments are included in selling and administrative expenses ($15 millions), we suppose it is $1 million (Saito Industries has large property and plant already ($176 millions)). Then, fixed charges = 1M + 0.5M = 1.5M
    Income before fixed charges and taxes = EBIT + lease payments = 14M + 1M = 15M.

Over the course of the business cycle, sales and profitability may expand and contract, and ratio analysis for any single year may not present an accurate picture of the firm, in particular for a start-up company. Therefore, we look at the trend of performance over a period of years.

6.2.2 Financial Forecasting

Forecasting for the future is not easy, but, as a small business executive, one must try. Without financial forecasting the business plan cannot be completed, nor can we persuade investors or venture capitalists.

6.2.2.1 Pro Forma Statements

A comprehensive means of financial forecasting is developing a series of pro forma or projected financial statements. Particular attention is paid to the *pro forma income statement*, the *cash budget*, and the *pro forma balance sheet*.

In developing the pro forma income statement, we follow these four steps:

1. Establish a sales projection.
2. Determine a production schedule and the associated use of new material, direct labor, and overhead to arrive at gross profit.

3. Compute other expenses.
4. Determine profit by completing the actual pro forma statement.

Table 6.12 shows an example of revenue projection for MMI. First- and second-year revenues are actual and are used to project the sales in the following years. While statistical techniques such as regression and time series analysis may be employed for this purpose, it is the company executive who is best able to predict future sales of specific products in the case of a small high-tech firm. To calculate the sales growth, MMI company executive Barb Shay used a 30% increase per year based on the second year's $150,000. The growth rate can be changed from 10% (pessimistic) to 30% (optimistic), depending on the executive's confidence in the business. Regarding the R&D revenue, Barb estimated a drastic jump from $147,000 to $500,000 in the third year, because of her confidence in the research proposals submitted in the second year. Due to the average acceptance rate (10–20%) of SBIR programs, 10–20% of the total requested research budget can be expected. After the third year, she used a 20% increase rate per year.

Once the income projection is finished, the production schedule and manufacturing cost are calculated. Monthly cash flow is computed using the method explained in Table 6.10, including the collection and payment schedule. Finally, the pro forma balance sheet will be prepared.

When the company grows and the budget size becomes large enough, an alternative method called the *percent-of-sales method* can be adopted, where an assumption is made that accounts on the balance sheet will maintain a given percentage relationship to sales. To determine the need for new funds associated with the sales, the following formula can be used:

$$\text{Required new funds} = \frac{A}{S}(\Delta S) - \frac{L}{S}(\Delta S) - PS_2(1-D) \tag{6.34}$$

where

$\frac{A}{S}$ = Percentage relationship of variable assets to sales
(ΔS) = Change in sales
$\frac{L}{S}$ = Percentage relationship of variable liabilities to sales
P = Profit margin
S_2 = New sales level
D = Dividend payout ratio.

6.2.2.2 Linear Regression

Once the firm has been operating for a long time, a time series of financial data has been collected. We can then forecast the company's future status by extrapolating the present trend over time statistically [5].

Table 6.12 MMI Sales and R&D Revenues Projection (Unit: US$)

Budget Period	*1st Year*	*2nd Year*	*3rd Year*	*4th Year*	*5th Year*
Sales revenue	166K	150K	195K	253K	330K
R&D revenue	0	147K	500K	600K	720K

Let us start from the fundamental linear regression analysis with the *least-squares method*, where the model parameters of a linear regression equation are estimated by finding the line that minimizes the sum of the squared distances from each sample data point to the model regression line.

Table 6.13 and Figure 6.8 show the sales revenue change from 2001 until 2007 for Cheng Kung Corporation (unit: US$K), which is used as an example. As you can clearly see, the data scattering is larger for scenario B than for scenario A.

As introduced in Table 6.12, a sales increase may be a power law with an annual increase rate of r; that is, the sales of a successive year (T) is provided by multiplying $(1 + r)$ on the sales of the focusing year ($T - 1$):

$$\begin{aligned} R_T &= (1+r)R_{(T-1)} \\ &= (1+r)^T R_0 \end{aligned} \quad (6.35)$$

or, by taking the logarithmic expression

$$\log R_T = T \log(1+r) + \log R_0 \quad (6.36)$$

This expression corresponds to a *log-linear regression* analysis. When r is small, Equation 6.35 is equivalent to a *linear regression* analysis:

$$R_T = (1+Tr)R_0 \quad (6.37)$$

In the simplest linear regression model, the dependent variable Y is related to only one variable X as

$$Y = a + bX$$

Table 6.13 Time Series of Sales Revenues in Cheng Kung Corporation (Unit: US$)

Budget Period	*2001*	*2002*	*2003*	*2004*	*2005*	*2006*	*2007*
Scenario A	71K	73K	79K	87K	92K	94K	100K
Scenario B	74K	69K	72K	95K	99K	87K	100K

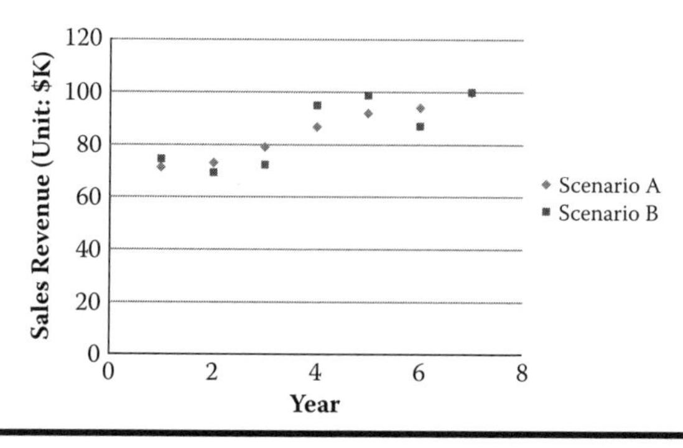

Figure 6.8 Sales revenue change from 2001 through 2007 for Cheng Kung Corporation for two scenarios.

The best-fitting estimates of a and b, denoted by $\hat{a}$ and $\hat{b}$, can be obtained from the data series (X_i, Y_i) as

$$\hat{b} = \frac{\sum_i (X_i - \overline{X})(Y_i - \overline{Y})}{\sum_i (X_i - \overline{X})^2}$$

$$\hat{a} = \overline{Y} - \hat{b}\overline{X}$$

where $\overline{X}$ and $\overline{Y}$ are the sample means of the dependent variable Y_i and independent variable X_i.

Microsoft Office Excel offers automatic calculation. Figure 6.9 shows how to obtain the regression trendline by using Excel. First, make the data table as shown in Figure 6.9a. Second, highlight the necessary columns and rows to the correlation, then click Insert, Charts, and then Scatter. You obtain the scattered plot, as shown in Figure 6.9b. Click the data points on the graph, and highlight them as in Figure 6.9b, then right-click to choose Add Trendline. In the Trendline option (Figure 6.9c), you may choose "Display equation," and "Display R squared value." Finally, you can receive the trendlines and their equations and R^2 values, as shown in Figure 6.9d. The *coefficient of determination* (R^2) measures the fraction of the total variation in Y that is explained by the variation in X. R^2 ranges in value from 0 to 1. A high R^2 indicates a high degree of correlation of

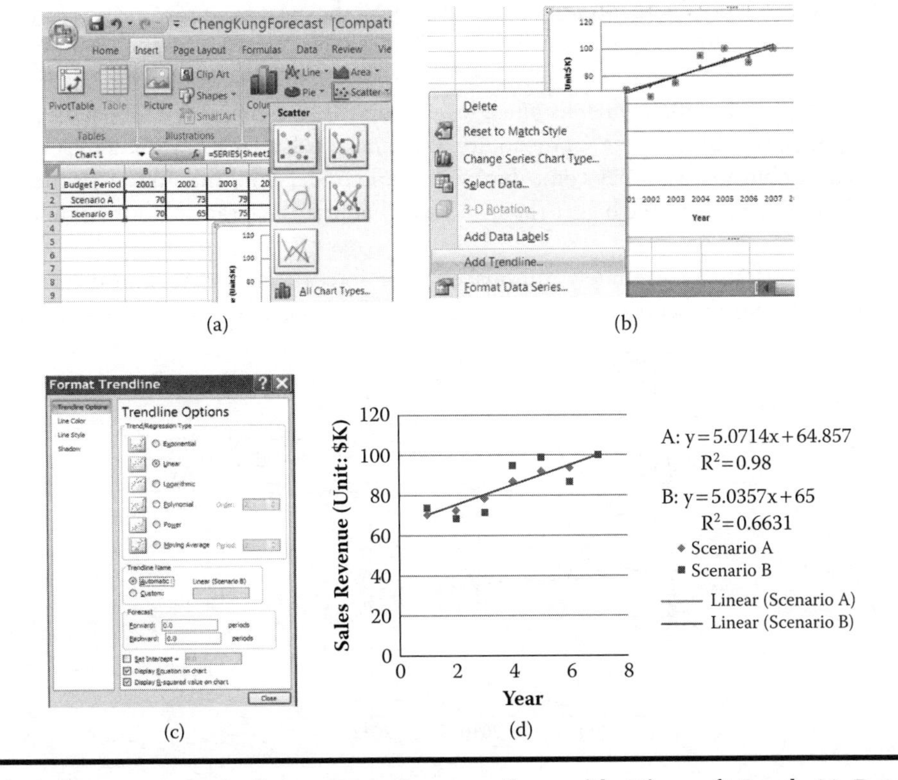

Figure 6.9 How to obtain linear regression equations with Microsoft Excel. (a) Date plot, (b) trendline adding, with (c) trendline option selection, and (d) final results for the sales trend of Cheng Kung Corporation (for two scenarios).

Y with X. Notice that the R^2 value is 0.98, very close to 1 for scenario A, while R^2 is only 0.74 for scenario B (widely scattered data).

6.2.2.3 Standard Deviation and Risk

The obtained equations for scenarios A and B are $Y = 5.0714X + 64.857$ and $Y = 5.0357X + 65.000$. By putting $X = 8$, we can forecast the sales amount for year 2008. We obtain 105.4 and 105.3, respectively, which are almost the same values. Is this amount highly probable to realize in 2008? Is there a possibility to dramatically increase or decrease the sales? A dramatic increase is an unexpected lucky situation, which does not cause a problem. However, in the event of a dramatic decrease, a cash flow problem may arise because the investment/purchase might have already been made according to the above sales projection, leading to the bankruptcy of the firm. Thus, the estimation should be as accurate as possible.

Based on the previous two scenarios, we can further calculate the standard deviation of the data points from the estimated linear regression curves. We can obtain $\sigma^2 = 2.1$ and 52, or $\sigma = 1.5$ and 7.2 for A and B, respectively. This difference suggests that we should describe the sales estimation as

$$68\%\ \text{probability: scenario A} = 105.4 \pm 1.5 \text{ and scenario B} = 105.3 \pm 7.2$$

$$95\%\ \text{probability: scenario A} = 105.4 \pm 3.0 \text{ and scenario B} = 105.3 \pm 15$$

When the historical sales fluctuation is small (scenario A), the most pessimistic sales estimation for 2008 is $105.4 - 3.0 = 102.4$, which is still higher than the sales amount in 2007. However, when the historical sales fluctuation is relatively large (scenario B), the most pessimistic sales estimation for 2008 is $105.3 - 15 = 90.3$, which is much lower than the sales amount in 2007. This low sales number happens, very likely, at 95% probability. If financial risk relates to the inability of the firm to meet its debt obligations as they come due, you should understand that the larger the standard deviation (data fluctuation), the bigger the firm's risk, because the financial estimation includes a large ambiguity.

Our ability to forecast accurately diminishes as we forecast farther out in time. Figure 6.10 depicts the relationship between risk and time. As the time horizon becomes longer, more

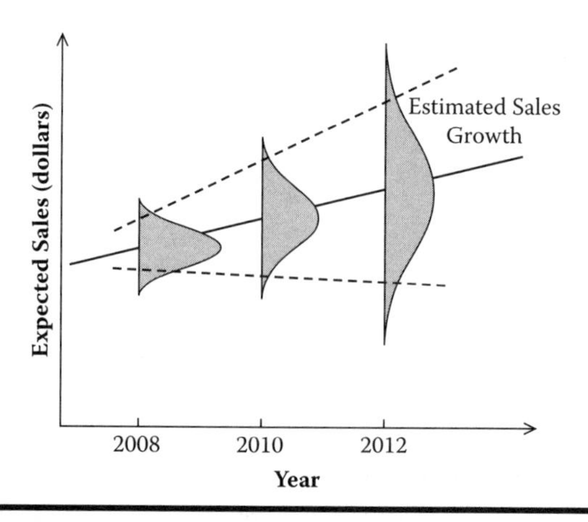

Figure 6.10 Standard deviation over time horizon.

uncertainty enters the forecast. Even though a forecast of sales shows a constant growth rate, the range of outcomes and probabilities increases as we move from year 8 to year 12 in Figure 6.10. The standard deviations increase for each forecast of sales growth.

6.2.3 Time Value of Money

When I was born, my parents started to accumulate $100 per month (annuity) for preparing my tuition coverage for a university. By the time I was 18 years old, 216 payments of $100 were accumulated, leading to approximately $50,000, which was more than sufficient to cover the university entrance and tuition fees for 4 years in my generation. Although the total principle was $21,600 (216 times of $100), the resultant amount accrued appears to be more than double. What is the trick? The trick is the *interest rate*!

The decision to purchase new equipment, a new plant or to start a new product requires additional capital allocation. Without identifying future benefits, which will surpass the current outlays, these expenditures are useless for the firm. Because there will be some time lag from the initial investment until profit creation in most of these cases, we have to learn the mathematical tools of the *time value of money*.

6.2.3.1 Future Value—Single Amount

A so-called greenhouse (incubator supporting institute) at the State University of Pennsylvania (SUP) offers a loan for a small start-up firm, lending for 5 years with an interest rate of 13% per year. Suppose that MMI borrows $100,000 from the greenhouse, how much should MMI pay back after 5 years? The calculation may not be difficult for you. The answer is

$$\$100{,}000 \times (1+0.13)^5 = \$184{,}200$$

We interpret this situation as "$100,000 now is equivalent to $184,200 in the future." The entrepreneur must understand that the firm needs to pay back roughly double (including various application fees) the initially loaned amount after 5 years. Be cautious of the compound interest system.

The following formula is essential:

$$\mathrm{FV} = \mathrm{PV} \times \mathrm{FV}_{\mathrm{IF}}, \quad \mathrm{FV}_{\mathrm{IF}} = (1+i)^n \tag{6.38}$$

where FV is the future value, PV is the present value, $\mathrm{FV}_{\mathrm{IF}}$ is the interest factor to interpret the present value to the future value, i is the interest rate, and n is the number of periods. We can also use a reverse form:

$$\mathrm{PV} = \mathrm{FV} \times \mathrm{PV}_{\mathrm{IF}}, \quad \mathrm{PV}_{\mathrm{IF}} = \frac{1}{(1+i)^n} = (1+i)^{-n} \tag{6.39}$$

where $\mathrm{PV}_{\mathrm{IF}}$ is the interest factor to interpret the future value to the present.

6.2.3.2 Future Value—Annuity

The annuity values are generally assumed to occur at the end of each period. When you invest amount A at the end of each year for n years, and your funds grow at the interest rate of i, what is the future value of this annuity? The first year's fund A will grow by the factor of $(1 + i)^{n-1}$ due to the $(n - 1)$ years stay in the bank until the closing, the second year's fund A will grow by the factor of $(1 + i)^{n-2}$ due to the $(n - 2)$ year stay, $(n - 1)$th year's fund A will grow by a factor of $(1 + i)$ due to just a one year stay in the bank, and the final fund A will not grow, but stay as it is. The future value of annuity FV_A thus can be calculated as

$$FV_A = A \times FV_{IFA} \tag{6.40}$$

$$FV_{IFA} = (1+i)^{n-1} + (1+i)^{n-2} + \cdots\cdots + (1+i)^1 + (1+i)^0$$

$$= \frac{[(1+i)^n - 1]}{i}$$

where FV_A is the future value of annuity, FV_{IFA} is the interest factor of future value of annuity, i is the interest rate, and n is the number of periods. In my parents' case, they accumulated \$100 for 216 months (12 × 18) at a monthly interest rate of 0.67% (annual 8%/12 months), so we get $FV_{IFA} = [(1 + 0.0067)^{216} - 1]/0.0067 = 482$.

The present value of an annuity, where every amount of A will be paid at the end of each year for n years (like a pension payment), can be calculated as

$$PV_A = A \times PV_{IFA} \tag{6.41}$$

$$PV_{IFA} = \left[\frac{1}{(1+i)^1}\right] + \left[\frac{1}{(1+i)^2}\right] + \cdots\cdots + \left[\frac{1}{(1+i)^{n-1}}\right] + \left[\frac{1}{(1+i)^n}\right]$$

$$= \frac{[1-(1+i)^{-n}]}{i}$$

where PV_A is the present value of annuity and PV_{IFA} is the interest factor of present value of annuity. Note that the interest factors for future and present values in the annuity's case are not just an inverse relationship as in the single amount case, but $FV_{IFA} = PV_{IFA} \times (1 + i)^n$.

Refer to Practical Exercise Problem 6.3, Pension Calculation, to further understand the present and future annuity concept.

6.2.4 Short-Term Financing

Your firm may need to borrow money for compensating a short-term cash flow deficit. This scenario often happens when an SBIR program is awarded to the firm, or a new purchase order comes for their product. There are several short-term financing types, which the financial company may offer to your firm:

1. Cash discount policy
2. Compensating balances
3. Commercial bank financing
4. Installment loans

Note that most of these interest rates are higher than the *prime rate*, which is the rate a bank charges its most creditworthy customers. The rate usually increases as a customer's credit risk gets higher.

Let us take the example of MMI. MMI plans to borrow $20,000 for a year. Four financial companies/banks provided the following four offers. The stated interest rate is 12% because of the following:

1. The interest is discounted.
2. There is a 20% compensating balance requirement.
3. Assume the interest is only $600 and the loan is for 90 days.
4. It is a 12-month installment loan.

You need to compute the effective interest rate per year under each of these assumptions, and to provide the advice to Barb Shay.

6.2.4.1 Discounted Loan

A "2/10, net 30 cash discount" means that we can deduct 2% if we remit out funds 10 days after billing, but failing this, we must pay the full amount by the 30th day. The standard formula for this example is

$$\text{Cost of failing to take a cash discount} = \frac{\text{Discount \%}}{(100\% - \text{Discount\%})} \times \frac{360}{(\text{Final due date} - \text{Discount period})}$$

$$= \frac{2\%}{(100\% - 2\%)} \times \frac{360}{(30 - 10)} = 36.72\%$$

Similarly, discounted loan means we will borrow (principle – interest), and return principle:

$$\text{Effective rate on discounted loan} = \frac{\text{Interest}}{(\text{Principal} - \text{Interest})} \times \frac{360 \text{ days}}{(\text{Days loan is outstanding})}$$

$$= \frac{12\%}{(100\% - 12\%)} \times \frac{360}{(360)} = 13.64\% \quad (6.42)$$

6.2.4.2 Compensating Balances

A bank may require that the business customer maintain a minimum average account balance, referred to as a *compensating balance*. Although MMI will borrow $20,000, 20% of this amount must be in the company's bank account. This means only (principal – compensating balance) can be loaned:

$$\begin{array}{c}\text{Effective rate with}\\ \text{compensating balances}\end{array} = \frac{\text{Interest}}{(\text{Principal} - \text{Compensating balance})} \times \frac{360 \text{ days}}{(\text{Days loan is outstanding})}$$

$$= \frac{12\%}{(100\% - 20\%)} \times \frac{360}{(360)} = 15\% \quad (6.43)$$

6.2.4.3 Commercial Bank Financing

The effective interest rate is based on the loan amount, the dollar interest, the length of the loan, and the method of repayment:

$$\text{Effective rate} = \frac{\text{Interest}}{(\text{Principal})} \times \frac{360 \text{ days}}{(\text{Days loan is outstanding})}$$
$$= \frac{\$600}{(\$20{,}000)} \times \frac{360}{90} = 12.0\% \tag{6.44}$$

6.2.4.4 Installment Loans

The most confusing borrowing arrangement to a customer is the installment loan, which calls for a series of equal payments over the life of the loan. MMI borrows \$20,000 first, paying back this principal and the interest of \$2400 (12%) by the end of 1 year. However, MMI needs to pay back the equal amount of (\$20,000 + \$2400)/12 months every month. Though MMI will pay a total of \$2400 in interest, as you can easily recognize, MMI does not have the use of \$20,000 for a whole year, rather the average outstanding loan may be approximately \$10,000. Thus, the effective rate is

$$\text{Effective rate on installment loan} = \frac{2(\text{Annual no. of payments})(\text{Interest})}{(\text{Total no. of payments} + 1)(\text{Principal})} \tag{6.45}$$
$$= \frac{2(12)(12\%)}{(12+1)(100\%)} = 22.15\%$$

6.2.5 Investment Decisions

A decision to build a new plant or to purchase new equipment is made according to the future profit that will be generated after the decision. There are two main factors to be considered: (1) how much in new earnings will be generated, and (2) how much cash flow will be created from the tax savings due to the depreciation.

We will again use an example of the tape-casting equipment at MMI. Table 6.14 provides annual depreciation and cash flow benefit related to the purchase of equipment, which was prepared by extending Table 6.9 by introducing new earnings R_1, R_2, …, R_8 for each year by this equipment (i.e., the ML product income) and the time value of money. The present value of total inflows (from year 1 to year 8) is provided:

$$\text{Present value of inflows} = \sum_n \frac{[(1-T)R_n + TD_n]}{(1+i)^n} \tag{6.46}$$

where R_n and D_n are the equipment-related earnings and annual depreciation in the nth year, T is the tax rate, and i is the discount (or interest) rate. Because present value of outflows (cost) is equal to the equipment price \$300,000, the *net present value* can be calculated as

Table 6.14 Annual Depreciation and Cash Flow Benefit Related to the Purchase of Equipment*

Items	*Year*							
	1	*2*	*3*	*4*	*5*	*6*	*7*	*8*
Earnings before depreciation and taxes	R_1	R_2	R_3	R_4	R_5	R_6	R_7	R_8
Depreciation base (tape-cast equipment)	$300,000	$300,000	$300,000	$300,000	$300,000	$300,000	$300,000	$300,000
Depreciation rate	0.143	0.245	0.175	0.125	0.089	0.089	0.089	0.045
Annual depreciation	$42,900	$73,500	$52,500	$37,500	$26,700	$26,700	$26,700	$13,500
Earnings before taxes	$R_1 - 42,900$	$R_2 - 73,500$	$R_3 - 52,500$	$R_4 - 37,500$	$R_5 - 26,700$	$R_6 - 26,700$	$R_7 - 26,700$	$R_8 - 13,500$
Tax rate (at 40%)	0.40	0.40	0.40	0.40	0.40	0.40	0.40	0.40
Earnings after taxes	$0.6(R_1 - 42,900)$	$0.6(R_2 - 73,500)$	$0.6(R_3 - 52,500)$	$0.6(R_4 - 37,500)$	$0.6(R_5 - 26,700)$	$0.6(R_6 - 26,700)$	$0.6(R_7 - 26,700)$	$0.6(R_8 - 13,500)$
New cash flow	$0.6R_1 +$ $17,160	$0.6R_2 +$ $29,400	$0.6R_3 +$ $21,000	$0.6R_4 +$ $15,000	$0.6R_5 +$ $10,680	$0.6R_6 +$ $10,680	$0.6R_7 +$ $10,680	$0.6R_8 + 5,400$
PV_{IF}	$\frac{1}{(1+i)^1}$	$\frac{1}{(1+i)^2}$	$\frac{1}{(1+i)^3}$	$\frac{1}{(1+i)^4}$	$\frac{1}{(1+i)^5}$	$\frac{1}{(1+i)^6}$	$\frac{1}{(1+i)^7}$	$\frac{1}{(1+i)^8}$
PV of cash flow	$(0.6R_1 +$ $17,160)/ $(1+i)^1$	$(0.6R_2 +$ $29,400)/ $(1+i)^2$	$(0.6R_3 +$ $21,000)/ $(1+i)^3$	$(0.6R_4 +$ $15,000)/ $(1+i)^4$	$(0.6R_5 +$ $10,680)/ $(1+i)^5$	$(0.6R_6 +$ $10,680)/ $(1+i)^6$	$(0.6R_7 +$ $10,680)/ $(1+i)^7$	$(0.6R_8 +$ $5,400)/ $(1+i)^8$

*Extended from Table 6.9 by introducing new earnings by this equipment and the time value of money.

$$\text{Net present value} = \text{Present value of inflows} - \text{Present value of outflows}$$
$$= \sum_n \frac{[(1-T)R_n + TD_n]}{(1+i)^n} - \sum_n D_n \qquad (6.47)$$

Note that $\sum_n D_n$ is equal to the equipment cost.

Instead of computing Table 6.14 further, we will consider a much simpler model to obtain intuitive analytical sense to the net present value. Let us adopt an assumption that all R_ns and D_ns are the same: $R_n = R_0$, $D_n = D_0$ = equipment cost/8. In this case

$$\text{Net present value} = [(1-T)R_0 + (T)D_0]\sum_n \frac{1}{(1+i)^n} - \sum_n D_n \qquad (6.48)$$
$$= \frac{[(1-T)R_0 + (T)D_0][1-(1+i)^{-8}]}{i8D_0}$$
$$= [(1-T)R_0 + (T)D_0]\mathrm{PV}_{\mathrm{IF}A}(8,i)8D_0$$

In the case of $R_0 = D_0$, $\mathrm{PV}_{\mathrm{IF}A} = 8$ for $i = 0$, it is obvious that $R_0 \geq D_0$ is required to obtain a positive net present value. Only when the net present value is positive does this investment (the purchase of the tape-casting equipment) become meaningful. Figure 6.11 shows a calculation example for the case of $R_0 = 2D_0$, where the net present value (unit of D_0) is plotted as a function of the discount rate (%). This figure indicates that when the discount rate is 8% and $R_0 = 2D_0 =$ \$75,000 (= 2 × \$300,000/8), the net present value is about $1.2D_0$ = \$45,000. A *profitability index* is obtained:

$$\text{Profitability index} = \frac{\text{Present value of the inflows}}{\text{Present value of the outflows}} \qquad (6.49)$$
$$= \frac{\$300{,}00 + \$45{,}000}{\$300{,}000} = 1.15$$

The crossover point with the horizontal axis, where the net present value becomes zero, is about 12% of the discount rate, which is called the *internal rate of return* (IRR). The IRR determines the yield of an investment, which equates the cash outflows (cost) with the subsequent cash inflows.

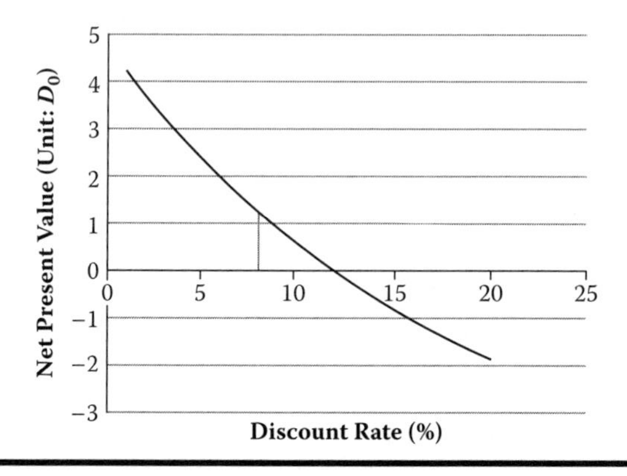

Figure 6.11 Net present value profile plotted as a function of discount rate ($R_0 = 2D_0$).

Chapter Summary

6.1 Profit = Sales revenue − Cost
Sales revenue = Unit cost × Number of products sold

6.2 Cost = Product cost + Period cost
Period cost includes research and development, selling, and administrative costs.

6.3 Product cost (= Cost of goods sold) is the cost of purchases plus transportation costs for a retailer, while it is equal to manufacturing costs for a manufacturer.

6.4 Manufacturing cost = Direct materials + Direct labor + Manufacturing overhead

6.5 Income statement—base format

Sales
Less: Cost of goods sold
Gross margin
Less: Selling and administrative expenses
Less: Depreciation (building, etc.)
Earnings before interests and taxes (EBIT)
Less: Interest charges
Earnings before taxes (EBT)
Less: Income tax expenses (at 40%)
After-tax net income (EAT)

6.6 Balance sheet—base format

Assets
Current assets
Cash
Marketable securities
Accounts and notes receivable
Inventories
Total current assets
Property, plant, and equipment
Buildings, machines, and equipment
Less: accumulated depreciation
Land
Total property, plant, and equipment

Other assets
Receivables due after 1 year
Other
Total assets
Liabilities and shareholders' equity
Current liabilities
Accounts payable
Accrued liabilities
Current maturity of long-term debt
Federal income and other taxes
Dividends payable
Total current liabilities
Other liabilities
Long-term debt
Total liabilities
Shareholders' equity
Preferred stock
Common stock
Additional paid-in capital
Retained earnings
Total shareholders' equity
Total liabilities and shareholders' equity

6.7 Price elasticity of demand

$$E = \frac{d(\log Q_d)}{d(\log P)} = \left(\frac{dQ_d}{dp}\right)\left(\frac{P}{Q_d}\right) = \frac{\text{Percent change in quantity of demand}}{\text{Percent change in price}}$$

6.8 Break-even analysis:

$$\text{BE} = \frac{\text{FM}}{\text{CM}} = \frac{\text{Fixed cost}}{\text{Contribution margin}}$$

(CM = Price − Variable cost per unit)

6.9 Profit maximization requires the output quantity at which marginal revenue equals marginal cost: MR = MC.

6.10 Depreciation of properties and equipment saves taxes and generates cash flows. The total cash flows generated should equal the product of equipment price and tax rate (by neglecting the time value of money).

6.11 Price-earnings ratio:

$$\frac{P}{E} = \frac{\text{Current price of the common stock}}{\text{Earnings per share}}$$

$$\text{Earnings per share} = \frac{\text{Earnings available to common stockholders}}{\text{Number of shares outstanding}}$$

6.12 Financial forecasting often uses a linear/log-linear regression model. Microsoft Excel provides easy simulation for this purpose. The larger the standard deviation (data fluctuation) in the financial data, the bigger the firm's risk.

6.13 Time value of money: *present value* and *future value*, single amount and annuity:

$$FV = PV \times FV_{IF}, \quad FV_{IF} = (1+i)^n$$

$$PV = FV \times PV_{IF}, \quad PV_{IF} = \frac{1}{(1+i)^n} = (1+i)^{-n}$$

$$FV_A = A \times FV_{IFA}, \quad FV_{IFA} = \frac{[(1+i)^n - 1]}{i}$$

$$PV_A = A \times PV_{IFA}, \quad PV_{IFA} = \frac{[1-(1+i)^{-n}]}{i}$$

6.14 Short-term financing includes
1. Cash discount
2. Compensating balances
3. Commercial bank financing
4. Installment loans

6.15 Investment decision: net present value should be positive.

$$\text{Net present value} = \text{Present value of inflows} - \text{Present value of outflows}$$
$$= \sum_n \frac{[(1-T)R_n + TD_n]}{(1+i)^n} - \sum_n D_n$$

where R_n and D_n are the equipment/property-related earnings (return) and annual depreciation in the nth year. Note that $\Sigma_n D_n$ is equal to the equipment/property cost.

$$\text{Profitability index} = \frac{\text{Present value of the inflows}}{\text{Present value of the outflows}}$$

The IRR determines the yield of an investment, which equates the cash outflows (cost) with the subsequent cash inflows.

Practical Exercise Problems

P6.1 Depreciation as a Tax Shield

Corporations A and B created the same EBDT of \$400,000. Corporation A charges off \$100,000 in depreciation, while Corporation B charges off none. Compute the cash flow for both companies, and discuss the cash flow difference. Use the tax rate of 40%.

Answers

\$280,000, \$240,000; \$40,000

P6.2 Demand Elasticity

Suppose that the demand function is provided by $Q_d = A/P$ (P = price; A = constant); compute the price elasticity of demand, total revenue under the condition that all goods manufactured are sold (i.e., equilibrium), and marginal revenue.

Answers

−1, A, 0

Comments

We will consider a general case; $Q_d = \mathrm{A}\ \mathrm{P}^{-\lambda}$ $(0 \leq \lambda \leq 1)$.

Total revenue: $\mathrm{R} = \mathrm{P} \cdot Q_d = \mathrm{A}\ \mathrm{P}^{1-\lambda}$
Marginal revenue: $\mathrm{MR} = d\mathrm{R}/dQ_d = [(\lambda - 1)/\lambda]\mathrm{P}$
Price elasticity of demand: $\mathrm{E} = (dQ_d/d\mathrm{P})\ (\mathrm{P}/Q_d) = -\lambda$

$\lambda = 0$ means "perfectly inelastic"; $E = 0$ and $MR = -\infty$, while $\lambda = 1$ corresponds to this question's case; unitary $E = -1$; and $MR = 0$. This negative exponent model corresponds to an inelastic case, and MR is always negative or zero.

P6.3 Pension Calculation

Barb Shay wishes to retire in 20 years, at which time she wants to have accumulated enough money to receive a yearly annuity of \$50,000 for 25 years after retirement. During the period before retirement, suppose that she can earn 8% annually, while after retirement she can earn 10% annually. What annual contributions to the retirement fund will allow her to receive the \$50,000 annuity?

Solution

We will calculate first the amount of money Barb will need at the retirement point for paying \$50,000 annual annuity for 25 years at an interest rate of 10%. The present value of annuity (at the retirement point) should be

$$\mathrm{PV}_A = A_1 \times \mathrm{PV}_{\mathrm{IFA}} = \frac{A_1 \times [1-(1+i)^{-n}]}{i} = \frac{50{,}000 \times [1-(1+0.10)^{-25}]}{0.10} = 50{,}000 \times 9.077$$
$$= \$453{,}850$$

The above amount should be accumulated for 20 years from now until retirement with the yearly annuity of A at an interest rate of 8%:

$$FV_A = A_2 \times FV_{IFA} = \frac{A_2 \times [(1+i)^n - 1]}{i} = \frac{A \times [(1+0.08)^{20} - 1]}{0.08} = A \times 45.762$$

$$= \$453,850$$

Thus, $A = \$9918$.

P6.4 Research Fund Forecasting

Forecasting methods are initially categorized into two types: *deterministic* models and *probabilistic* (or *stochastic*) models. The time series–linear regression estimation in Section 6.2.2 corresponds to the deterministic model. The probabilistic model estimation is described here.

MMI submitted 10 R&D proposals to federal SBIR programs and private corporations last year; these are listed in Table 6.15 with the requested budget amount of the research fund. During the proposal preparation period, Barb Shay, CTO, contacted each program officer to obtain detailed information on what the program officer would like to seek through the program. This lobby work is essential to a successful award. According to the reply/response from the program officer, Barb scored a success probability for each proposal in five ranks: 0, 25, 50, 75, and 100%. These probabilities are also inserted in the table. Estimate the R&D budget forecast for the next year on behalf of Barb, taking into account the following two possible scenarios:

1. All proposal reviews are conducted independently, and the awards are decided without restrictions.
2. The federal institutes regulate award duplication; that is, there can only be one award at one time from the same institute. In this list, three piezo proposals are submitted to Institute X, four motor proposals to Institute Y, and three medical proposals to Institute Z.

Table 6.15 MMI's Proposals Submitted Last Year with the Proposed Budget and Probability

Proposal Name	*Proposed Budget*	*Probability Estimation*
Piezo A	$80,000	25%
Piezo B	$80,000	75%
Piezo C	$300,000	50%
Motor D	$80,000	75%
Motor E	$150,000	75%
Motor F	$300,000	25%
Motor G	$800,000	25%
Medical H	$80,000	50%
Medical J	$150,000	25%
Medical K	$300,000	75%

Table 6.16 MMI's Proposals Submitted Last Year with the Expected Amount of Funds

Proposal Name	*Proposed Budget*	*Probability Estimation*	*Expected Funds*
Piezo A	$80,000	25%	$20,000
Piezo B	$80,000	75%	$60,000
Piezo C	$300,000	50%	$150,000
Motor D	$80,000	75%	$60,000
Motor E	$150,000	75%	$112,500
Motor F	$300,000	25%	$75,000
Motor G	$800,000	25%	$200,000
Medical H	$80,000	50%	$40,000
Medical J	$150,000	25%	$37,500
Medical K	$300,000	75%	$225,000
		Total	$980,000

Solution

1. When all proposal reviews are conducted independently and the award decided without any restrictions, the total expected funds can be obtained by ∑Proposal budget × Probability. The answer is $980,000, as shown in Table 6.16.
2. However, when there is one award at one time from the same institute, the situation is different. Among three piezo proposals to Institute X, four motor proposals to Institute Y, and three medical proposals to Institute Z, we need to select one each from these three institutes. How can we select them? We use a hypothesis that the proposal with the highest probability score should be selected. Because Barb got a good feeling during her communication with the program officer, that proposal must have obtained higher evaluations even from the proposal reviewers, leading to the award. MMI may be awarded the three proposals: Piezo B, Motor D or E (50:50 chance of selection), and Medical K. From this selection,

$$\text{Expected fund} = \$60,000 + \frac{\$60,000 + \$112,500}{2} + \$225,000 = \$371,250$$

References

1. Hilton, R. W. *Managerial accounting.* New York: McGraw-Hill Irwin, 2008.
2. Sobel, M. *The 12-hour MBA program.* Englewood Cliffs, NJ: Prentice Hall, 1994.
3. Block, S. B., and G. A. Hirt. Financial management, 12th ed. New York: McGraw-Hill/Irwin, 2008.
4. Wikipedia, "Tax Credit," http://en.wikipedia.org/wiki/Tax_credit.
5. Thomas, C. R., and S. C. Maurice. Managerial econo*mics*, 8th ed. New York: McGraw-Hill-Irwin, 2005.

Chapter 7

Quantitative Business Analysis—Beneficial Tools for Business

You will learn management science in this chapter, including the following topics:

- Analyzing and building mathematical models of complex business situations
- Solving and refining the mathematical models typically using Excel spreadsheets or other software programs to gain insight into the business situation
- Communicating and implementing the resulting insights and recommendations based on these models

From my experience, the mathematical approaches used in this chapter are already known by most readers with intensive engineering backgrounds. Thus, our focus will be on how to use Excel software more efficiently. We will use Excel Solver extensively to solve most of the problems throughout this chapter. This chapter owes extensively to Lawrence and Pasternack [1].

Add-In Process for Microsoft Excel 2003: If you are not familiar with (or have not uploaded) Solver in Microsoft Excel 97 or above, please do it now. Figure 7.1 shows the starting page of Solver. If, when you click Data, you cannot find the Solver icon, your Solver function is inactive. From the Excel Option page, click Add-Ins, then click Solver for activating.

Add-In Process for Microsoft Excel 2007:

1. Click Office button
2. Click Excel option
3. Click Add-Ins
4. Click Solver Add-In

Refer to Figure 7.2 for the Solver Add-In process.

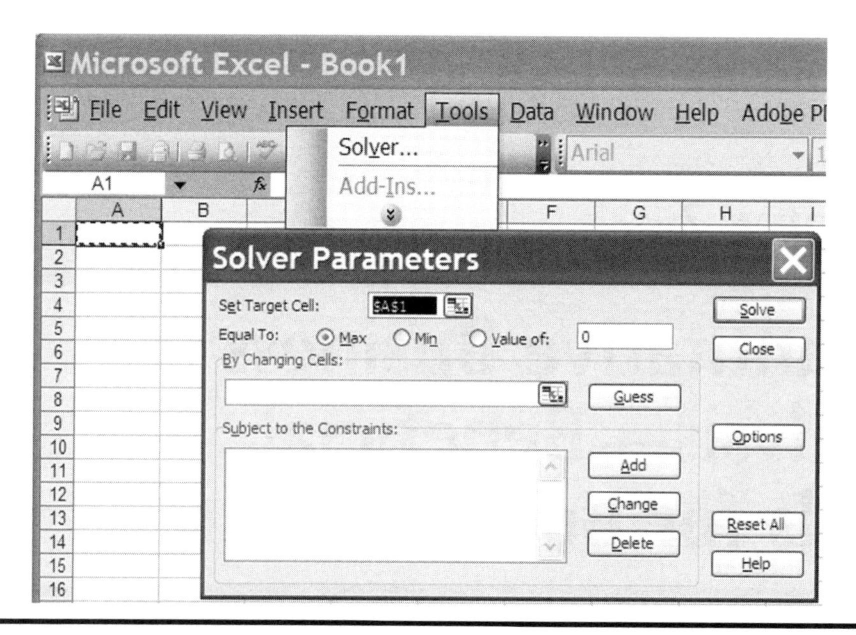

Figure 7.1 Microsoft Excel 2003 with Solver.

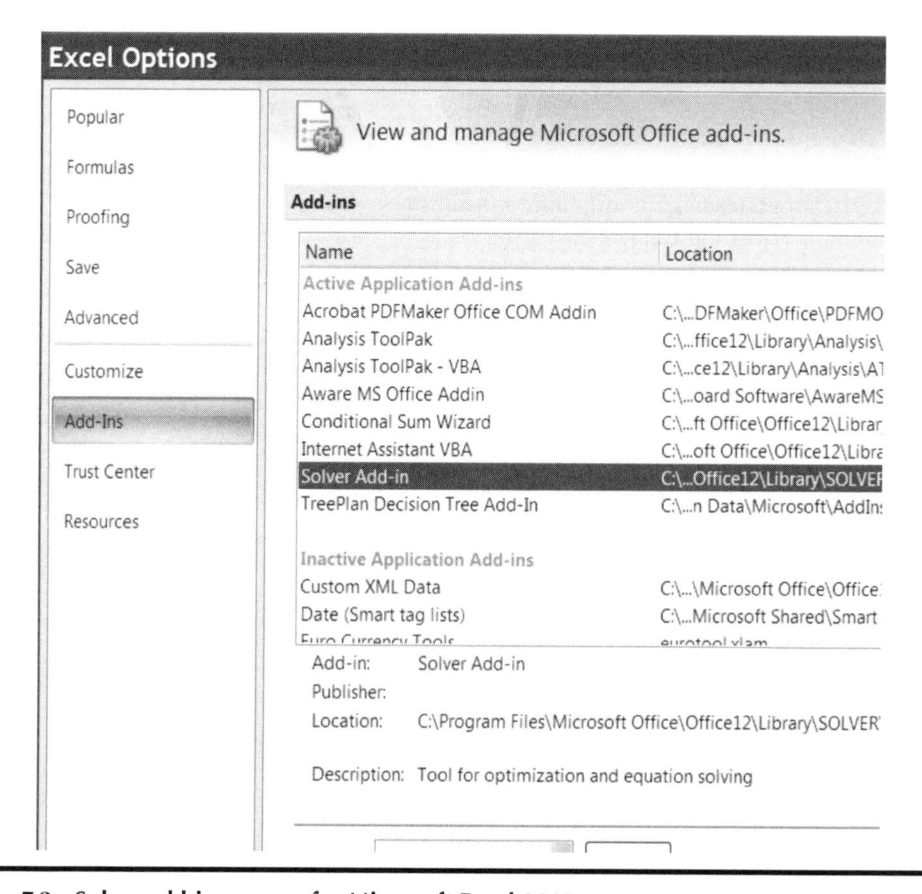

Figure 7.2 Solver add-in process for Microsoft Excel 2007.

7.1 Linear Programming

Let us use a simple production schedule problem to explain the solving procedure in detail.

MMI Example—Part I

Micro Motors Inc. (MMI) produces multilayer actuators (MLAs) and ultrasonic motors (USMs) every month. Table 7.1 summarizes the necessary data for producing MLAs and USMs. Though USMs generate a larger gross profit per unit with a smaller amount of raw materials per unit than MLAs, USMs require a working period five times longer than MLAs. MMI has two major constraints: (1) labor—total 500 h/month maximum (= 3 engineers × 40 h/week × 4.17 weeks/month), and (2) raw materials available—total 300 g/month maximum. MMI would like to determine the optimum production quantities (unit number) X for MLAs and Y for USMs in order to maximize the gross profit.

Part I seeks the solutions for nonnegative continuous numbers of X and Y.

7.1.1 Mathematical Modeling

The production quantities (unit number) X for MLAs and Y for USMs are called *decision variables*. The total gross profit is provided by the sum of MLA profit \$$25X$ and USM profit \$$100Y$:

$$25X + 100Y$$

which MMI would like to maximize. This is called *objective function*. Two major physical constraints at MMI are expressed with functional constraints, ≤, ≥, or =:

Labor: $5X + 25Y \leq 500$
Raw materials: $5X + 2Y \leq 300$

Now we have the following constrained mathematical model for this problem:

$$\text{MAXIMIZE:} \quad 25X + 100Y \quad \text{(Total profit)}$$

$$\text{SUBJECT TO:} \quad 5X + 25Y \leq 500 \quad \text{(Labor)}$$

$$5X + 2Y \leq 300 \quad \text{(Raw materials)}$$

$$X, Y \geq 0 \quad \text{(Nonnegativity)}$$

(The condition that "X and Y are integers" is not added in Part I.)

Table 7.1 Production Data for Multilayer Actuators (MLAs) and Ultrasonic Motors (USMs)

Product	*Gross Profit/Unit*	*Required Labor/Unit*	*Gross Profit/Labor*	*Required Material/Unit*
MLA	\$25	5 h	\$5/h	5 g
USM	\$100	25 h	\$4/h	2 g

7.1.2 Graphical Solution

Two constraints can be transformed as

$$Y \leq -\left(\frac{1}{5}\right)X + 20$$

and

$$Y \leq -2.5X + 150$$

Taking into account X and Y nonnegative, the constraints exhibit a quadrilateral region shown in gray in Figure 7.3. The coordinates for four corner points are obtained as (0,0), (0,20), (60,0), and M. The red-star point M can be obtained from the equation

$$Y = -\left(\frac{1}{5}\right)X + 20 = -2.5X + 150$$

as (56.52, 8.696). This point of the feasible region is called the *extreme point*, which is the intersection of the limit on labor time and the limit for materials quantity.

Equating the total profit to K (where $25X + 100Y = K$), we obtain

$$Y = -\left(\frac{1}{4}\right)X + \frac{K}{100}$$

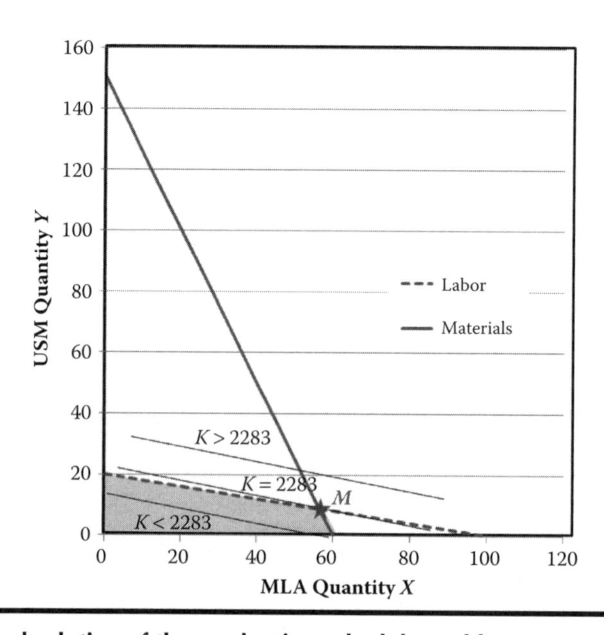

Figure 7.3 Graphical solution of the production schedule problem.

By scanning this line upward and downward, we obtain the maximum K value. When $K > 2283$, this line will not overlap with the quadrilateral region, while for $K < 2283$, the line overlaps with the quadrilateral area. As you can clearly see, the K value becomes the maximum when this line crosses the star point M (56.52, 8.696), that is, the extreme point. Now we can obtain the maximum profit $2283 for the MLA production of 56.52 units and the USM quantity of 8.696 units.

7.1.3 Excel Spreadsheet Solver

We will now trace the same solution by using the Solver function of the Excel spreadsheet. Solver is an option found in the Tools menu. If you do not see Solver listed, you must check the Solver Add-In box of the Add-Ins option under the Tools menu.

We will start from Figure 7.4, which shows how to assign the cells. To use Solver, I recommend you designate cells to contain the following:

- Values of the decision variables (changing cells)—cells C5 and D5
- Value of the objective function (target cell)—cell E7
- Total value of the left-hand side of the constraints—cells E9 and E10

Also familiarize yourself with Excel's SUM and SUMPRODUCT functions for this setting. Cell E7 is total profit, which is provided by the sum of MLA and USM profits. It should be equal to SUMPRODUCT(C5:D5,C7:D7), which means that this cell is computed by

$$(C5 \times C7 + D5 \times D7)$$

As you may recognize, if this cell E7 is copied to E9, the row numbers are automatically shifted to

$$(C7 \times C9 + D7 \times D9)$$

MMI-ProductionSched... M

E7 =SUMPRODUCT($C$5:$D$5,C7:D7)

	A	B	C	D	E	F	G
1		MMI					
2		Product Schedule					
3							
4			MLA	USM			
5	Decision variables:	Units	1	1			
6					Total		Limit
7	Objective function:	Profit	25	100	125		
8							
9	Constraints:	Labor	5	25		<=	500
10		Materials	5	2		<=	300
11							

Graphical | Continuous | Integer

Ready 100%

Figure 7.4 Microsoft Excel 2007 with Solver for an MMI production schedule.

Because we should not shift the row number 5, which is the variable unit number (X and Y) row, we must fix the row number. For this purpose, we will use the "$" symbol. Thus, instead of SUMPRODUCT(C5:D5,C7:D7), we type here:

SUMPRODUCT($C$5:$D$5,C7:D7)

The F4 function key at the top of the keyboard is easy to use for this exchange. Highlight only the first array (C5:D5) of the formula in the formula bar and press the F4 key; then you can insert dollar signs.

Next, copy E7 into E9 and E10, so that you obtain the numbers 30 and 7 in these cells, because the initial unit numbers in C5 and D5 are tentatively 1 for both.

We now click Solver in the Tools menu. This gives the dialog box shown in Figure 7.1. Note that the dialog box is the same for both Excel 2003 and 2007.

Step 1: Set Target Cell. This is the cell for the objective function. With the cursor in the Set Target Cell box, click on cell E7.

Step 2: Equal To. Problem type (maximization or minimization) is set. Leave the button for Max highlighted. (When you need to compute the minimum, choose "Min.")

Step 3: Changing Cells. Changing cells are the cells that contain the decision variables. With the cursor in the By Changing Cells box, highlight cells C5 and D5.

Step 4: Subject to the Constraints. To introduce the constraints, click the Add button in Subject to the Constraints. You will get the Add Constraint dialog box shown in Figure 7.5. The constraint options include ≤, =, ≥, int, and bin. The last two options restrict a cell to be integer-valued or binary (0 or 1), respectively. With the cursor in the Cell Reference box, highlight cells in E9 and E10. Click the direction as ≤ in this case. With the cursor in the Constraint box, highlight cells G9 and G10. Click Add, if we have more constraints. If finished, click OK.

Step 5: Options. Click Options to open the dialog box shown in Figure 7.6. It is important to check both Assume Linear Model and Assume Non-Negative. Click OK.

Step 6: Solve. Figure 7.7 shows the completed Solver dialog box for the MMI product schedule. To solve for the optimal solution, click Solve.

Step 7: Reports. Once an optimal solution is found, the Solver Results dialog box shows up (Figure 7.8). At the same time, Figure 7.4 is automatically changed to Figure 7.9. You can find that the solutions (quantity $X = 56.52$, quantity $Y = 8.696$, and the maximum profit $2283) are exactly the same in the calculation error range.

Two further reports are available: Answer Report and Sensitivity Report. Highlight one of them and click OK. Note that these two reports are created on separate spreadsheets. No new dialog boxes show up.

Answer Report: Click on the Answer Report tab at the bottom of the spreadsheet to access the window shown in Figure 7.10. The Answer Report includes three parts: Target Cell, Adjustable Cells, and Constraints.

Target Cell provides the optimal value of the objective function in the Final Value column, with the initial input ($125 in our trial). Similarly, the optimum values for the decision variables are shown in the Final Value column of the Adjustable Cells section. In the Constraints section, the Cell Value gives the total values of the left side of the constraints. The information entered in the Constraint dialog box of Solver appears in the Formula column. The Slack column shows the amount of slack for each constraint. Note that if the slack is 0, the word "Binding" is printed in the Status column; "Not Binding" is printed when the slack is positive.

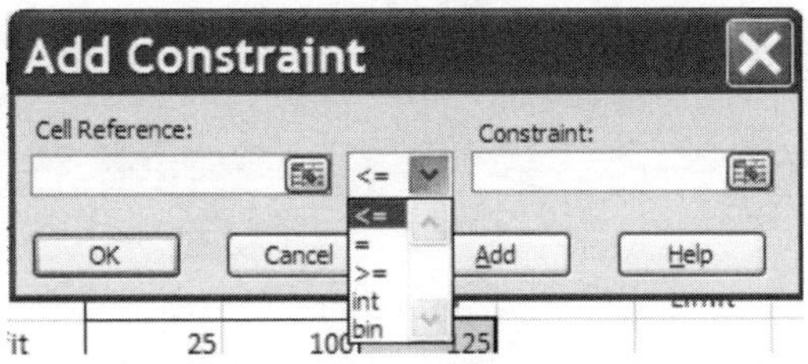

Figure 7.5 Adding the Constraint dialog box in Solver.

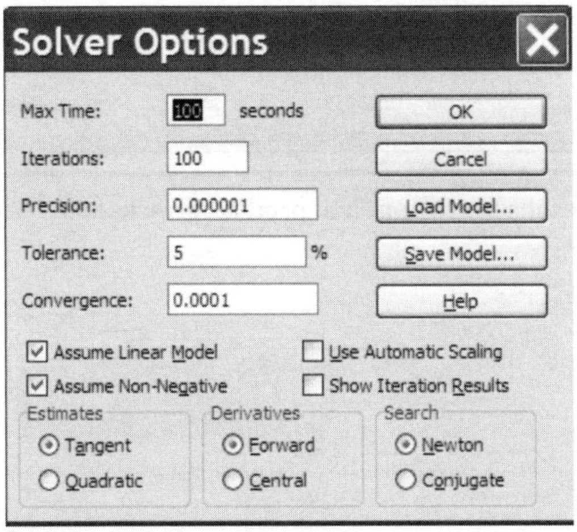

Figure 7.6 Solver Options dialog box.

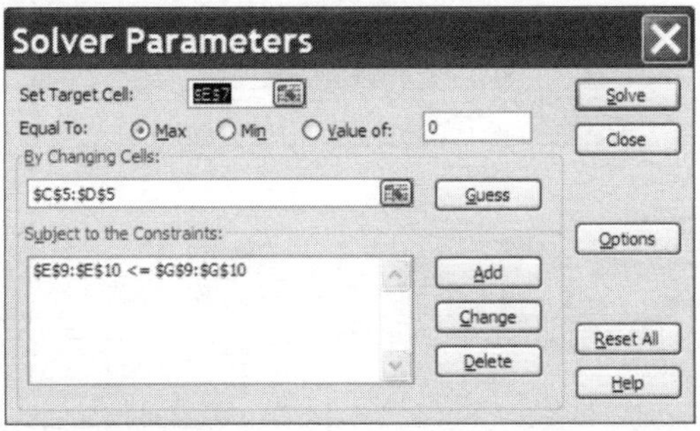

Figure 7.7 Solver Parameters dialog box.

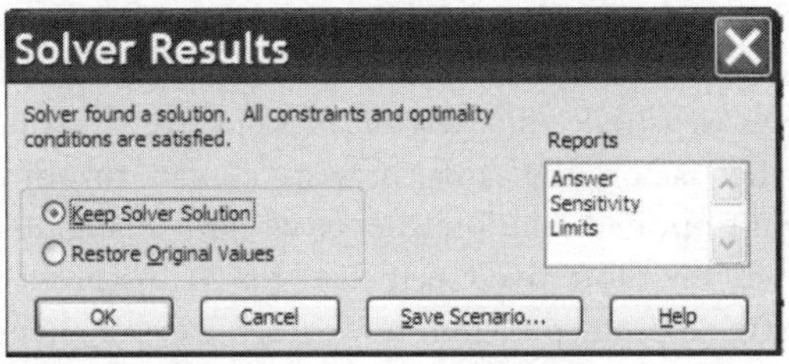

Figure 7.8 Solver Results dialog box.

MMI-ProductionSche...

E7 =SUMPRODUCT($C$5:$D$5,C7:D7)

	A	B	C	D	E	F	G
1		MMI					
2		Product Schedule					
3							
4			MLA	USM			
5	Decision variables:	Units	56.52174	8.695652			
6					Total		Limit
7	Objective function:	Profit	25	100	2282.609		
8							
9	Constraints:	Labor	5	25	500	<=	500
10		Materials	5	2	300	<=	300

Continuous | Integer | Binary

Figure 7.9 Computed results for the optimal production schedule.

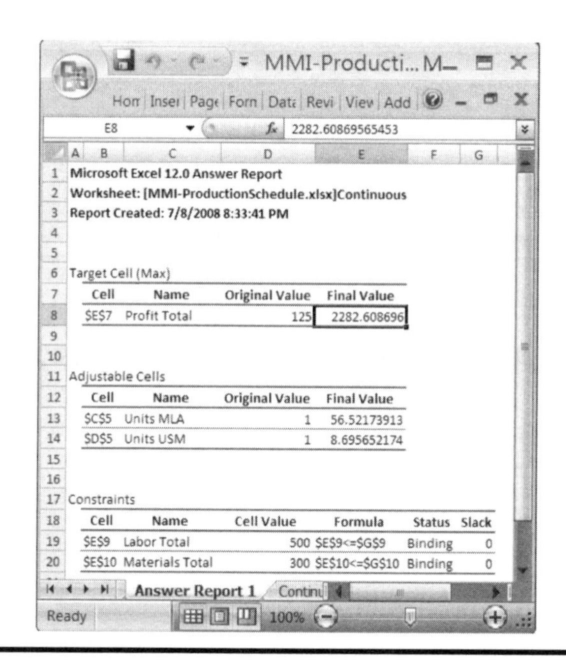

E8 2282.60869565453

Microsoft Excel 12.0 Answer Report
Worksheet: [MMI-ProductionSchedule.xlsx]Continuous
Report Created: 7/8/2008 8:33:41 PM

Target Cell (Max)

Cell	Name	Original Value	Final Value
$E$7	Profit Total	125	2282.608696

Adjustable Cells

Cell	Name	Original Value	Final Value
$C$5	Units MLA	1	56.52173913
$D$5	Units USM	1	8.695652174

Constraints

Cell	Name	Cell Value	Formula	Status	Slack
$E$9	Labor Total	500	$E$9<=$G$9	Binding	0
$E$10	Materials Total	300	$E$10<=$G$10	Binding	0

Answer Report 1 | Continu

Figure 7.10 Answer report.

Sensitivity Report: The Sensitivity Report (Figure 7.11) exhibits important information concerning the effects of changes to either an objective function coefficient or a constraint value.

In the Adjustable Cells, sensitivity analysis of an objective function coefficient answers the question, "Keeping all other factors the same, how much can an objective function coefficient change without changing the optimal solution (that is, without changing the production quantities of X and Y)?" For example, the Objective Coefficient for MLA (profit per unit) is initially \$25. Without changing the extreme point position, how much can we change this number? In other

Microsoft Excel 12.0 Sensitivity Report
Worksheet: [MMI-ProductionSchedule.xlsx]Continuous
Report Created: 7/8/2008 8:35:58 PM

Adjustable Cells

Cell	Name	Final Value	Reduced Cost	Objective Coefficient	Allowable Increase	Allowable Decrease
$C$5	Units MLA	56.52173913	0	25	225	5
$D$5	Units USM	8.695652174	0	100	25	90

Constraints

Cell	Name	Final Value	Shadow Price	Constraint R.H. Side	Allowable Increase	Allowable Decrease
$E$9	Labor Total	500	3.913043478	500	3250	200
$E$10	Materials Total	300	1.086956522	300	200	260

Figure 7.11 Sensitivity report.

words, how much can we swing the slope of the objective function curve in Figure 7.3? The answer is provided by Allowable Increase and Allowable Decrease. The MLA profit can be changed from \$20 (= 25 − 5) to \$250 (= 25 + 225). This range is called *range of optimality*. Similarly, the gross profit for USM can be changed from \$10 (= 100 − 90) to \$125 (100 + 25). Note that the value of the objective function (total profit) will, of course, change, according to the unit profit change.

Shadow price is a useful number from a practical operation viewpoint. The shadow for a constraint is the change to the objective function value per unit increase to its constraint value. For example, MMI set the maximum labor (working time) as 500 h. If MMI asks the researcher to work one more hour (up to 501 h), how much will total profit increase (a sort of "what-if" question)? The answer can be found in the Shadow Price column as \$3.91. If this amount (additional profit) is higher than the additionally required cost (such as an extra portion of the overwork payment, after subtracting the regular salary rate, utility, etc.), MMI should move in this direction.

The *range of feasibility* is the range of values for a constraint value in which the shadow prices for the constraints remain unchanged. The range is provided by Allowable Increase and Allowable Decrease. In the labor constraint in MMI, the range will be from 300 h (= 500 − 200) to 3750 h (= 500 + 3250), while in the materials constraint, from 40 g (= 300 − 260) to 500 g (= 300 + 200).

7.1.4 Integer Model

MMI Example—Part II

As described in Part I, MMI produces MLAs and USMs every month. Table 7.1 summarized the necessary data for producing MLAs and USMs. Though USMs generate a larger gross profit per unit, with a smaller amount of raw materials per unit than MLAs, USMs require a working period five times longer than MLAs. MMI has two major constraints: (1) labor—total 500 h/month maximum (= 3 engineers × 40 h/week × 4.17 weeks/month), and (2) raw materials available—total 300 g/month maximum. MMI would like to determine the optimum production quantities (unit number) X for MLAs and Y for USMs in order to maximize the gross profit.

Part II seeks the solutions for nonnegative integer numbers of X and Y.

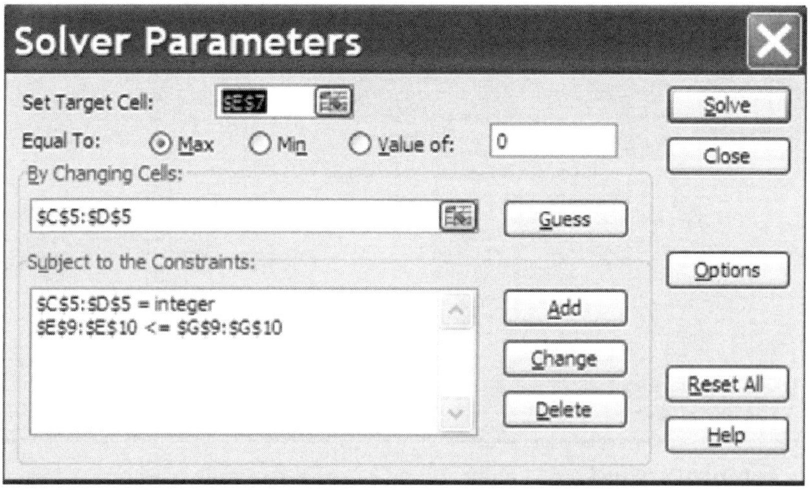

Figure 7.12 Solver Options dialog box (integer analysis).

Solution—Part II

In Part I, we found that one solution (quantity $X = 56.52$, quantity $Y = 8.696$, and the maximum profit \$2283) is the optimal solution. However, the product unit should be an integer, since we cannot sell 0.696 unit of USM, in practice.

If we initially consider the graphical solution first, the integer optimum solution should exist around the MLA production of 56.52 units and the USM quantity of 8.696 units. Let us take into account the following positions: $(X,Y) = (56,8)$, (57,8), (58,8), (59,8); (53,9), (54,9), (55,9), and (56,9). Only (56,8), (53,9), (54,9), and (55,9) pass the criteria: labor $5X + 25Y \leq 500$, and raw materials $5X + 2Y \leq 300$. Thus, it is obvious the maximum profit can be obtained for the position (55,9).

Let us now solve the same question on Excel Solver by adding the integer constraint to the decision variables (quantities X and Y). We will add one more condition to "Subject to the Constraints":

$$\$C\$5{:}\$D\$5 = \text{integer}$$

as shown in Figure 7.12 (find the difference from Figure 7.7). The solution can easily be found (Figure 7.13). The Answer Report is shown in Figure 7.14, where the optimum production schedule—55 units MLA, 9 units USM, and total profit \$2275—is the same as we analyzed graphically. Note that there is Slack in Materials Constraint (7 g material remained), leading to a "Not Binding" note in the Status column. Also note that there is a slight decrease in the total profit (\$2275) in comparison with the profit obtained from the continuous parameter model.

It is important to know that the Excel Solver does not generate sensitivity analysis for the integer analysis. Thus, I strongly recommend you use the continuous parameter analysis in parallel to the integer analysis in order to obtain the sensitivity analysis.

7.1.5 Binary Model

MMI Example—Part III

MMI produces MLAs and USMs every month. Table 7.1 summarizes the necessary data for producing MLAs and USMs. MMI has two major constraints: (1) labor—total 500 h/month

MMI-Producti... M

E7 | =SUMPRODUCT($C$5:$D$5,C7:

	A	B	C	D	E	F	G
1		MMI					
2		Product Schedule					
3							
4			MLA	USM			
5	Decision variables:	Units	55	9			
6					Total		Limit
7	Objective function:	Profit	25	100	2275		
8							
9	Constraints:	Labor	5	25	500	<=	500
10		Materials	5	2	293	<=	300

Sensitivity Report 1 | Int

Ready 100%

Figure 7.13 Computed results for the optimal production schedule (integer analysis).

E7 | 2275

1 Microsoft Excel 12.0 Answer Report
2 Worksheet: [MMI-ProductionSchedule.xlsx]Integer
3 Report Created: 7/9/2008 11:02:55 AM

5 Target Cell (Max)

Cell	Name	Original Value	Final Value
$E$7	Profit Total	2275	2275

9 Adjustable Cells

Cell	Name	Original Value	Final Value
$C$5	Units MLA	55	55
$D$5	Units USM	9	9

14 Constraints

Cell	Name	Cell Value	Formula	Status	Slack
$E$9	Labor Total	500	$E$9<=$G$9	Binding	0
$E$10	Materials Total	293	$E$10<=$G$10	Not Binding	7
$C$5	Units MLA	55	$C$5=integer	Binding	0
$D$5	Units USM	9	$D$5=integer	Binding	0

Integer | Answer Report 2

Figure 7.14 Answer report (integer analysis).

maximum (= 3 engineers × 40 h/week × 4.17 weeks/month), and (2) raw materials available—total 300 g/month maximum. MMI would like to determine the optimum production quantities (unit number) X for MLAs and Y for USMs in order to maximize the gross profit.

Part III seeks the solution for choosing only one product, either MLAs or USMs.

Appropriate use of binary variables can help us in expressing comparative relationships. To illustrate, suppose Z_1, Z_2, and Z_3 are binary variables representing whether each of three production lines should be set ($Z_i = 1$) or not set ($Z_i = 0$). The following relationships can then be expressed by these variables:

- At least two production lines must be set: $Z_1 + Z_2 + Z_3 \geq 2$
- If line 1 is set, line 2 must not be set: $Z_1 + Z_2 \leq 1$
- If line 1 is set, line 2 must be set: $Z_1 - Z_2 \leq 0$

- One but not both of lines 1 and 2 must be set: $Z_1 + Z_2 = 1$
- Both or neither of lines 1 and 2 must be set: $Z_1 - Z_2 = 0$
- Total line construction budget cannot exceed $\$K$, and the setting costs for lines 1, 2, and 3 are $\$A$, $\$B$, and $\$C$, respectively: $AZ_1 + BZ_2 + CZ_3 \leq K$

Binary variables can also be used to indicate restrictions in certain conditional situations. Suppose X denotes the production amount in MLA line ($X \geq 0$). If the MLA line is set, there is no other restriction on the value of X, but if it is not set, X must be 0. This relation can be expressed by

$$X \leq M Z_1$$

In this equation, M denotes an extremely large number such as 10^{20} (or 1E + 20) that does not restrict the value of X if $Z_1 = 1$. Because M is large, $X \leq M$ will not restrict the number of MLA products. But if the MLA line is not set ($Z_1 = 0$), the constraint becomes $X \leq 0$, leading to $X = 0$; that is, no product is made in the MLA line. In summary, the binary parameters can be used to set On/Off or used for the selection of projects.

Solution—Part III

Now, we will apply the binary variables to the MMI production schedule. Because MMI production line space is limited, Barb Shay decided to set only one production line; that is, either MLAs or USMs, not both. The problem is which to set—MLA or USM lines? We introduce here Z_1 and Z_2 for On/Off parameters for the MLA and USM production lines (C5 and D5, respectively; see Figure 7.15). Then, introducing an extremely large number M (1E+20) in G5, the products MZ_1 and MZ_2 are prepared in H5 and I5 cells. Subject to the Constraints in the Solver Parameter dialog box is set as in Figure 7.16. Three different points from Figure 7.12 are as follows: (1) \$C\$5 and \$D\$5 are binary parameters; (2) \$C\$5 and \$D\$5 are equal to or smaller than $M \times X$ and $M \times Y$, respectively; and, most importantly, (3) $Z_1 + Z_2 = 1$; that is, one but not both production lines must be set. The results are shown in Figure 7.15 and its Answer Report is in Figure 7.17. The computation "chose" the USM production (rather than the MLA production), if MMI needs to set only one production line. Twenty units of USMs generate the total gross profit of \$2000.

E8 =SUMPRODUCT($C$6:$D$6,C8:D8)

	A	B	C	D	E	F	G	H	I
4			MLA	USM	Z1+Z2		M	MZ1	MZ2
5	Decision variables:	On/Off	0	1	1		1E+20	0	1E+20
6		Units	-1.8E-15	20					
7					Total		Limit		
8	Objective function:	Profit	25	100	2000				
9									
10	Constraints:	Labor	5	25	500	<=	500		
11		Materials	5	2	40	<=	300		

Binary / Answer Report 3

Figure 7.15 Computed results for the optimal production schedule (binary model).

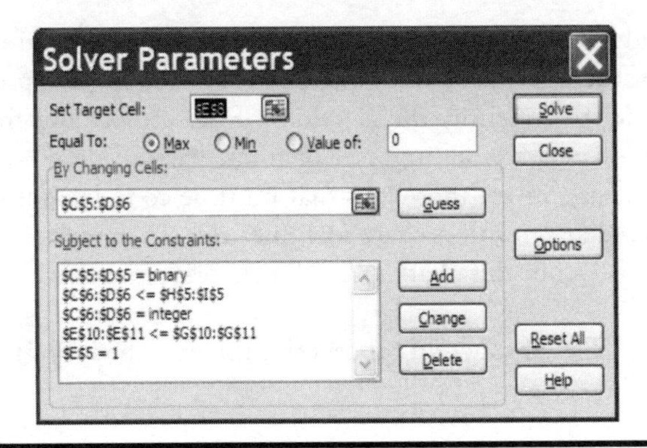

Figure 7.16 Solver Options dialog box (binary model).

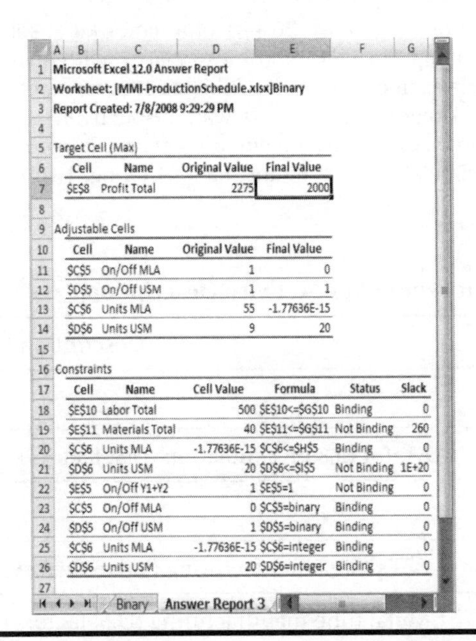

Microsoft Excel 12.0 Answer Report
Worksheet: [MMI-ProductionSchedule.xlsx]Binary
Report Created: 7/8/2008 9:29:29 PM

Target Cell (Max)

Cell	Name	Original Value	Final Value
$E$8	Profit Total	2275	2000

Adjustable Cells

Cell	Name	Original Value	Final Value
$C$5	On/Off MLA	1	0
$D$5	On/Off USM	1	1
$C$6	Units MLA	55	-1.77636E-15
$D$6	Units USM	9	20

Constraints

Cell	Name	Cell Value	Formula	Status	Slack
$E$10	Labor Total	500	$E$10<=$G$10	Binding	0
$E$11	Materials Total	40	$E$11<=$G$11	Not Binding	260
$C$6	Units MLA	-1.77636E-15	$C$6<=$H$5	Binding	0
$D$6	Units USM	20	$D$6<=$I$5	Not Binding	1E+20
$E$5	On/Off Y1+Y2	1	$E$5=1	Not Binding	0
$C$5	On/Off MLA	0	$C$5=binary	Binding	0
$D$5	On/Off USM	1	$D$5=binary	Binding	0
$C$6	Units MLA	-1.77636E-15	$C$6=integer	Binding	0
$D$6	Units USM	20	$D$6=integer	Binding	0

Binary | Answer Report 3

Figure 7.17 Answer report (binary model).

7.2 Program Evaluation and Review Technique

In Section 7.1, we learned about the production schedule in terms of how multiple projects share the activity time and budget to maximize the organization's profit. In this section, we will learn about the production schedule in terms of time series or sequence. The objectives will include the following:

- Determining a schedule of earliest and latest start and finish times for each activity that leads to the earliest completion time for the entire project
- Calculating the likelihood that a project will be completed within a certain time period
- Finding the minimum cost schedule that will complete a project by a certain date

Program evaluation and review technique (PERT) was developed in the mid-1950s for treating the completion time of the activities as random variables with specific probability distributions. PERT requires the modeler to identify the activities of the project and the precedence relations between them. This involves determining a set of immediate predecessors for each activity. An activity's immediate predecessors are those jobs that must be completed just prior to the activity's commencement. A precedence relation chart identifies the separate activities of the project and their precedence relations. From this chart a PERT network representation of the project can be constructed.

Let us use again the USM production process (for project scheduling) at MMI to explain the PERT concepts.

MMI Example

MMI received a purchase order (PO) for 20 sets of a motor kit (a set of the metal tube motor and its driver electronic chip) from Saito Industries in Japan with delivery due in 90 days. Lead zirconate titanate (PZT) plates, metal tubes, and drivers are manufactured in outsourcing factories located in Japan, the United States, and Taiwan, respectively, so additional transportation periods should be taken into account. The entire project can be represented by the 10 activities given in Table 7.2: one simulation, five motor manufacturing, two driver manufacturing, and two final testing.

Table 7.2 MMI Motor Kit Production Activity Description

Activity	*Description*
Design Simulation	
A	Motor design simulation (U.S. factory)
Motor Manufacturing	
B	PZT plate manufacturing (Japan factory)
C	Metal tube manufacturing (U.S. factory)
E	Transportation 1 (Japan to United States)
F	Motor assembly (U.S. factory)
G	Motor characterization (U.S. factory)
Driver Manufacturing	
D	Driver chip manufacturing (Taiwan factory)
H	Transportation 2 (Taiwan to United States)
Testing	
I	Test and adjustment of a motor and a driver chip (U.S. factory)
J	Transportation 3 (United States to Japan)

7.2.1 PERT Network

The PERT network is shown in Figure 7.18, where the nodes designate activities and their duration, and the arrows define the precedence relations between the activities. Table 7.3 summarizes all activities with immediate predecessors and the expected times to perform each activity. Three "times"—optimistic, pessimistic, and most likely—are listed, which will be used in Section 7.2.3. Note that m is not just an average of a and b; that is, we consider a "skewed" distribution curve of the probability. The pessimistic time tails off more than the optimistic time. The duration in Figure 7.18 shows only the value m.

7.2.2 PERT Approach

Two of the primary objectives of PERT analyses are (1) to determine the minimum possible completion time for the project, and (2) to determine a range of start and finish times for each activity so that the project can be completed in minimum time.

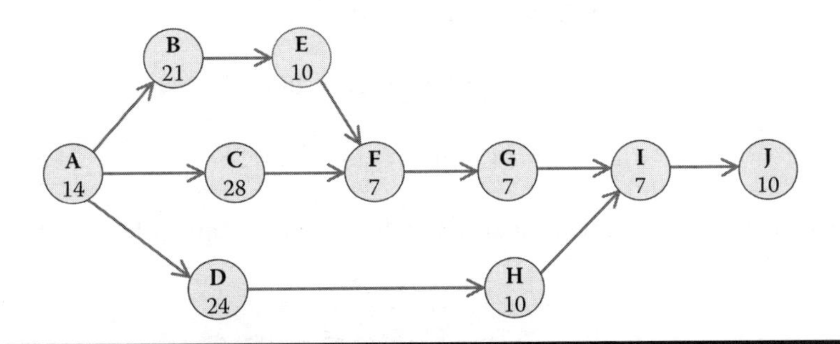

Figure 7.18 PERT network for the MMI motor kit project.

Table 7.3 MMI Motor Kit Production Precedence Relation Chart

Activity	*Immediate Predecessors*	*Optimistic Time* a	*Pessimistic Time* b	*Most Likely Time to Perform the Activity* m
A	—	12	16	14
B	A	18	24	21
C	A	24	32	28
E	B	7	17	10
F	C, E	5	11	7
G	F	5	9	7
D	A	21	27	24
H	D	7	17	10
I	G, H	5	9	7
J	I	7	17	10

Because each activity has some time flexibility for duration, the PERT analyses introduce the earliest start time (ES), earliest finish time (EF), latest start time (LS), and the latest finish time (LF).

7.2.2.1 Earliest Start/Finish Times

To determine the ES and EF times for the activities, a forward pass is made through the network. First, we solve the problem on manual. Let us start from activity A. The ES with no predecessors is set as 0, and its EF is simply the activity's completion time; EF = ES + Completion time = 0 + 14 days = 14. The first approach adopts "most likely time" only, without considering the probabilistic distribution of the time to perform the activity.

We proceed to the next nodes, activity B, C, or D. Because all of these activities' immediate predecessors must be completed before the activity can begin, B, C, or D's ES is the maximum of the EFs of its immediate predecessors. Then, its EF equals ES + its completion time:

$$\text{ES(B)} = 14,\ \text{EF(B)} = 14 + 21 = 35$$

$$\text{ES(C)} = 14,\ \text{EF(C)} = 14 + 28 = 42$$

$$\text{ES(D)} = 14,\ \text{EF(D)} = 14 + 24 = 38$$

Because activity B is the immediate predecessor for activity E and EF(B) = 35, we now can conclude:

$$\text{ES(E)} = 35, \text{ and EF(E)} = 35 + 10 = 45$$

Now we consider activity F. Both activity C (with EF(C) = 42) and activity E (with EF(E) = 45) are immediately predecessors of activity F. As both must be completed before starting activity F, the ES for activity F is the maximum of the EF for activities C and E:

$$\text{ES(F)} = \text{MAX(EF(C), EF(E))} = \text{MAX } (42, 45) = 45$$

and

$$\text{EF(F)} = \text{ES(F)} + 7 = 45 + 7 = 52$$

We repeat this process until all nodes have been evaluated, finalizing a schedule of ES and EF times for each activity: EF(G) = 52 + 7 = 59, EF(I) = 59 + 7 = 66, EF(J) = 66 + 10 = 76. The maximum EF time, 76, is the estimated completion time of the entire project. In summary:

$$\text{ES} = \text{Maximum EF of all its immediate predecessors}$$

$$\text{EF} = \text{ES} + \text{(Activity completion time)}$$

Second, we use Excel Solver. We define the variables:

$$X_A = \text{ES}(A), \quad X_B = \text{ES}(B), \ \ldots, \quad X_J = \text{ES}(J)$$

The constraint set consists of the nonnegativity constraints and one constraint for each immediate predecessor relationship in the project:

ES for an activity ≥ ES for the immediate predecessor activity + immediate predecessor's completion time

$$\text{MINIMIZE:} \quad X_A + X_B + X_C + \ldots + X_J$$

$$\text{SUBJECT TO:} \quad X_B \geq X_A + 14, X_C \geq X_A + 14, X_D \geq X_A + 14$$

$$X_E \geq X_B + 21$$

$$X_F \geq X_E + 10, X_F \geq X_C + 28$$

$$X_G \geq X_F + 7$$

$$X_H \geq X_D + 24$$

$$X_I \geq X_G + 7, X_I \geq X_H + 10$$

$$X_J \geq X_I + 7$$

$$\text{All } X\text{'s} \geq 0.$$

Although the actual value of the objective function ($X_A + X_B + X_C + \cdots + X_J$) is meaningless, the resulting set of X's will give the ES times for each of the activities. The EF times can then be calculated by adding the activity completion times to these ES times. The overall project completion time is then the maximum of these earliest finish times. Figure 7.19 shows the PERT Excel spreadsheet, and the final results for the ES and EF values on the forward pass. Figure 7.20 shows the Solver Parameters dialog box setting for the PERT minimization process. The project completion time (= maximum EF time) of 76 days is calculated, which is the same as the manually obtained value and less than the 90 days, which was Saito Industries' delivery due date.

7.2.2.2 Latest Start/Finish Times

To determine the LS and LF times for the activities, which allow the project to be completed by its minimum completion date of 76 days, a backward pass is made through the network. First, we begin by evaluating all activities that have no successor activities. In our case study, this is activity J, which has an LF = 76 and a completion time of 10 days. The LS for activity J is, thus, LS = LF – Completion time = 76 – 10 = 66 days.

We then go backward to the next nodes, activity I. Because activity J is the only successor activity to activity I, LF(I) should be LS(J) = 66 days. Subtracting the corresponding activity duration, we obtain LS(I):

$$\text{LS(I)} = \text{LF(I)} - 7 = 66 - 7 = 59$$

PERT - Microsoft Excel non-comm...

Home Insert Page Layout Formulas Data Review View Add-Ins

D21 =SUM(J9:J18)

	A	B	C	D	E	F	G	H	I	J	K	L	M	N
2	MMI													
3														
4	MEAN													
5	STANDARD DEVIATION													
6	VARIANCE													
7														
8	Activity	Mode	Immediate Predecessors	a	b	m	mu	sigma	variance	ES	EF	LS	LF	Slack
9	Simulation	A		12	16	14				0	14			
10	PZT Plate	B	A	18	24	21				14	35			
11	Metal Tube	C	A	24	32	28				14	42			
12	Driver	D	A	21	27	24				14	38			
13	Transport 1	E	B	7	17	10				35	45			
14	Motor Assembly	F	C,E	5	11	7				45	52			
15	Characterization	G	F	5	9	7				52	59			
16	Transport 2	H	D	7	17	10				38	48			
17	Test	I	G,H	5	9	7				59	66			
18	Transport 3	J	I	7	17	10				66	76			
19														
20	Project Completion Time =			76										
21	Min Σ ES =			337										

Answer Report 1 Sensitivity Report 1 PE

Ready 100%

Figure 7.19 PERT preparation sheet for the MMI motor kit.

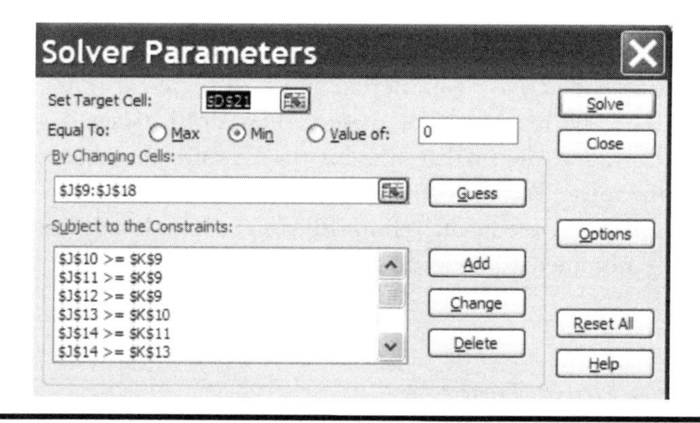

Figure 7.20 Solver Parameters dialog box setting for the PERT minimization process (forward pass).

Continuing the backward pass through the network, we now consider activities G and H:

$$LF(G) = LS(I) = 59, LS(G) = LF(G) - 7 = 52$$

$$LF(H) = LS(I) = 59, LS(H) = LF(H) - 10 = 49$$

We will further continue the procedure on the most time consuming pass, activities F, E, and B:

$$LF(F) = LS(G) = 52, LS(F) = LF(F) - 7 = 52 - 7 = 45$$

$$LF(E) = LS(F) = 45, LS(E) = LF(E) - 10 = 45 - 10 = 35$$

$$LF(B) = LS(E) = 35, LS(B) = LF(B) - 21 = 35 - 21 = 14$$

Similarly

$$LF(C) = LS(F) = 45, LS(C) = LF(C) - 28 = 45 - 28 = 17$$

$$LF(D) = LS(H) = 49, LS(D) = LF(D) - 24 = 49 - 24 = 25$$

Finally, we consider activity A, the predecessors from which are three activities B, C, and D. Activity A should be finished prior to the starting time of all activities B, C, and D, thus the LF time for activity A is the minimum of the LS times for activities B, C, and D:

$$LF(A) = MIN(LS(B), LS(C), LS(D))$$

$$= MIN\ (14, 17, 25) = 14$$

and finally

$$LS(A) = LF(A) - 14 = 14 - 14 = 0$$

In summary, LF is the minimum LS of all its immediate successor activities, and LS = LF – activity completion time.

Second, we will use Excel Solver. We define the variables:

$$Y_A = LF(A), \quad Y_B = LF(B), \ldots, \quad Y_J = LF(J)$$

The constraint set consists of the nonnegativity constraints and one constraint for each immediate successor relationship in the project:

LF for an Activity ≤ LF for the immediate Successor Activity
– Immediate Successor's Activity Completion Time

$$\text{MAXIMIZE:} \quad Y_A + Y_B + Y_C + \ldots + Y_J$$

$$\text{SUBJECT TO:} \quad Y_A \leq Y_B - 21$$

$$Y_A \leq Y_C - 28$$

$$Y_A \leq Y_D - 24$$

$$Y_B \leq Y_E - 10$$

$$Y_C \le Y_F - 7$$

$$Y_D \le Y_H - 10$$

$$Y_E \le Y_F - 7$$

$$Y_F \le Y_G - 7$$

$$Y_G \le Y_I - 7$$

$$Y_H \le Y_I - 7$$

$$Y_I \le Y_J - 10$$

$$Y_J = \text{EF(J)} = 76$$

$$\text{All } Y\text{'s} \ge 0$$

Although the actual value of the objective function ($Y_A + Y_B + Y_C + \ldots + Y_J$) is meaningless, the resulting set of *Y*s will give the LF times for each of the activities. The LS times can then be calculated by subtracting the activity completion times from these LF times. Figure 7.21 shows the Solver Parameters dialog box setting for the PERT maximization process (backward process). Figure 7.22 shows the PERT Excel spreadsheet, and the final results for the LS and LF values on the backward pass, in addition to ES and EF values.

7.2.2.3 Critical Path and Slack Times

In the course of completing a project, both planned and unforeseen delays can affect activity start or completion times. If delivery of either PZT plates or metal tubes is delayed, the following activity F, motor assembly, may not be started. Some of the activity delays affect the overall completion date of the project, while others may not. To analyze the impact of such delays on the project, we determine the slack time for each activity. *Slack time* is the amount of time an activity can be delayed from its ES without delaying the project's estimated completion time. It is calculated by

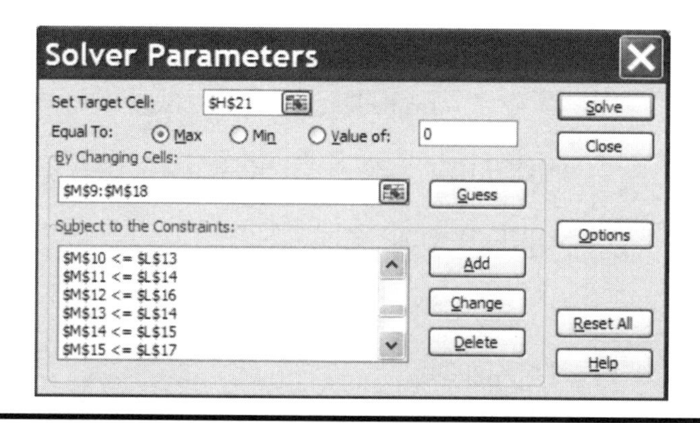

Figure 7.21 Solver Parameters dialog box setting for the PERT maximization process (backward pass).

	A	B	C	D	E	F	G	H	I	J	K	L	M	N
1	Motor Set													
2	MMI													
3														
4	MEAN													
5	STANDARD DEVIATION													
6	VARIANCE													
7														
8	Activity	Mode	Immediate Predecessors	a	b	m	mu	sigma	variance	ES	EF	LS	LF	Slack
9	Simulation	A		12	16	14				0	14	0	14	0
10	PZT Plate	B	A	18	24	21				14	35	14	35	0
11	Metal Tube	C	A	24	32	28				14	42	17	45	3
12	Driver	D	A	21	27	24				14	38	25	49	11
13	Transport 1	E	B	7	17	10				35	45	35	45	0
14	Motor Assembly	F	C,E	5	11	7				45	52	45	52	0
15	Characterization	G	F	5	9	7				52	59	52	59	0
16	Transport 2	H	D	7	17	10				38	48	49	59	11
17	Test	I	G,H	5	9	7				59	66	59	66	0
18	Transport 3	J	I	7	17	10				66	76	66	76	0
19														
20	Project Completion Time =			76										
21	Min Σ ES =			337		Max Σ LF =		500						

Figure 7.22 PERT preparation sheet for the MMI motor kit project (backward pass).

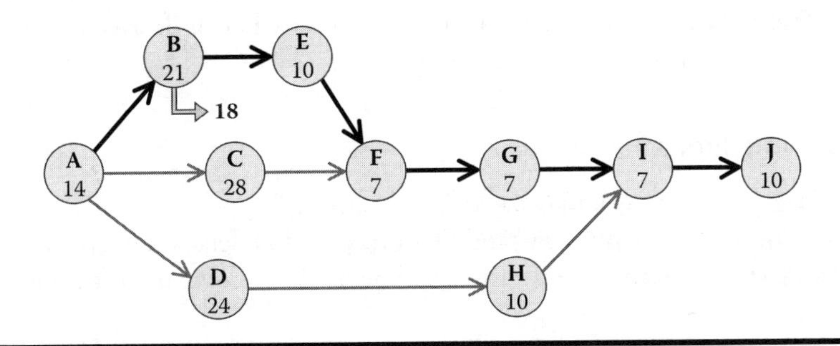

Figure 7.23 PERT critical path for the MMI motor kit project.

subtracting an activities ES from its LS (or its EF from its LF). This value for an activity's slack time assumes that only the completion time of this single activity has been changed and that there are no other delays to activities in the project. Figure 7.22 also includes the slack time, calculated from LS – ES.

When an activity has slack time, the manager has some flexibility in scheduling and may be able to distribute the workload more evenly throughout the project's duration without affecting its overall completion date. This is especially important in projects with limited staff or resources.

Activities that have no slack time are called *critical activities.* These activities must be rigidly scheduled to start and finish at their specific ES and EF times, respectively. Any delay in completing a critical activity will delay completion time of the entire project beyond 76 days by the corresponding amount. Figure 7.23 illustrates a critical path in thick arrows, formed from the critical activities. The sum of the completion times of the activities on the critical path is the minimum completion time for the project ($14 + 21 + 10 + 7 + 7 + 7 + 10 = 76$). Because it consists of a sequence of activities that cannot be delayed without affecting the earliest project completion date, the critical path is actually the longest path in the directed network.

In the present MMI project, the critical path is A-B-E-F-G-I-J. It is possible to have more than one critical path in a PERT network. For example, if the completion time of activity B (PZT plates manufacturing at a Japanese partner) had been 18 days (rather than 21 days), the total completion time would be reduced to 73 days, and the slack time for activity C would be 0. Thus, a second critical path A-C-F-G-I-J would show up.

In summary:

1. The critical activities (activities with 0 slack) form at least one critical path in the network.
2. A critical path is the longest path through the network.
3. The sum of the completion times for the activities on the critical path gives the minimum completion time of the project.

7.2.2.4 Analysis of Possible Delays

7.2.2.4.1 Single Delays

A delay in a single critical activity will result in an equivalent delay in the entire project. On the other hand, a delay in a noncritical activity will only delay the project by the amount the delay exceeds the activity's slack; a delay less than the slack time of the activity will not affect the project completion time. For example, delaying activity E (transportation from Japan to the United States) by 3 days directly causes a 3-day delay of the entire project, while delaying activity C by 3 days does not delay the entire project at all. However, if activity C is delayed by 5 days, 5 – 3 = 2, so a 2-day delay will occur throughout the entire project.

7.2.2.4.2 Multiple Delays

Case 1: 3-day delay in activity C and 6-day delay in activity D

There is no path between activities C and D. Both C and D delay times are not more than the slacks (3 days and 11 days, respectively). There will be no delay in the project completion.

Case 2: 6-day delay in activity D and 6-day delay in activity H

Both D and H delay times are not more than slacks (11 days and 11 days, respectively). However, as these activities are connected, attention is needed. Because of a 6-day delay in activity D, new EF(D) = original EF(D) + 6 = 38 + 6 = 44. Thus, the successor activity H would have new ES(H) = new EF(D) = 44, then new EF(H) = 44 + 10 = 54. Comparing with LF(H) = 59, the slack of activity H is reduced only to 5 days (= 59 – 54), not 11 days anymore. If activity H (transportation from Taiwan to the United States) is actually delayed by 6 days, 6 – 5 = 1 day delay will occur in the entire project schedule.

In Excel Solver, merely changing the completion durations of the delayed activities by adding those delayed times, and a re-solve provides new results on the project completion time and the critical path.

7.2.3 Gantt Charts

One responsibility of the project managers is to track the progress of the project. A popular tool used to display activities and monitor their progress is the Gantt chart (named for Henry Gantt, who introduced such a chart in 1918). In a Gantt chart, time is measured on the horizontal axis, each activity is listed on the vertical axis, and a bar is drawn corresponding to its expected completion time. In one of the earliest time Gantt charts, the bar begins at the ES time of the activity (which is when all the activity's immediate predecessors are expected to be completed). The end of the bar represents the earliest completion time for that specific activity [2].

Figure 7.24 shows how to draw the Gantt chart with Excel: (a) Prepare three columns of data: activity, ES, and m (completion time). Note that activity sequence is prepared up-side-down, because of the final chart arrangement. (b) Highlight these three columns, then convert to 2-D or 3-D bar chart. (c) Eliminate the bar color of the ES portion bar. The result is shown in Figure 7.25.

Motor Set

MMI

Node	Activity	ES	m
J	Transport 3	66	10
I	Test	59	7
H	Transport 2	38	10
D	Driver	14	24
G	Characterization	52	7
F	Motor Assembly	45	7
C	Metal Tube	14	28
E	Transport 1	35	10
B	PZT Plate	14	21
A	Simulation	0	14

(a)

PERT - Microsoft Excel

Home Insert Page Layout Formulas

PivotTable Table Picture Clip Art Shapes SmartArt Column Line Pie Bar

Tables

2-D Bar

3-D Bar

Cylinder

(b)

Figure 7.24 Gantt chart preparation procedure: (a) Prepare the data in three columns, including activity, ES and *m* (completion time). (b) Convert to a chart via a 3-D bar.

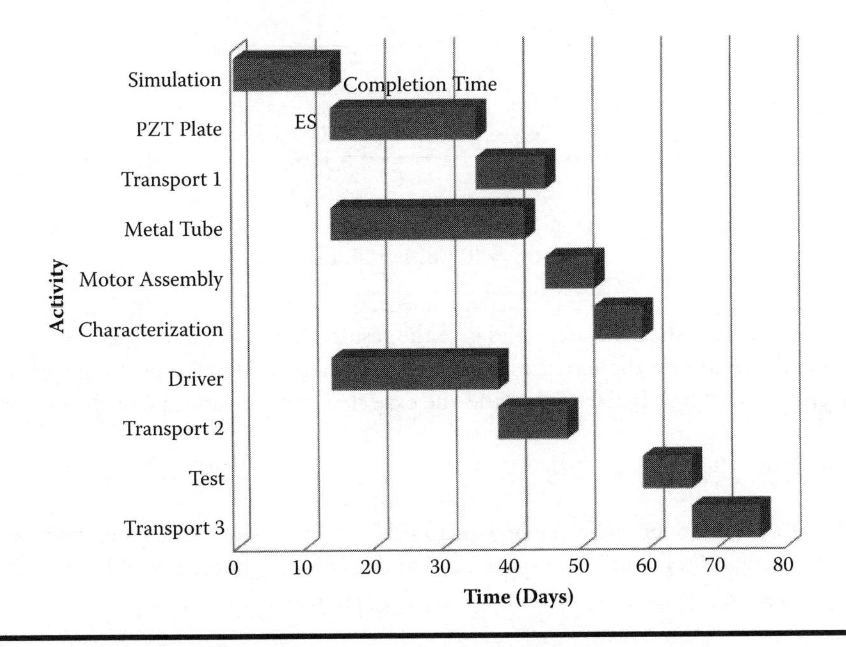

Figure 7.25 Earliest time Gantt chart for a motor kit production at MMI.

A crucial step in meeting a target completion date and containing costs is management's ability to monitor a project's progress. We can use a Gantt chart as a visual aid for tracking the progress of project activities by shading an appropriate percentage of the corresponding bar to document the completed work. A manager then needs only to glance at the chart on a given date to see if the project is being completed on schedule with respect to the earliest possible completion times of the activities.

7.2.4 Probabilistic Approach to Project Scheduling

We have used definitive completion times so far in the PERT analysis. However, no manager has such confidence in the activity completion times. As was shown in Table 7.3 in Section 7.2.1, we introduced three "times"—optimistic a, pessimistic b, and most-likely m—which are maintained in Figure 7.22. Here, m is not just an average of a and b, which is known as a beta distribution. Statisticians have found that a *beta distribution* is useful in approximating distributions with limited data and fixed end points. The assumption that activity times follow a beta distribution has only a modest effect on the analysis of the completion time of the entire project. Of greater concern are approximations for the average, or mean, activity completion time, μ, and its standard deviation, σ, which are based on the time estimates a, m, and b. The estimates for the average time μ and its standard deviation σ are [3]

$$\mu = \frac{a + 4m + b}{6} \tag{7.1}$$

$$\sigma = \frac{b - a}{6} \tag{7.2}$$

The mean, variance, and standard deviation for activity A can be found by

$$\mu_A = \frac{a + 4m + b}{6} = \frac{12 + 4 \times 14 + 16}{6} = 14$$

$$\sigma_A = \frac{b - a}{6} = \frac{16 - 12}{6} = 0.667$$

$$\sigma_A^2 = (0.667)^2 = 0.444$$

Similar calculations for the other activities give the results shown in Figure 7.26.

Note that the means for the activities are the same as those used in the previous PERT analysis. Thus, the critical path is A-B-E-F-G-I-J, and the expected completion time of the project, μ, is

$$\mu = \mu_A + \mu_B + \mu_E + \mu_F + \mu_G + \mu_I + \mu_J = 14 + 21 + 10.67 + 7.33 + 7 + 7 + 10.67 = 77.67$$

This value is 1.67 days longer than the previous PERT analysis, which originated from the skewed distribution for activities E and J (transportation) and activity F (motor assembly). Other activities having a symmetrical distribution exhibit $\mu = m$. The variance, σ^2, is

C4 =SUM(G9,G10,G13,G14,G15,G17,G18)

	A	B	C	D	E	F	G	H	I	J	K	L	M	N
1	Motor Set													
2	MMI													
3														
4	MEAN (A-B-E-F-G-I-J)		77.6666667											
5	STANDARD DEVIATION		2.98142397											
6	VARIANCE		8.88888889											
7														
8	Activity	Node	Immediate Predecessors	a	b	m	mu	sigma	variance	ES	EF	LS	LF	Slack
9	Simulation	A		12	16	14	14	0.667	0.44444	0	14	0	14	0
10	PZT Plate	B	A	18	24	21	21	1	1	14	35	14	35	0
11	Metal Tube	C	A	24	32	28	28	1.333	1.77778	14	42	17	45	3
12	Driver	D	A	21	27	24	24	1	1	14	38	25	49	11
13	Transport 1	E	B	7	17	10	10.67	1.667	2.77778	35	45	35	45	0
14	Motor Assembly	F	C,E	5	11	7	7.333	1	1	45	52	45	52	0
15	Characterization	G	F	5	9	7	7	0.667	0.44444	52	59	52	59	0
16	Transport 2	H	D	7	17	10	10.67	1.667	2.77778	38	48	49	59	11
17	Test	I	G,H	5	9	7	7	0.667	0.44444	59	66	59	66	0
18	Transport 3	J	I	7	17	10	10.67	1.667	2.77778	66	76	66	76	0
19														
20	Project Completion Time =			76										
21	Min Σ ES =			337		Max Σ LF =		500						

PERT1 PERT2 GanttChart

Figure 7.26 Mean, variance, and standard deviation are added to Figure 7.22.

$$\sigma^2 = \sigma_A^2 + \sigma_B^2 + \sigma_E^2 + \sigma_F^2 + \sigma_G^2 + \sigma_I^2 + \sigma_J^2$$
$$= 0.444 + 1 + 2.778 + 1 + 0.444 + 0.444 + 2.778 = 8.889$$

Note that two transportations (Japan–U.S. and U.S.–Japan) contribute largely to this variance. The standard deviation for the entire project, σ, is

$$\sigma = \sqrt{\sigma^2} = 2.98$$

We can say that the expected completion time of the project is 77.67 ± 2.98 days (in about 2/3 probability). Because the due delivery time is 90 days, the probability of the entire project completion in less than 90 days can be calculated as

$$P(X \le 90 \text{ days}) = P\left(Z \le \frac{90 - 77.67}{2.98}\right) = P(Z \le 4.1) = 99.9\%$$

Because this is higher than 4σ, we can say that the probability of failing to meet the due date is negligible.

7.2.5 Critical Path Method

In previous discussions, MMI's regular manufacturing process of metal tube motors took a minimum of 76 days, according to the PERT analysis, which is shorter than Saito Industries' original delivery due period of 90 days. However, Saito requested that MMI decrease manufacturing time to 60 days to meet their new product release schedule. Is it possible to shrink the delivery period from 76 days to 60 days? We will consider this sort of problem. The critical path method (CPM) is a deterministic approach to project planning, based on the assumption that an activity's completion time can be determined with certainty. This time depends only on the amount of money allocated to the activity. The process of reducing an activity's completion time by committing additional monetary resources is known as *crashing*. CPM assumes that there are two crucial time points for each activity: (1) its normal completion time (T_N), achieved when the usual or normal cost (C_N) is spent to complete the activity; and (2) its crash completion time (T_C), the minimum possible completion time for the activity. The T_C is attained when a maximum crash cost (C_C) is spent. The assumption is that spending an amount greater than C_C on an activity will not significantly reduce the completion time any further. Think of a PZT plate manufacturing process, for example. The ceramic sintering period cannot be shrunk so easily, because the time is determined by the equipment, such as ball-milling time and furnace sintering time, which cannot be changed by the monetary investment. However, because ceramic cutting and polishing are manual processes, adding to the workforce dramatically reduces the manufacturing period. In this sense, the minimum T_C at the maximum C_C for an activity is defined. CPM analyses are based on the following linearity assumption:

$$M = \frac{E}{R} = \frac{C_C - C_N}{T_N - T_C} \tag{7.3}$$

where $R = T_N - T_C$ = maximum possible time reduction (crashing) of an activity
$E = C_C - C_N$ = maximum additional (crash) cost required to achieve the maximum time reduction
M = marginal cost of reducing an activity's completion time by one unit time

Figure 7.27 summarizes crashing possibilities for each activity: crashing time *R*, crash cost *E*, and marginal cost *M*. We need to assume that Saito is willing to commit additional monetary resources to meet a deadline (*D*) of 60 days. An approach to determine the amount of time each activity should be crashed can be developed by taking into account the following:

1. The project time is reduced only when activities on the critical path are reduced.
2. The maximum time reduction for each activity is limited.
3. The amount of time an activity on the critical path can be reduced before another path also becomes a critical path is limited.

7.2.5.1 Linear Programming Approach to Crashing

A simple modification to the linear program given in Section 7.2.2 is required. We now define two variables for each activity, *j*:

X_j = ES time for the activity
$Z_j = T_N - T_C$ = amount by which the activity is to be crashed

N17 fx =SUMPRODUCT(H5:H14,I5:I14)

	A	B	C	D	E	F	G	H	I	J	K	L	M	N	O
1	Motor Set														
2	MMI														
3															
4	Activity	Node	Immediate Predecessors	TN	CN	R	E	M=E/R	TN-TC	TC	ES	EF	LS	LF	Slack
5	Simulation	A		14	$ 6,000	4	$5,000	$1,250	0	14	0	14	-0	14	0
6	PZT Plate	B	A	21	$ 2,000	5	$2,000	$ 400	5	16	14	30	14	30	0
7	Metal Tube	C	A	28	$ 4,000	7	$4,000	$ 571	5	23	14	37	14	37	0
8	Driver	D	A	24	$ 5,000	7	$4,000	$ 571	0	24	14	38	17	41	3
9	Transport 1	E	B	10	$ 60	3	$ 140	$ 47	3	7	30	37	30	37	0
10	Motor Assembly	F	C,E	7	$ 3,000	2	$2,000	$1,000	2	5	37	42	37	42	0
11	Characterization	G	F	7	$ 3,000	2	$2,000	$1,000	1	6	42	48	42	48	0
12	Transport 2	H	D	10	$ 60	3	$ 140	$ 47	3	7	38	45	41	48	3
13	Test	I	G,H	7	$ 3,000	2	$2,000	$1,000	2	5	48	53	48	53	0
14	Transport 3	J	I	10	$ 60	3	$ 140	$ 47	3	7	53	60	53	60	0
15				Total	$ 26,180										
16	Project Completion Time =			60		Due Date		60							
17	Min Σ ES =			290		Max Σ LF =		410			Min Σ M Z =			10277	
18															

PCM / PCM (2) / PCM (3) / PCM (4)

Figure 7.27 Linear programming approach to crashing (MMI's motor manufacturing).

Because the normal cost must always be paid, the objective is to minimize the the additional investment spent to reduce the completion times of activities. The cost per unit reduction time for an activity is M_j, and the amount of time the activity is reduced is the decision variable, Z_j. Therefore, the total extra amount spent crashing the activity is M_jZ_j. Because we want to minimize the total additional investment spent to crash the project, the objective function is the sum of all such costs:

$$\text{MIN} \sum_j M_j \, Z_j$$

There are three constraints in this approach:

1. No activity can be reduced more than its maximum time reduction.

$$Z_j \leq R_j$$

2. The start time for an activity must be at least as great as the finished time of all immediate predecessor activities.

   X_j (ES for an activity) ≥ EF for an immediate predecessor of the activity

   Note, however, that the activity finish times are reduced by the amount of time each activity is crashed.

   EF for an immediate predecessor of the activity = X_j (ES for a predecessor activity) + T_N (normal completion time of the predecessor activity) – Z_j (crashed time of the predecessor activity).

For example, in Figure 7.27,

$$\text{EF(C)} = \text{ES(C)} + T_N - (T_N - T_C) = X_C + T_N - Z_C = 14 \text{ days} + 28 \text{ days} - 5 \text{ days} = 37 \text{ days}$$

3. The project must be completed by its deadline D.

Because project completion time is determined by the maximum of the EF times of the terminal (final) activities (activity J in our case) in the project, we add constraints of the form:

$$\text{EF for a terminal activity} \leq D$$

In our example, $X_J + T_N(\text{J}) - Y_J \leq 60$ days.

The complete linear programming model for MMI is then:

$$\text{MINIMIZE: SUMPRODUCT}(M_j, Z_j)$$

$$\text{SUBJECT TO: } Z_j \leq R_j$$

$$X_B \geq X_A + (14 - Z_A)$$

$$X_C \geq X_A + (14 - Z_A)$$

$$X_D \geq X_A + (14 - Z_A)$$

$$X_E \geq X_B + (21 - Z_B)$$

$$X_F \geq X_C + (28 - Z_C), X_F \geq X_E + (10 - Z_E)$$

$$X_G \geq X_F + (7 - Z_F)$$

$$X_H \geq X_D + (24 - Z_D)$$

$$X_I \geq X_G + (7 - Z_G), X_I \geq X_H + (10 - Z_H)$$

$$X_J \geq X_I + (7 - Z_I)$$

$$\text{All } X\text{'s and } Y\text{'s} \geq 0$$

I5:I14 ($T_C - T_N$) and K5:K14 (ES) were initially determined as a minimization problem of SUMPRODUCT. T_C and EF were then calculated. LS, LF, and Slack were calculated as a maximization problem of the total sum of LF, as we learned in Section 7.2.2. The results are shown in Figure 7.27.

Compared to Figure 7.26, Figure 7.27 suggests that, except for activities A (simulation) and D (driver manufacturing), all other activities were crashed. Because the marginal cost for crashing activity A is most expensive ($1250 per day), this activity crash was not used. Because activity D crash did not affect any change in the entire project completion time, it was not spent (actually, there still remain 3 days as slack). The slack in activity C in Figure 7.26 disappeared in Figure 7.27, which means that A-C-F-G-I-J also became a critical path after the activity crashing. This crashing (16-day shrinkage from 76 days to 60 days) required an additional resource of $10,277 on top of the normal cost of $26,180.

7.3 Game Theory

Game theory was first proposed by von Neumann in 1928, and its applications to economics were widely discussed in a paper coauthored with O. Morgenstern in 1944 [4]. In the following sections, we will learn the relationship of game theory to linear programming.

7.3.1 Two-Person Zero-Sum Game

As the simplest case, a two-person *zero-sum* game is introduced. Let us start with the rock-paper-scissors game. We will consider two players, player I and player II, for this *janken* game. Table 7.4 shows a payoff matrix for the *janken* game under the condition that +1 is obtained for a win, −1 for a loss, and 0 for a draw. Note that this payoff matrix is generated from player I's standpoint. When player II gestures scissors, player I will receive +1, −1, and 0, when he gestures rock, paper, and scissors, respectively. From player II's standpoint, the matrix should be the opposite. When player I receives +1, −1, and 0, player II should obtain −1, +1, and 0, respectively, as such, a zero-sum game.

Everybody knows the best strategy for this game, by learning empirically from his or her childhood; that is, we had better use 1/3, 1/3, and 1/3 probability for rock, paper, and scissors, and apply them randomly with an unfamiliar opponent (without knowing his or her playing tendencies).

Next, let us change a rule slightly: +5 is obtained for a paper win, +2 for a scissors win, and +1 for a rock win, their equivalent negative scores for loss, and 0 for a draw. The payoff matrix is described in Table 7.5. In this scenario, what probabilities are suitable to rock, paper, and scissors as the best strategy? Player I obtains +5 and −2 for win and loss with paper, while he obtains +1 and −5 for win and loss with rock, respectively. It is natural for player I to be reluctant to gesture rock. However, if player I uses too much paper for winning +5, player II may use scissors frequently to defeat player I (player I loses by −2). So, what probabilities should player I use for rock, paper, and scissors? This is not obvious, and this type of decision is the objective of this section. (For your information, the probabilities should be 0.25, 0.125, and 0.625 for rock, paper, and scissors. You will find that paper usage is surprisingly low!)

Table 7.4 Payoff Matrix for a *Janken* Game for Two Players—Even Condition

		Player II		
		Rock	Paper	Scissors
Player I	Rock	0	−1	1
	Paper	1	0	−1
	Scissors	−1	1	0

Table 7.5 Payoff Matrix for a *Janken* Game for Two Players—Biased Condition

		Player II		
		Rock	Paper	Scissors
Player I	Rock	0	–5	1
	Paper	5	0	–2
	Scissors	–1	2	0

7.3.2 Game Theory Outline

We consider only two players, player I and player II. Player I uses one strategy $i \in M$ from a finite number of strategies:

$$M = \{1, 2, \ldots, m\}$$

and player II uses one strategy $j \in N$ from a finite number of strategies:

$$N = \{1, 2, \ldots, n\}$$

For this occasion, player I receives payoff a_{ij}, and player II pays this amount (loss). A payoff matrix $A = [a_{ij}]$ can be defined as

$$A = \begin{bmatrix} a_{11} & a_{12} & \cdots & a_{1n} \\ a_{21} & a_{22} & \cdots & a_{2n} \\ \vdots & \vdots & \ddots & \vdots \\ a_{m1} & a_{m2} & \cdots & a_{mn} \end{bmatrix}$$

M, N, and A determine the game rule, and you remember that the payoff of player II is $-a_{ij}$ and

$$(\text{Payoff of Player I}) + (\text{Payoff of Player II}) = a_{ij} + (-a_{ij}) = 0$$

Player I wants to maximize his payoff from the payoff matrix, while player II wants to minimize his loss (remember again that the payoff matrix is based on player I's standpoint). In this sense, we refer to player I as a *maximum chaser*, and player II as a *minimum chaser*.

We consider a multiple number of games, and the probability of the strategy i by player I is denoted by p_i:

$$p_i \geq 0, \sum_{i=1}^{m} p_i = 1$$

Player I is supposed to apply the strategy randomly from M. On the other hand, the probability of the strategy j by player II is denoted by q_j:

$$q_j \geq 0, \sum_{j=1}^{n} q_j = 1$$

Player II is supposed to apply the strategy randomly from N. A mixed strategy by player I is described by

$$p = \begin{bmatrix} p_1 & p_2 & \cdots & p_m \end{bmatrix}$$

where p is a probability distribution on M. Similarly

$$q = \begin{bmatrix} q_1 \\ q_2 \\ \vdots \\ q_n \end{bmatrix}$$

is called a *mixed strategy* by player II, and shows a probability distribution on N. In particular, a *pure strategy* is defined for either player I or player II, who can choose a unique strategy i or j, described as follows:

$$p = [0 \; \cdots \; 0 \; \overset{i}{1} \; 0 \; \cdots \; 0]$$

$$q = \begin{bmatrix} 0 \\ \vdots \\ 0 \\ 1 \\ 0 \\ \vdots \\ 0 \end{bmatrix} j$$

When player I uses a mixed strategy p and player II uses a mixed strategy q, the expected payoff for player I is represented by

$$pAq = \begin{bmatrix} p_1 \, p_2 \cdots p_m \end{bmatrix} \begin{bmatrix} a_{11} & a_{12} & \cdots & a_{1n} \\ a_{21} & a_{22} & \cdots & a_{2n} \\ \vdots & \vdots & \ddots & \vdots \\ a_{m1} & a_{m2} & \cdots & a_{mn} \end{bmatrix} \begin{bmatrix} q_1 \\ q_2 \\ \vdots \\ q_n \end{bmatrix} = \sum_{i=1}^{m} \sum_{j=1}^{n} p_i \, a_{ij} \, q_j \qquad (7.4)$$

which is called a *payoff function*. When the payoff function is defined by this simplest bilinear form (Equation 7.4), the game is called a *rectangular game*. For example, payoff function for a payoff matrix

$A = \begin{bmatrix} 3 & -2 & 1 \\ -1 & 4 & -5 \end{bmatrix}$ is provided by

$$pAq = \begin{bmatrix} p_1 & p_2 \end{bmatrix} \begin{bmatrix} 3 & -2 & 1 \\ -1 & 4 & -5 \end{bmatrix} \begin{bmatrix} q_1 \\ q_2 \\ q_3 \end{bmatrix} = 3p_1q_1 - 2p_1q_2 + p_1q_3 - p_2q_1 + 4p_2q_2 - 5p_2q_3$$

The minimax principle is the key to find the optimized mixed strategy for each player. We will use S and T for representing probability distribution groups on M and N, respectively:

$$S = \left\{ p \mid p = \begin{bmatrix} p_1 & p_2 & \cdots & p_m \end{bmatrix}, p_i \geq 0, \sum_{i=1}^{m} p_i = 1 \right\}$$

$$T = \left\{ q \mid q = \begin{bmatrix} q_1 \\ q_2 \\ \vdots \\ q_n \end{bmatrix}, q_j \geq 0, \sum_{j=1}^{n} q_j = 1 \right\}$$

1. Under the assumption that player I will choose a mixed strategy p in order to maximize the payoff function, player II will choose a mixed strategy q to minimize opponent I's payoff. The final payoff function value is represented as

$$\operatorname*{Min}_{q \in T} \operatorname*{Max}_{p \in S} pAq$$

2. Under the assumption that player II will choose a mixed strategy q in order to minimize the payoff function of the opponent, player I will choose a mixed strategy p to maximize his payoff. The final payoff function value is represented as

$$\operatorname*{Max}_{p \in S} \operatorname*{Min}_{q \in T} pAq$$

The above processes 1 and 2 can be analogous to a saddle point problem, as visualized in Figure 7.28. In general, the point showing $v_1 = \operatorname*{Min}_{q \in T} \operatorname*{Max}_{p \in S} pAq$ does not necessarily coincide with the point showing $v_2 = \operatorname*{Max}_{p \in S} \operatorname*{Min}_{q \in T} pAq$. However, if the payoff function has a bilinear form, $v_1 = v_2$ can be verified, and the game becomes strictly determined. Thus, when player II chooses a mixed strategy q_0 via process (a), and player I chooses a mixed strategy p_0 via process (b), the following minimax theorem is derived:

$$\text{Minimax theorem: } \operatorname*{Min}_{q \in T} \operatorname*{Max}_{p \in S} pAq = \operatorname*{Max}_{p \in S} \operatorname*{Min}_{q \in T} pAq = p_0 A q_0 = v$$

Here, q_0 is called *minimax strategy*, while p_0 is called *maximin strategy*. And $v = p_0Aq_0$ is called its *game value*, which corresponds to an average payoff per game, following multiple games between these two players with their best maximin and minimax processes.

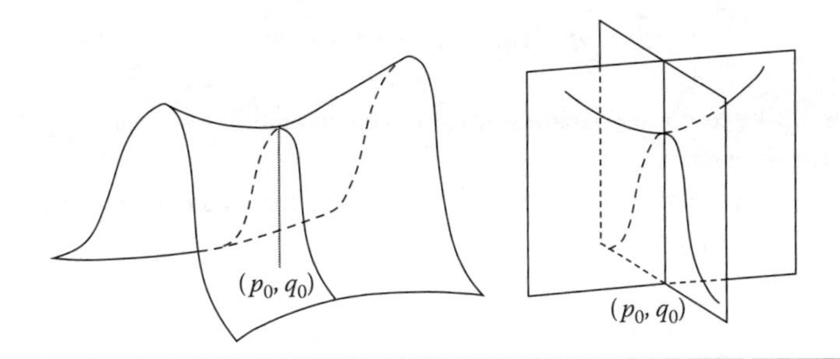

Figure 7.28 A saddle point of a payoff function in a rectangular game, showing the minimax and maximin strategies.

7.3.3 Rock-Paper-Scissors

Now we will reconsider the two players' *janken* game. Table 7.5 shows a payoff matrix with a special skewed condition: +5 is obtained for a paper win, +2 for a scissors win, and +1 for a rock win, their negative scores for a loss, and 0 for a draw.

First, we calculate the payoff function:

$$pAq = \begin{bmatrix} p_1 & p_2 & p_3 \end{bmatrix} \begin{bmatrix} 0 & -5 & 1 \\ 5 & 0 & -2 \\ -1 & 2 & 0 \end{bmatrix} \begin{bmatrix} q_1 \\ q_2 \\ q_3 \end{bmatrix}$$

$$= (0\,p_1 + 5\,p_2 - 1\,p_3)\,q_1 + (-5\,p_1 + 0\,p_2 + 2\,p_3)\,q_2 + (1\,p_1 - 2\,p_2 + 0\,p_3)\,q_3$$

Because we do not know the strategy of the opponent (player II), we will consider here three pure strategies:

$$q_1 = \begin{bmatrix} 1 \\ 0 \\ 0 \end{bmatrix},\ q_2 = \begin{bmatrix} 0 \\ 1 \\ 0 \end{bmatrix},\ q_3 = \begin{bmatrix} 0 \\ 0 \\ 1 \end{bmatrix}$$

Accordingly, we obtain three payoff functions:

$$q_1\text{: } pAq = (0p_1 + 5p_2 - 1p_3)$$

$$q_2\text{: } pAq = (-5p_1 + 0p_2 + 2p_3)$$

$$q_3\text{: } pAq = (1p_1 - 2p_2 + 0p_3)$$

The payoff value depends on player I's strategy, that is, a combination of probabilities (p_1, p_2, p_3). When we denote player II's expected minimum value as v, we obtain

$$q_1\text{: } (0p_1 + 5p_2 - 1p_3) \geq v$$

$$q_2\text{: } (-5p_1 + 0p_2 + 2p_3) \geq v$$

$$q_3\text{: } (1p_1 - 2p_2 + 0p_3) \geq v$$

Now player I would like to choose the maximum value available for v, as suggested in process (b), the so-called maximin process.

Microsoft Excel Solver can be introduced for solving this maximization problem. We introduce three variables, X_1, X_2, and X_3:

$$X_1 = \text{probability of using rock}$$

$$X_2 = \text{probability of using paper}$$

$$X_3 = \text{probability of using scissors}$$

with an important constraint, $X_1 + X_2 + X_3 = 1$.

In summary:

$$\text{MAXIMIZATION:} \quad V$$

$$\text{SUBJECT TO:} \quad 0X_1 + 5X_2 - 1X_3 - V \geq 0$$

$$-5X_1 + 0X_2 + 2X_3 - V \geq 0$$

$$1X_1 - 2X_2 + 0X_3 - V \geq 0$$

$$X_1 + X_2 + X_3 = 1$$

$$X_1, X_2, X_3 \geq 0, \ V \text{ unrestricted}$$

Figure 7.29 shows the setting for the Excel Solver Parameters dialog box, and the obtained results. The maximum V expected is zero, the same zero payoff average gain for any rock-paper-scissors strategy. The probabilities for rock, paper, and scissors by player I should be 0.25, 0.125, and 0.625, respectively (a mixed strategy). You will find that paper usage is surprisingly low, despite the initial discussion!

Because Table 7.5 exhibits a skew-symmetric (i.e., ${}^tA = -A$), the opponent's (player II) strategy can be easily recognized to be the same as that of player I. We will consider more general cases in the next section.

7.3.4 Case Study for Bidding on the Multilayer Actuator

Saito Industries has another supplier of MLAs: Piezo Tech in Taiwan. MMI has confidence that their MLAs are higher quality than those of Piezo Tech. However, as schematically illustrated in Figure 7.30, because the fixed cost is lower for Piezo Tech, they can reach the break-even (BE) point more easily than MMI. In other words, MMI is inferior to Piezo Tech in terms of the MLA price.

Table 7.6 summarizes the current MLA production analyses for MMI and Piezo Tech. Note a big difference in the fixed cost with the same variable cost per unit, which provides a great advantage to Piezo Tech with regard to price reduction.

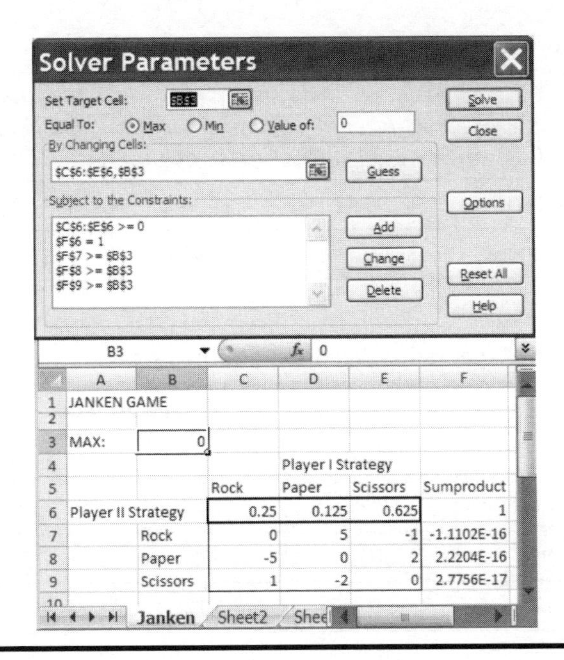

Figure 7.29 Excel Solver for the *janken* game with a biased payoff matrix.

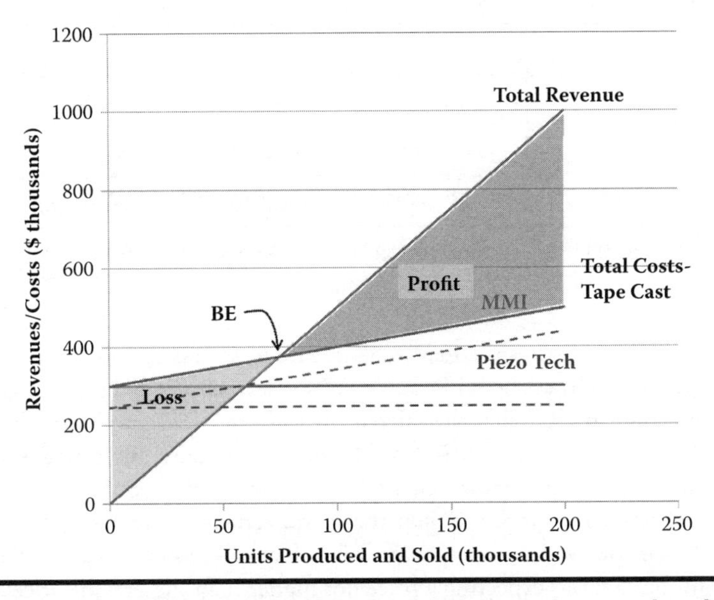

Figure 7.30 Break-even analysis results for MMI and Piezo Tech. Note that the BE point for MMI is higher than for Piezo Tech.

Saito is planning to purchase 200,000 pieces of the MLAs in total, and has asked for quotes from both MMI and Piezo Tech. Saito is expecting the price reduction, of course. For MMI, if we set BE equal to 100,000 (supposing 50% share of the total 200,000 units), the contribution margin can be reduced to $3, leading to a price of $4 (lowest price for MMI), while for Piezo Tech, the contribution margin can be reduced to $2.50, leading to a price of $3.50 (the lowest price for Piezo Tech).

Table 7.6 Manufacturing Cost, MLA Price, BE Breakdown

Company	*Fixed Cost*	*ML Current Price per Unit*	*Manufacturing Cost per Unit*	*Contribution Margin per Unit*	*Break-Even Point (Units)*
MMI	$300,000	$5.00	$1.00	$4.00	75,000
Piezo Tech	$250,000	$5.00	$1.00	$4.00	62,500

Table 7.7 Market Share Expectation According to the Price

		Piezo Tech			
		$3.50	$4.00	$4.50	$5.00
MMI	$4.00	40%	70%	90%	100%
	$4.50	30%	40%	70%	90%
	$5.00	20%	30%	40%	70%

Table 7.8 Payoff Matrix for MLAs (Unit: US$)

		Piezo Tech			
		$3.50	$4.00	$4.50	$5.00
MMI	$4.00	–60K (50K)	120K (–70K)	240K (–180K)	300K (–250K)
	$4.50	–90K (100K)	–20K (110K)	190K (–40K)	330K (–170K)
	$5.00	–140K (150K)	–60K (170K)	20K (170K)	260K (–10K)

Barb Shay of MMI anticipates the market share rate of a total of 200,000 pieces by MMI (in comparison with Piezo Tech) as shown in Table 7.7 for several possible price offers by MMI and Piezo Tech to Saito Industries. When the offered prices from MMI and Piezo Tech are the same, Saito will take 70% from MMI and 30% from Piezo Tech, because the product quality of MMI is superior to that of Piezo Tech. However, when Piezo Tech offers $3.50 per unit, which MMI cannot offer because the price is lower than the manufacturing cost, Piezo Tech increases their share significantly. Of course, both companies would like to keep the price as high as possible to increase the profit, but Saito is expecting a price not higher than the present price of $5.00. From the market share expectation, Barb calculated the anticipated payoff matrix, as shown in Table 7.8. Top figures without parentheses (–60K, 120K, 240K, 300K ...) in each cell show profit/loss on the MMI side, and the bottom figures in parentheses (50K, –70K, –180K, –250K ...) in each cell show profit/loss on the Piezo Tech side. The calculation procedure is explained with an example of MMI $4.50/Piezo Tech $4.50: (1) MMI will share 70% of 200,000 pieces, i.e., 140,000 units. (2) Sales revenues will be $4.50 × 140,000 = $630,000. (3) Total cost = fixed cost ($300,000) + $1.00 × 140,000 = $440,000. (4) Profit = revenues – total cost = $190,000. Note first that this is not a simple "zero-sum" game, though the higher MMI's payoff, the lower Piezo Tech's payoff, in general. For example, MMI $5.00/Piezo Tech $4.50 can provide positive profits ($20,000 and $170,000) for both firms.

7.3.4.1 Decision-Making Criteria

We will start from decision making under uncertainty, because it is probable that MMI will not discuss anything with Piezo Tech about the bidding price. There are four approaches: maximin, minimax regret, maximax, and expected value criteria.

7.3.4.1.1 Maximin Criterion—Pessimistic or Conservative Approach

A pessimistic decision maker believes that the worst possible result will occur. Because this criterion is based on a worst-case scenario, the decision maker first finds the minimum payoff for each decision alternative. The optimal decision alternative is the one that maximizes the minimum payoff. A similar concept is found in the rock-paper-scissors game.

Excel Solver can be introduced for solving this maximization problem. We introduce three variables: X_1, X_2, and X_3:

$$X_1 = \text{MMI probability of bidding at \$4.00}$$

$$X_2 = \text{MMI probability of bidding at \$4.50}$$

$$X_3 = \text{MMI probability of bidding at \$5.00}$$

with an important constraint, $X_1 + X_2 + X_3 = 1$.

In summary, on the MMI side:

$$\text{MAXIMIZATION:} \quad V_1$$

$$\text{SUBJECT TO:}$$

$$-60X_1 - 90X_2 - 140X_3 - V_1 \geq 0$$

$$120X_1 - 20X_2 - 60X_3 - V_1 \geq 0$$

$$240X_1 + 190X_2 + 20X_3 - V_1 \geq 0$$

$$300X_1 + 330X_2 + 260X_3 - V_1 \geq 0$$

$$X_1 + X_2 + X_3 = 1$$

$$X_1, X_2, X_3 \geq 0, \; V_1 \text{ unrestricted}$$

On the Piezo Tech side:

$$Y_1 = \text{PT probability of bidding at \$3.50}$$

$$Y_2 = \text{PT probability of bidding at \$4.00}$$

$$Y_3 = \text{PT probability of bidding at \$4.50}$$

Y_4 = PT probability of bidding at \$5.00

MAXIMIZATION: V_2

SUBJECT TO:

$$50Y_1 - 70Y_2 - 180Y_3 - 250Y_4 - V_2 \geq 0$$

$$100Y_1 + 110Y_2 + -40Y_3 - 170Y_4 - V_2 \geq 0$$

$$150Y_1 + 170Y_2 + 170Y_3 + -10Y_4 - V_2 \geq 0$$

$$Y_1 + Y_2 + Y_3 + Y_4 = 1$$

$$Y_1, Y_2, Y_3, Y_4 \geq 0, V \text{ unrestricted}$$

Figure 7.31 shows the results obtained via Excel Solver. The "pure" strategy (only one quote price) for MMI is \$4.00 with negative profit (loss) of –\$60,000, and the pure strategy for Piezo Tech is \$3.50 with positive profit of \$50,000. This point (p, q) = (\$4.00, \$3.50) seems to be the maximin point for this bidding game.

7.3.4.1.2 Minimax Regret Criterion—Pessimistic or Conservative Approach

This approach is identical to the minimax approach based on lost opportunity, or "regret." The decision maker incurs regret by failing to choose the best decision (the one with the highest profit or payoff). The regret table is determined from the payoff matrix by calculating the regret for each decision alternative as the difference between its payoff value and the best value. Let us calculate the regret table for MMI from Table 7.8. When MMI chooses \$4.00, the best profit value \$300,000 can be obtained for \$5.00 of Piezo Tech's selection. The regret can be calculated by the difference from this \$300,000: –\$60,000, \$120,000, and \$240,000 can be converted to \$360,000, \$180,000, and \$60,000, respectively (see the first column of Table 7.9). Find the maximum regret for each decision alternative. That is, \$360,000, \$420,000, and \$400,000 for \$4.00, \$4.50, and \$5.00 choices by MMI. Select the decision alternative that has the minimum among the maximum regrets, that is, \$360,000 for MMI \$4.00/Piezo Tech \$3.50 combination—the same result as the maximin criterion discussed in Section 7.3.4.1.1.

7.3.4.1.3 Maximax Criterion—Optimistic or Aggressive Approach

In contrast to a pessimistic decision maker, an optimistic decision maker feels that luck is always imminent, and the best possible outcome corresponding to the decision will occur. In this scenario, first determine the maximum payoff for each decision alternative. As shown in Table 7.8, MMI can use \$300,000, \$330,000, and \$260,000 for the bidding prices \$4.00, \$4.50, and \$5.00, respectively, hoping that Piezo Tech will bid by \$5.00. Then, select the decision alternative that has the maximum payoff, that is, \$330,000 for MMI \$4.50/Piezo Tech \$5.00.

Even though Barb Shay is rather optimistic, she may not use a maximax approach—it is too risky!

	A	B	C	D	E	F	G
1	MMI Payoff Maximization						
2							
3	MAX:	-60					
4				MMI Strategy			
5			$4.00	$4.50	$5.00	Sumproduct	
6	Piezo Tech Strategy		1	0	0	1	
7		$3.50	-60	-90	-140	-60	
8		$4.00	120	-20	-60	120	
9		$4.50	240	190	20	240	
10		$5.00	300	330	260	300	
11							
12	Piezo Tech Payoff Maximization						
13							
14	Max:	50					
15				Piezo Tech Strategy			
16			$3.50	$4.00	$4.50	$5.00	Sumproduct
17	MMI Strategy		1	0	0	0	1
18		$4.00	50	-70	-180	-250	50
19		$4.50	100	110	-40	-170	100
20		$5.00	150	170	170	-10	150
21							

Figure 7.31 Excel Solver for the MMI bidding game with a payoff matrix.

Table 7.9 Regret Table for MMI (Unit: US$)

		Piezo Tech			
		$3.50	$4.00	$4.50	$5.00
MMI	$4.00	360K	180K	60K	0
	$4.50	420K	350K	140K	0
	$5.00	400K	320K	240K	0

7.3.4.1.4 Expected Value Criterion—Statistical Approach or Expected Value

This is a statistical approach to calculate the expected value for each decision alternative. If a probability estimate for the occurrence of each strategy is available, it is possible to calculate an expected value associated with each decision alternative.

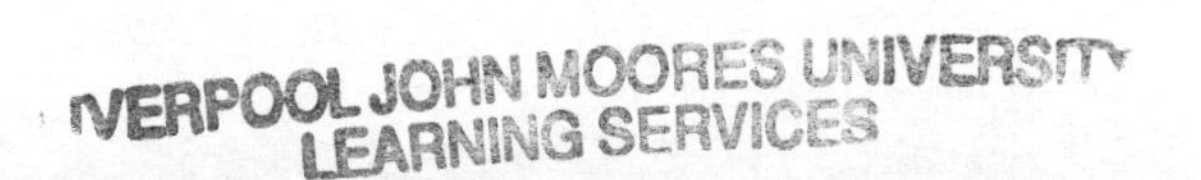

If MMI does not know anything about the personality of Piezo Tech's manager, Barb may adopt an even probability among their choices for Piezo Tech's strategy, that is, 25% each for $3.50, $4.00, $4.50, and $5.00 bidding. By using this even probability distribution, the expected payoff for each of MMI's decision alternatives can be calculated:

$4.00: 0.25(−$60,000) + 0.25($120,000) + 0.25($240,000) + 0.25($300,000) = $150,000

$4.50: 0.25(−$90,000) + 0.25(−$20,000) + 0.25($190,000) + 0.25($330,000) = $102.5,000

$5.00: 0.25(−$140,000) + 0.25(−$60,000) + 0.25($20,000) + 0.25($260,000) = $20,000

Then, select the decision alternative that has the best expected payoff, that is, $4.00.

In conclusion, decision making is not easy; depending on the manager's personality, the game solution will change. Both the lowest bids (p, q) = ($4.00, $3.50) and the highest bids (p, q) = ($5.00, $5.00) have reasons to be supported. Among the four uncertain criteria for Piezo Tech's decision, three approaches suggest that the best strategy for MMI is $4.00 bidding. However, as the minimax approach suggests, if Piezo Tech bids at $3.50, MMI's profit is actually negative (−$60,000). Thus, MMI should drop out of the bidding for this product supply.

The solution can be found only in the game theory under cooperative negotiation. The equilibrium point (p, q) = ($4.00, $3.50) can generate payoffs of −$60,000 for MMI, and $50,000 for Piezo Tech. As long as the new strategic point (p, q) can generate profit for Piezo Tech higher than $50,000, Piezo Tech may be interested in cooperating with MMI. Initially consider the maximization of the total payoff of the two firms (MMI and Piezo Tech), which is shown in Table 7.10. The maximum of the total profit $250,000 can be found for MMI $5.00/Piezo Tech $5.00. However, as initially considered, Saito is not favorable to the price of $5.00, equal to the present sales cost. The second choice is MMI $4.50/Piezo Tech $4.50, leading to the total profit $150,000. If MMI guarantees to pay back the sum of $40,000 (Piezo Tech's deficit) and $75,000 (a half of the total profit), Piezo Tech may agree. Another alternative strategy is a combination of MMI $4.50/Piezo Tech $4.50 and MMI $4.50/Piezo Tech $4.00. In this scenario, if they combine these two strategies using a 0.36:0.64 ratio, both MMI and Piezo Tech can obtain the same profit of $56,000, higher than $50,000 of the original equilibrium point. In practice, the bidding prices by MMI and Piezo Tech may be $4.50 and $4.18, respectively. If we combine MMI $4.50/Piezo Tech $4.50 and MMI $5.00/Piezo Tech $4.50, with a ratio of 0.39:0.61, both will obtain the same profit of $87,000, much higher than $50,000 for Piezo Tech. In practice, the bidding prices by MMI and Piezo Tech may be $4.80 and $4.50, respectively.

Table 7.10 Payoff Matrix for MLAs. Total Profit of Two Firms (Top) and Profit of Piezo Tech (Bottom in Parentheses) (Unit: US$)

		Piezo Tech			
		$3.50	$4.00	$4.50	$5.00
MMI	$4.00	−10K (50K)	50K (−70K)	60K (−180K)	50K (−250K)
	$4.50	10K (100K)	90K (110K)	150K (−40K)	160K (−170K)
	$5.00	10K (150K)	110K (170K)	190K (170K)	250K (−10K)

Chapter Summary

7.1 Excel Solver is a useful tool for solving a maximization or minimization problem.

7.2 Linear programming: manufacturing schedule of multiple products, X and Y (units):

$$\text{MAXIMIZE: } aX + bY \quad \text{(Total profit)}$$

$$\text{SUBJECT TO: } cX + dY \leq L \quad \text{(Labor)}$$

$$eX + fY \leq M \quad \text{(Raw materials)}$$

$$gX + hY \leq S \quad \text{(Space)}$$

$$X,\ Y \geq 0 \quad \text{(Nonnegativity).}$$

7.3 Program evaluation and review technique (PERT) was developed in the mid-1950s for treating the completion time of activities as random variables with specific probability distributions. PERT requires the modeler to identify the activities of the project and the precedence relations between them. This involves determining a set of immediate predecessors for each activity. An activity's immediate predecessors are those jobs that must be completed just prior to the activity's commencement. A precedence relation chart identifies the separate activities of the project and their precedence relations. From this chart a PERT network representation of the project can be constructed.

7.4 Earliest start time (ES): ES for an activity ≥ ES for the immediate predecessor activity + immediate predecessor's activity completion time.

7.5 A Gantt chart is a visual aid for tracking the progress of project activities by shading an appropriate percentage of the corresponding bar to document the completed work. A manager can see on a given date if the project is being completed on schedule.

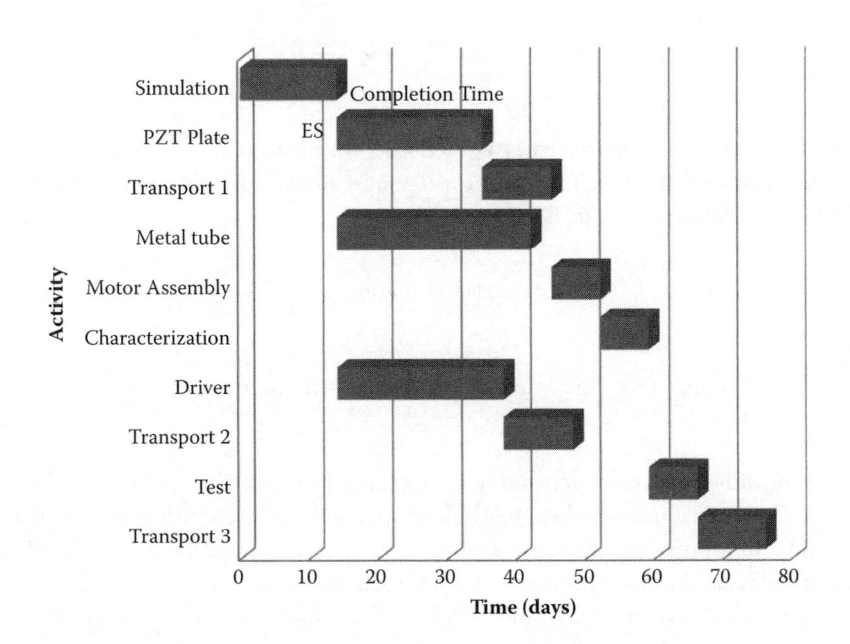

7.6 Critical path method (CPM) assumes that there are two crucial time points for each activity: (1) its normal completion time (T_N), achieved when the usual or normal cost (C_N) is spent to complete the activity; and (2) its crash completion time (T_C), the minimum possible completion time for the activity. The T_C is attained when the maximum crash cost (C_C) is spent.

7.7 Game theory: The mixed strategies followed by player I and player II are described by

$$p = [p_1 \; p_2 \; \cdots \; p_m]$$

$$q = \begin{bmatrix} q_1 \\ q_2 \\ \vdots \\ q_n \end{bmatrix}$$

where p and q are probability distribution.
The payoff function is given by

$$pAq = [p_1 \; p_2 \; \cdots \; p_m] \begin{bmatrix} a_{11} & a_{12} & \cdots & a_{1n} \\ a_{21} & a_{21} & \cdots & a_{2n} \\ \vdots & \vdots & \ddots & \vdots \\ a_{m1} & a_{m2} & \cdots & a_{mn} \end{bmatrix} \begin{bmatrix} q_1 \\ q_2 \\ \vdots \\ q_n \end{bmatrix} = \sum_{i=1}^{m} \sum_{j=1}^{n} p_i a_{ij} q_j$$

7.8 Minimax theorem: Under the supposition that player I will choose a mixed strategy p in order to maximize the payoff function, player II will choose a mixed strategy q to minimize player I's payoff. The final payoff function value is represented as

$$\underset{q \in T}{\text{Min}} \; \underset{p \in S}{\text{Max}} \; pAq$$

Under the supposition that player II will choose mixed strategy q in order to minimize the payoff function of the opponent, player I will choose mixed strategy p to maximize his payoff. The final payoff function value is represented as

$$\underset{p \in S}{\text{Max}} \; \underset{q \in T}{\text{Min}} \; pAq$$

$$\text{Minimax theorem: } \underset{q \in T}{\text{Min}} \; \underset{p \in S}{\text{Max}} \; pAq = \underset{p \in S}{\text{Max}} \; \underset{q \in T}{\text{Min}} \; pAq = p_0 A q_0 = v$$

7.9 Decision making is not easy; depending on a manager's personality, the game solution will be changed. The maximin method is a pessimistic approach, or a very risk-averse approach, where we consider maximization of the possible worst profit (or loss). On the other hand, the maximax method is an optimistic approach. Both the game solutions (p, q) changed from the lowest bids ($4.00, $3.50) to the highest bids ($5.00, $5.00) have reasonable supports.

Practical Exercise Problems

P7.1 Linear Programming

Take one of the quantum business analysis models in this chapter (linear programming, PERT, game theory, etc.) and apply it to a situation in your life. Some examples may include the following:

- Production scheduling
- Allocating manpower or budget at a branch
- Proposal submission strategy competing with a rival firm

The report should be prepared in a context as if you were presenting results to your manager or a decision maker in your company. A typical outline of each type of report is listed below:

1. Application description, setting, background
2. Proposed model description
3. Data
4. Model output/analysis
5. Conclusions/suggestions for further study

MMI Example—Multilayer Actuator Production Plan

In October 20XX, MMI provided MLA components to Saito Industries, initially with only 1000 pieces for a test purpose. Once this test is finished, much larger production will be requested in January of the next year, finally reaching around 3 million pieces per year. MMI needs to develop the production plan for supplying actuators to Saito Industries.

The customer's demands include: (1) the supply quantity will be exponentially increased in two years, and (2) the unit price will be drastically decreased with the quantity. MMI has three manufacturing factories: MMI in Pennsylvania (no tape-cast facility), a partner company, Cheng Kung Corporation in Taiwan (one tape-cast facility), and Chiang Mai factory in Thailand (no tape-cast facility). Labor cost drastically decreases from Pennsylvania to Taiwan to Thailand. Taking into account the transportation period, which can be counted as a manufacturing delay, MMI will distribute the manufacturing load to these three places, in order to minimize the manufacturing cost. The manufacturing factory locations of these three facilities are shown in Figure 1.14 in Chapter 1.

P7.1.1 Production Schedule Modeling

The introduction of a tape-casting facility, which is an expensive automation production system ($300,000 per set), is usually recommended if the production amount exceeds 1 million pieces per year. Otherwise, the conventional cut-and-bond method should be employed, which requires hiring several manufacturing workers. For a tape-casting system, the labor fee is negligible but the initial investment is very high, while for the cut-and-bond method, the labor fees are proportional to the product quantity—even the equipment cost is negligible. There is a threshold product quantity above which the equipment installation starts to provide the profit benefit. This quantity is about 1 million pieces in MLA production.

We will consider the following two scenarios:

1. Setup of a manual labor manufacturing line composed of six workers, without the introduction of the tape-cast equipment. Taking into account the necessary labor cost S (the annual

salary for one product line of six workers), and the product line's manufacturing capability q (quantity of manufacturing actuators, which averages 50,000 pieces per year per product line), the total cost can be calculated as

$$\mathrm{TC}_{\mathrm{ProdLine}}(Q) = \left(P_I + \frac{S}{q}\right) \cdot Q$$

where P_I is the input cost, including raw material cost, such as PZT ceramic and Ag-Pd electrode, around $0.4 per piece of MLA.

2. Introduction of automatic tape-casting manufacturing equipment (typically the maximum capability of manufacturing is 1 million pieces per year) costs TFC (fixed cost, $300,000). Taking into account the input cost P_I, the total cost for the tape-casting is:

$$\mathrm{TC}_{\mathrm{TapeCast}}\,(Q) = P_I \cdot Q + \mathrm{TFC})$$

We will adopt the following values for each parameter: $P_I = \$0.40$, $q = 50{,}000$ pieces per year, and $S = \$150{,}000$ (in the United States) or $7500 (in Thailand). By setting the last two equations to be equal, we obtain

$$\mathrm{TFC} = \left(\frac{S}{q}\right) \cdot Q$$

In the United States, $S/q = \$150{,}000/50{,}000 = \$3.00$, while in Thailand, $S/q = \$7500/\ 50{,}000 = \$0.15$. Thus, we obtain the transition production quantity $Q = 100{,}000$ and 2 million for the United States and Thailand, respectively. Because of the higher salary, the automation equipment should be introduced much earlier in the United States. On the contrary, we may use manual labor manufacturing lines for a rather *large production quantity in Thailand.*

Note that Taiwan has a special situation. Because there are many inexpensive machinery companies in Taiwan, the tape-casting equipment is cheaper than in the United States or Thailand. We presume that TFC = $250,000 in Taiwan, which may provide a privilege for Taiwan Cheng Kung Corporation to purchase the equipment over MMI in the United States or from the Chiang Mai factory.

P7.1.2 Constraints from the Customer (Saito Industries)

Saito requests from MMI a 12-week delivery for each PO. Requested MLA quantity and ML product unit price are summarized in Table P7.1. Because MMI is much smaller than Saito, MMI is merely a *price-taker* with only a slight negotiation possibility.

P7.1.3 Manufacturing Capability

MMI has three factories: MMI in Pennsylvania, Taiwan Cheng Kung, and Chiang Mai factory. Raw materials, such as piezoelectric ceramic powder and Ag/Pd metal electrode paste, are synthesized at MMI's laboratory at Pennsylvania, and transported to the MMI factory in Pennsylvania (no delay), to Taiwan Cheng Kung (1 week for one way), and to Chiang Mai factory (2 weeks for one way). For both a tape-cast machine and a manual labor production line, one production batch can be finished in two weeks (biweekly). Thus, during the 12-week PO period, MMI (Pennsylvania)

Table P7.1 Expected Purchase Order from Saito Industries to MMI in These 96 Weeks (2 Years)

PO No.	*1*	*2*	*3*	*4*	*5*	*6*	*7*	*8*
Week	1–12	13–24	25–36	37–48	49–60	61–72	73–84	85–96
Quantity	10K	50K	250K	1000K	2000K	2400K	2700K	3000K
Requested unit price	$10.00	$5.00	$2.80	$1.80	$1.20	$1.10	$1.00	$1.00
Total payment	$100K	$250K	$700K	$1800K	$2400K	$2640K	$2700K	3000K

can run six production batches. However, due to the transportation time loss, Taiwan Cheng Kung and Chiang Mai factory can run five and four production batches during 12 weeks. Only Taiwan Cheng Kung has a tape-cast facility, so other factories need to purchase the necessary number of tape-cast apparatuses, increasing the production quantity: $300,000 per set in the United States and Thailand, but $250,000 in Taiwan. This sales situation may provide an advantage to Cheng Kung for purchasing the equipment. One tape-cast apparatus can manufacture a maximum of 42,000 ML pieces per biweekly batch. Regarding the workers, MMI in Pennsylvania, Taiwan Cheng Kung, and Chiang Mai factory can hire a maximum of 60 people (10 lines), 240 people (40 lines), and 1200 people (200 lines), respectively, due to the factory space. Six workers make one manufacturing line, which can manufacture an average of 2100 pieces per biweekly batch. The average biweekly salary for one manufacturing line (six workers) is $6250, $1562, and $312 for the United States, Taiwan, and Thailand, respectively. Table P7.2 summarizes the necessary data for all three manufacturing plants.

MMI should, in general, use the cheapest labor in Thailand as much as possible. However, we need to take into account the freight cost and the manufacturing time loss, due to the transportation.

P7.1.4 Transportation Cost

The piezoelectric MLA is very compact (typically $2 \times 3 \times 5$ mm^3) and lightweight (0.24 gram per chip). Thus, 100,000 pieces will make one freight carton with a package size of $20 \times 30 \times 5$ cm^3 and a weight of 24 kg. MMI will initially send raw materials (with an assumption that the weight is almost the same as the final MLAs, but rather bulky compared with the final products) and receive all MLAs (neglecting the product wastes) from Taiwan Cheng Kung and Chiang Mai factory. The freight cost is summarized in Table P7.3.

P7.1.5 The Mathematical Model

P7.1.5.1 Decision Variables

1. MMI will purchase tape-cast systems and/or hire workers:
   Number of systems purchased:
   - At MMI Y_{11}
   - At Taiwan Cheng Kung Y_{21}
   - At Chiang Mai factory Y_{31}

Table P7.2 Estimation Sheet for the Manufacturing Capability of Three MMI Facilities

	MMI, Pennsylvania	*Cheng Kung, Taiwan*	*Chiang Mai, Thailand*
Maximum number of working rotation	6	5	4
Tape-cast Apparatus			
Production capability per system (pieces biweekly)	42,000	42,000	42,000
Total production capability per system (pieces during one PO period)	252,000	210,000	168,000
Price per system	$300,000	$250,000	$300,000
Number of present apparatuses	0	1	0
Maintenance fee per system (biweekly)	$8000	$2000	$400
Total maintenance fee per system (during one PO period)	$48,000	$10,000	$1600
Manual Process			
Biweekly salary for one manufacturing line (6 workers)	$6250	$1562	$312
Total salary for one manufacturing line (during one PO period)	$37,500	$7810	$1248
Production capability per one line (pieces biweekly)	2100	2100	2100
Total production capability per one line (during one PO period)	12,600	10,500	8400
Maximum workers/maximum lines	60/10	240/40	1200/200
Transportation			
Transportation dates required (weeks)	0	2	4

Number of worker lines employed (one line composed of six workers):
At MMI Y_{12}
At Taiwan Cheng Kung Y_{22}
At Chiang Mai factory Y_{33}
All Ys ≥ 0 and integer.

2. Number of ML chips to be manufactured:
At MMI
With tape-cast apparatus X_{11}
With human worker line X_{12}

Table P7.3 Transportation (Freight) Cost of ML Raw Materials and Final Products (between MMI and Taiwan or Thailand)

ML Quantity	*1–100,000*	*500,000*	*1 Million*	*2 Million*	*4 Million*
Number of cartons	1	5	10	20	40
To Taiwan (raw materials)	$600	$3000	$6000	$12,000	$24,000
From Taiwan (ML chips)	$400	$2000	$4000	$8000	$16,000
Total freight cost for Taiwan	$1000	$5000	$10,000	$20,000	$40,000
To Chiang Mai (raw materials)	$900	$4500	$9000	$18,000	$36,000
From Chiang Mai (ML chips)	$600	$3000	$6000	$12,000	$24,000
Total freight cost for Chiang Mai	$1500	$7500	$15,000	$30,000	$60,000

At Taiwan Cheng Kung
With tape-cast apparatus X_{21}
With human worker line X_{22}
At MMI Chiang Mai factory
With tape-cast apparatus X_{31}
With human worker line X_{32}
All Xs ≥ 0 and integer.

3. Transportation:
Number of freight cartons to send and return:
To and from Taiwan Cheng Kung Z_{21}
To and form Chiang Mai factory Z_{31}
All Zs ≥ 0 and integer.

P7.1.5.2 Objective Function

The objective is to minimize the total cost under a certain quantity (Q) of MLAs requested in the PO from Saito Industries.

1. Cost for purchasing tape-cast systems and/or for hiring workers:
Based on the number of systems purchased:
At MMI 300,000 Y_{11}
At Taiwan Cheng Kung 250,000 Y_{21}
At Chiang Mai factory 300,000 Y_{31}
Based on the number of worker lines employed; the line is supposed to work under the maximum rotation times during one PO period):
At MMI 37,500 Y_{12}
At Taiwan Cheng Kung 7810 Y_{22}
At Chiang Mai factory 1248 Y_{33}

2. Other manufacturing costs:
   Raw materials cost ($0.4 for all places):
   - At MMI 0.4 $(X_{11} + X_{12})$
   - At Taiwan Cheng Kung 0.4 $(X_{21} + X_{22})$
   - At Chiang Mai factory 0.4 $(X_{31} + X_{32})$

   Tape-cast apparatus maintenance cost:
   - At MMI 48,000 Y_{11}
   - At Taiwan Cheng Kung 10,000 Y_{21}
   - At Chiang Mai factory 160 Y_{31}

3. Transportation costs:
   - To and from Taiwan Cheng Kung 1000 Z_{21}
   - To and form Chiang Mai factory 1500 Z_{31}

The total cost can be obtained by summing up all above the cost.

P7.1.5.3 Constraints

1. Production quantity:

   Summing up the production quantity at three places and via both machine and manual manufacturing methods, which needs to be more than the PO quantity Q:

$$X_{11} + X_{12} + X_{21} + X_{22} + X_{31} + X_{32} \geq Q$$

   Further, taking into account the maximum production capability at each location (under the maximum number of working rotation at each location),

   - At MMI $X_{11} \leq 252{,}000 \quad Y_{11}$
   - $X_{12} \leq 12{,}600 \quad Y_{12}$
   - At Taiwan Cheng Kung $X_{21} \leq 210{,}000 \quad (Y_{21} + 1)$
   - $X_{22} \leq 10{,}500 \quad Y_{22}$
   - At Chiang Mai factory $X_{31} \leq 168{,}000 \quad Y_{31}$
   - $X_{32} \leq 8400 \quad Y_{32}$

   Note that 1 is added initially at Taiwan Cheng Kung Corporation, $X_{21} \leq 210{,}000\ (Y_{21} + 1)$, because Cheng Kung has a tape-cast apparatus already. The calculation should be made starting from PO period 1 and continuing successively higher. Once a factory purchases N tape-cast apparatuses in PO period (i), we need to use $(Y + N)$ in PO period $(i + 1)$ for evaluating the necessary cost for manufacturing X: N apparatuses already exist!

2. Space constraints:
   - At MMI $Y_{12} \leq 10$
   - At Taiwan Cheng Kung $Y_{22} \leq 40$
   - At Chiang Mai factory $Y_{33} \leq 200$

3. Number of freight cartons:

   To reflect Table P7.3 (one carton can include a maximum of 100,000 pieces of ML chips), Z_{21} and Z_{31} can be correlated with the sum of $(X_{21}+X_{22})$ and $(X_{31}+X_{32})$ as follows:

   $$(X_{21}+X_{22}) \le 100{,}000 \quad Z_{21}$$
   $$(X_{31}+X_{32}) \le 100{,}000 \quad Z_{31}$$

P7.1.6 Excel Solver

Now we will solve the problem with Excel Solver. Figure P7.1 shows the Solver Parameter setting in Excel for the case of PO period 1.

Tables P7.4 and P7.5 are the calculation results for PO period 1 and period 7, respectively.

P7.1.6.1 Small Purchase Order Quantity Less Than One Equipment Capability (PO 1–5)

In Table P7.4, all actuator production will be made in Taiwan Cheng Kung, because there is already one tape-cast apparatus, and no particular cost is required in Taiwan. Manpower is used at the Chiang Mai factory first, then at Taiwan Cheng Kung, due to the significant labor cost difference. However, if the worker line can be used on a biweekly basis (not in 12-week periods), the U.S. manpower may be used for the smallest (such as 1000 pieces) prototypes.

P7.1.6.2 Large Purchase Order Quantity Higher Than One Equipment Capability (PO 6–8)

In Table P7.5, with increasing the production quantity we need to purchase multiple tape-cast apparatuses to accelerate the production capability. Because the equipment is slightly cheaper in Taiwan than in the United States or Thailand, the purchasing privilege goes to Taiwan.

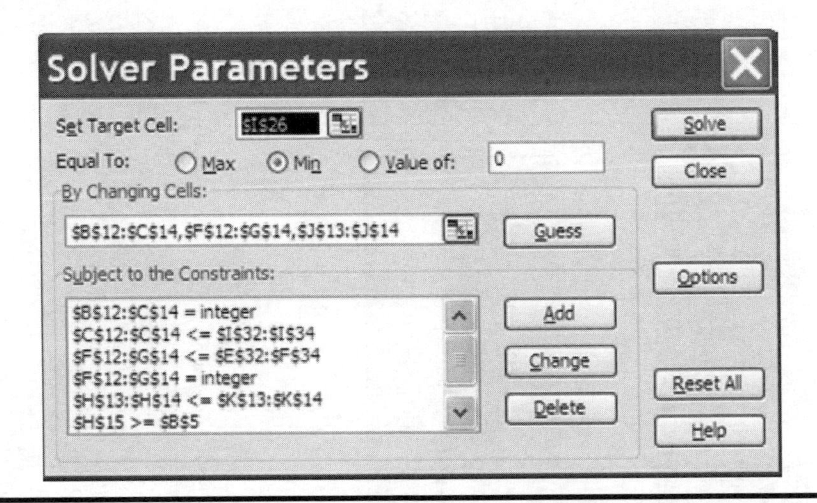

Figure P7.1 Solver parameter setting in Excel for the case purchase order period 1.

Table P7.4 Excel Solution for Purchase Order Period 1

Multilayer Actuator Production Plan *(Micro Motor Inc.)*

(1) Purchase Order from Saito

PO Number	*1*	*2*	*3*	*4*	*5*	*6*	*7*	*8*
Quantity Requested	10,000	50,000	250,000	1,000,000	2,000,000	2,400,000	2,700,000	3,000,000
Unit Price	$10.00	$5.00	$2.80	$1.80	$1.20	$1.10	$1.00	$1.00
Total Payment	$100.000	$250,000	$700,000	$1,800,000	$2,400,000	$2,640,000	$2,700,000	$3,000,000

(2) Decision Variables

# of new systems and worker lines

	Tape Cast	*Line Worker*
MMI at PA	0	0
Taiwan Cheng Kung	0	0
Chiang Mai	7.69327E-32	0

# of ML chips to be manufactured

	Tape Cast	*Line Worker*	*Total # of Production*		*# of Cartons*	*Quantity Carton #*
MMI at PA	0	0	0			
Taiwan Cheng Kung	10,000	0	10,000	<=	1	100,000
Chiang Mai	0	0	0	<=	1.29E-31	1.29E-26
			10,000			
			# of ML in a carton		100,000	

(3) Objective Function: Minimizing the Total Cost

Unit cost for a system and a worker line

	Tape Cast per System	*Line Worker*	*Total Cost*
MMI at PA	$ 300,000	$ 37,500	0
Taiwan Cheng Kung	$ 250,000	$ 7810	0
Chiang Mai	$ 300,000	$ 1248	2.308E-26
			2.308E-26

Other manufacturing cost / *Transportation cost*

	Raw Materials	*Apparatus Maintenance*	*Transportation cost*
MMI at PA	0.4	48,000	
Taiwan Cheng Kung	0.4	10,000	1000
Chiang Mai	0.4	1600	1500
Total Cost	4000	1.2309E-28	1000

The grand total cost of facilities, workers, manufacturing and transportation costs	$ 5000
Gross profit	$95,000

Table P7.4 Excel Solution for Purchase Order Period 1 *(Continued)*

(4) Constraints

Production quantity per unit

	Tape Cast	Line Worker
MMI at PA	252,000	12,600
Taiwan Cheng Kung	210,000	10,500
Chiang Mai	168,000	8400

Total production quantity

	Tape Cast	Line Worker
MMI at PA	0	0
Taiwan Cheng Kung	210,000	0
Chiang Mai	0	0

*Taiwan Cheng Kung: 1 Existing Apparatus

Space constraint

	Worker Lines
MMI at PA	10
Taiwan Cheng Kung	40
Chiang Mai	200

Table P7.5 Excel Solution for Purchase Order Period 7

Multilayer Actuator Production Plan *(Micro Motor Inc.)*

(1) Purchase Order from Saito

PO Number	1	2	3	4	5	6	7	8
Quantity Requested	10,000	50,000	250,000	1,000,000	2,000,000	2,400,000	2,700,000	3,000,000
Unit Price	$10.00	$5.00	$2.80	$1.80	$1.20	$1.10	$1.00	$1.00
Total Payment	$100.000	$250,000	$700,000	$1,800,000	$2,400,000	$2,640,000	$2,700,000	$3,000,000

(2) Decision Variables

# of new systems and worker lines

	Tape Cast	Line Worker
MMI at PA	5.82867E-16	0
Taiwan Cheng Kung	1	38
Chiang Mai	0	200

*Taiwan Sunnytec: Buy 1 Apparatus

# of ML chips to be manufactured

	Tape Cast	Line Worker	Total # of Production		# of Cartons	Quantity Carton #
MMI at PA	0	0	0			
Taiwan Cheng Kung	630,000	390,000	1,020,000	<=	11	1,100,000
Chiang Mai	0	0	1,680,000	<=	17	1,700,000
			2,700,000			

# of ML in a carton	100,000

(Continued)

Table P7.5 Excel Solution for Purchase Order Period 7 *(Continued)*

(3) Objective Function: Minimizing the Total Cost

Unit cost for a system and a worker line

	Tape Cast per System	*Line Worker*	*Total Cost*
MMI at PA	$ 300,000	$ 37,500	1.7486E-10
Taiwan Cheng Kung	$ 250,000	$ 7810	546,780
Chiang Mai	$ 300,000	$ 1248	249,600
			796,380

Other manufacturing cost / *Transportation cost*

	Raw Materials	*Apparatus Maintenance*	*Transportation cost*
MMI at PA	0.4	48,000	
Taiwan Cheng Kung	0.4	10,000	1000
Chiang Mai	0.4	1600	1500
Total Cost	1,080,000	10,000	36,500

The grand total cost of facilities, workers, manufacturing and transportation costs	$ 1,922,880
Gross profit	$ 777.120

(4) Constraints

Production quantity per unit

	Tape Cast	*Line Worker*
MMI at PA	252,000	12,600
Taiwan Cheng Kung	210,000	10,500
Chiang Mai	168,000	8,400

Total production quantity

	Tape Cast	*Line Worker*
MMI at PA	0	0
Taiwan Cheng Kung	630,000	399,000
Chiang Mai	0	1,680,000

*Taiwan Cheng Kung: 2 Existing Apparatuses

Space constraint

	Worker Lines
MMI at PA	10
Taiwan Cheng Kung	40
Chiang Mai	200

If the same purchasing price is set for Taiwan, the purchasing privilege goes to the United States, because no particular manpower cost is required, as well as due to the diminished transportation cost and lack of working time.

P7.1.7 Multilayer Actuator Production Plan

Figure P7.2 shows a graph of an expected PO from Saito Industries to MMI in these 96 weeks (12 weeks for one PO period), visualized from Table P7.1. The MLA unit price drops dramatically from the initial $10.00 (skimming price) down to $1.00, with an increase in the purchase quantity from 10,000 up to 3 million pieces. Thus total sales amount (revenue) increases more slowly than the quantity.

Figure P7.3 summarizes the number for new equipment purchase and line worker salary for three locations, MMI in Pennsylvania, Taiwan Cheng Kung and the Chiang Mai factory. Since Taiwan Cheng Kung has initially one set of tape-cast apparatus, MMI does not need to purchase additional apparatus until the sixth PO period. One, one, and two new sets should be purchased

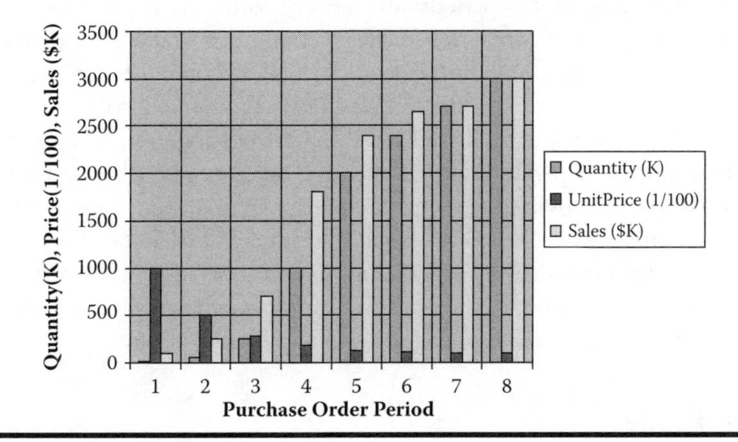

Figure P7.2 Visualization of expected purchase order from Saito Industries to MMI in these 96 weeks (two years).

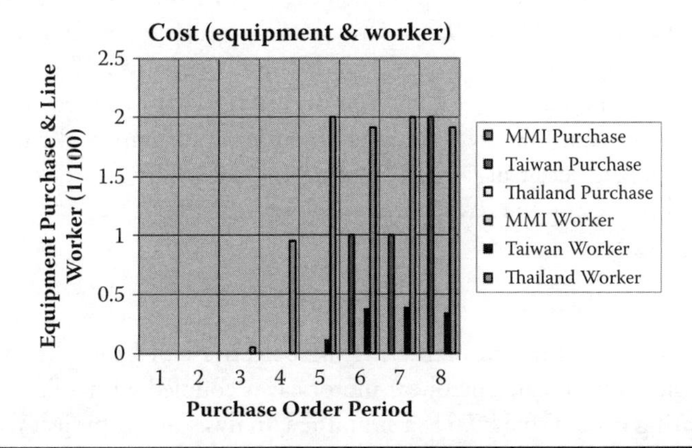

Figure P7.3 Numbers for new equipment purchases and line workers for three locations: MMI in Pennsylvania, Taiwan Cheng Kung, and the Chiang Mai factory.

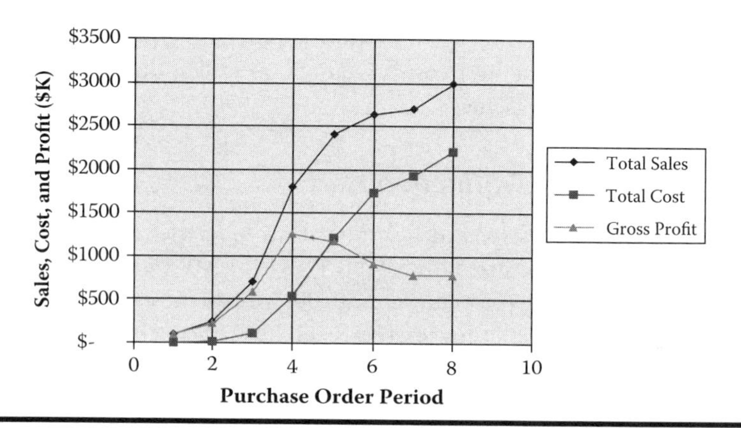

Figure P7.4 Total sales, total manufacturing cost, and gross profit (sales–cost) as a function of PO period sequence.

in the sixth, seventh, and eighth PO periods, respectively. Note that all purchases should be made through the Taiwan factory. As was mentioned already, this is due to the lower purchase cost of the tape-cast system in Taiwan. Regarding the line workers, Chiang Mai is the highest priority. Once all 200 workers there are almost spent out, the second choice, Taiwan, will start to hire workers.

Finally, Figure P7.4 shows the total sales, total manufacturing cost, and gross profit (sales – cost) as a function of the PO period sequence. Because we have a tape-cast apparatus in Taiwan Cheng Kung initially, we do not need to spend that particular manufacturing cost until PO period 3. Thus, the maximum gross profit will be reached around PO period 4. Because they need to spend $250,000 for the equipment as production quantity increases, the gross profit decreases after the maximum for PO period 4. Then the gross profit finally saturates around $800,000 per PO period (PO 7 and 8), which corresponds to a Profit/Sales rate of 26%.

P7.2 Janken Game

Rock-paper-scissors is played by player I and player II. Table 7.4 shows a payoff matrix for the *janken* game under the condition that +1 is obtained for a win, –1 for a loss, and 0 for a draw. Note that this payoff matrix is generated from player I's standpoint. When player II gestures scissors, player I will receive +1, –1, and 0 when he gestures rock, paper, and scissors, respectively. From the player II's standpoint, the matrix should be totally negative. When player I receives +1, –1, and 0, player II should obtain –1, +1, and 0, respectively, that is, a zero-sum game.

Everybody knows the best strategy for this game, by learning empirically from his or her small childhood; that is, we had better use 1/3, 1/3, and 1/3 probability for rock, paper, and scissors. Verify this empirical rule using the payoff matrix shown in Table 7.4.

P7.3 PERT

MMI is interested in developing and marketing piezoelectric USM toys. The preliminary idea is a sort of levitated linear motor car. The linear motor car is coupled with a flat road map, prepared with Teflon lubricating tape. Table P7.6 is a simplified analysis of the project components.

Table P7.6 MMI's Working Schedule

		Immediate Predecessors	*Optimistic Time*	*Most Likely Time*	*Pessimistic Time*
A	Focus group of likely buyers	-	1	2	2
B	Market analysis	A	3	5	7
C	Designs developed	A	3	6	8
D	Marketing blitz	B, C	2	3	4
E	Production run	C	3	5	9
F	Analysis of buyer response	D, E	2	4	5

1. What is the expected completion time of the project?
2. Which activities have the most flexibility in scheduling?
3. The company would like to complete the entire project within 3 months (13 weeks). What is the probability it will be able to meet this target date?

References

1. Lawrence, J. A., Jr., and B. A. Pasternack. *Applied management science*, 2nd ed. New York: John Wiley & Sons, 2002.
2. Gantt, H. L. "Work, Wages and Profit." The Engineering Magazine, 1910. Republished as *Work, Wages and Profits*. Easton, Pennsylvania: Hive Publishing Company, 1974.
3. Grubbs, F. E. Attempts to Validate Certain PERT Statistics or "Picking on PERT." *Operations Research*, 1962, 10(6): 912–915.
4. Neumann, J. V. and O. Morgenstern. Theory of Games and Economic Behavior. Princeton, NJ: Princeton University Press, 2007.

Chapter 8

Marketing Strategy—Fundamentals of Marketing

The high-tech entrepreneur needs to develop at least one unique and sophisticated product as soon as the firm starts up. However, if the product is not sold as originally planned, the firm will experience its first "death valley" due to a cash flow problem, which in the worst scenario will lead to bankruptcy.

In this chapter, you will learn the essentials of marketing for high-tech entrepreneurs: (1) marketing research, including the five Ps, (2) portfolio model, and (3) the marketing mix four Ps. I am indebted to *Marketing Management*, by Peter and Donnelly [1] for some of the content in this chapter.

8.1 Marketing Research

Marketing research is the process by which information about the market environment is generated, analyzed, and interpreted for use as an aid to marketing decision making [1]. You should recognize that marketing research: (1) can include errors and misdirections depending on the treatment of the research, (2) does not forecast with certainty what will happen in the future, and (3) does not provide all of the factors for decision making or the success of a strategy.

8.1.1 The Five Ps of Marketing Research

The marketing research process includes five steps: (1) purpose of the research, (2) plan of the research, (3) performance of the research, (4) processing of research data, and (5) preparation of a research report.

8.1.1.1 Purpose of the Research

The purpose of the research is to clarify the following:

- The current situation involving the problem to be researched
- The nature of the problem
- The specific questions the research is designed to investigate

8.1.1.2 Plan for the Research

Some of the key terminologies are as follows:

- *Primary data*—data collected specifically for the research problem under investigation.
- *Secondary data*—data previously collected for other purposes, but can be used for the problem at hand.
- *Qualitative research*—typically involving face-to-face interviews with respondents (focus groups and long interviews).
- *Quantitative research*—systematic procedures designed to obtain and analyze numerical data.
  - *Observational research*—watching people
  - *Survey research*—questionnaire by mail, phone, or in person
  - *Experimental research*—manipulating one variable and examining its impact on other variables
  - *Mathematical modeling research*—typically involving secondary data, such as scanner data collected and stored in computer files from retail checkout counters

8.1.1.3 Performance of the Research

This process involves preparing for data collection as well as actually collecting data. Questionnaire preparation and actual data collection should be done carefully.

8.1.1.4 Processing of the Research Data

Qualitative research data consist of interview records that are content-analyzed for ideas or themes. Quantitative research data may be analyzed in a variety of ways depending on the objectives of the research.

8.1.1.5 Preparation of the Research Report

The limitations of the research should be carefully noted. Test market areas may not be representative of the market in general, or sample size and design may be incorrectly formulated, partially because of budget constraints.

8.1.2 Target Marketing

A company's product may not be for universal usage. For instance, the micromotor by MMI may be used for mobile phone camera modules. The final camera phone may be used by most people, but the direct customers of the motor are mobile phone module manufacturing companies (business-to-business [B2B] relationship). Let us now consider market segmentation. Widely used criteria or dimensions for market segmentation include the items in the following sections [2].

8.1.2.1 Demographic

These factors include age, gender, race, education, marital status, income, and so forth.

8.1.2.2 Psychographic

Attitude, interests, and opinions (AIO) comprise the psychographic dimensions. These may be sociocultural, religious/spiritual, philosophical, ethical/moral, political, economic, technological/scientific, and so on.

8.1.2.3 Usage Related

This category deals with how the product is actually used.

- *Quantity:* A large bottle of wine is a product for heavy drinkers, and a half-size lunch is a product for weight watchers. Quantity is an element to differentiate the marketing target.
- *Timing:* Brunch is a typical menu only on the weekend, and most clothes are seasonal products.
- *Application:* The specific purpose of usage is critical. MMI's micromotor can be utilized for mobile phone applications, for which the product life seems to be very short (a half year maximum) and the price must be incredibly low. However, the same motor can be utilized for medical catheter/endoscope applications, for which the product life seems to be rather long, and luckily medical doctors and their patients do not hesitate to purchase expensive devices. The product life cycle means the period in which a product is favorably accepted in the market. For the mobile phone, this may be only half a year, due to the customer's attitude; that is, customers try to upgrade to cellular phones with better functions. The product life cycle is different from the reliability lifetime of the device. The mobile phone may function well for more than 5 years, but this is not essential.

Most consumer products are planned, designed, and manufactured by targeting a special market (need-pull). However, a product by a high-tech entrepreneur is a bit mysterious. The product sometimes comes first, and then applications are found (seed-push). MMI originally developed the micromotor for mobile phone applications, taking into account two key factors: compact size (including its drive circuit) and low cost. However, new medical applications have been requested, which creates an unexpected marketing segment.

There are at least three market domains for high-tech entrepreneurs to seek even though they are a B2B relationship:

1. *Information technology/robotics industries*—mature industries, requiring mass production and low cost
2. *Biomedical engineering industries*—rapid growth industries requiring high quality (price is not the primary factor)
3. *Ecological/energy industries*—booming industries in the development stage, without fixing the price or needs quantity. Only preliminary market research is available.

8.1.3 Market Research Examples

The following sections have been extensively cited from my popular textbook *Micromechatronics* [3].

8.1.3.1 Secondary Data

8.1.3.1.1 Direct Citing from the Publication—Camera Module Market

Reiter's Report on cellular phone trends was accessed from their Web site (Figure 8.1) [4]. The number of mobile phones in the world will reach 1 billion by 2010. It is amazing to find that 87% of them will feature a compact camera.

8.1.3.1.2 Modified from the Publication—Patent Search

A chart depicting the breakdown of 508 patent disclosures filed in Japan from 1972 to 1984 with respect to technical content appears in Figure 8.2a. Another chart representing the breakdown for 550 patents disclosed from 1988 to 1990 is shown in Figure 8.2b. Four major categories characterize the technical content: (1) materials development, (2) device design, (3) drive/control systems, and (4) applications. Specific areas within these major categories include new compositions, fabrication processes, multilayer actuators, bimorph structures, displacement magnification mechanisms, drive methods, control methods, servo displacement transducers, and pulse drive and ultrasonic motor applications. Note that device application concepts constitute most of the patents summarized in Figure 8.2a, while only about one-quarter of the patents classified in Figure 8.2b are related to this area. This tendency is generally observed in the development of any device. Once the basic design is proposed, various prototype devices are fabricated and application patents are filed. However, as the actual commercialization is promoted, the focus shifts to the optimization of designs, cheap and efficient manufacturing processes, and new drive/control methods.

The application patents filed from 1972 to 1984 were primarily concerned with servo displacement transducers (43%) and pulse drive motors (40%), while ultrasonic motor applications (5%) were of only marginal interest at that time. However, the proportions shift quite noticeably during the 1988 to 1990 time frame, when application patents generally decreased to about 27%, but

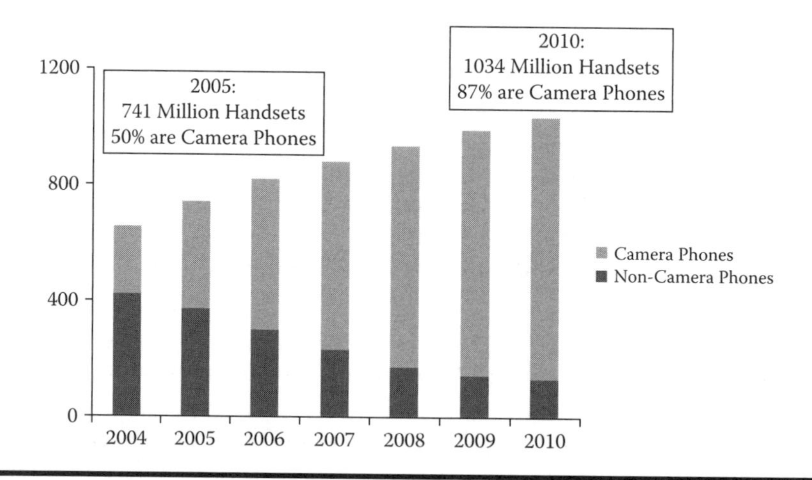

Figure 8.1 Worldwide mobile phone and camera phone shipments. (From Reiter's 2006 Camera Phone Report, http://www.cameraphonereport.com. With permission.)

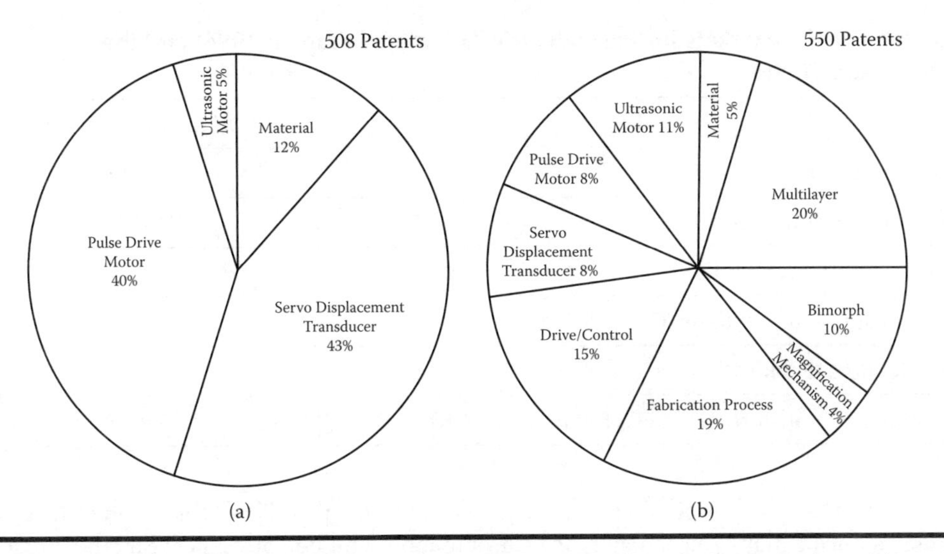

Figure 8.2 Breakdown of patent disclosures filed in Japan between 1972 and 1990 with respect to technical content: (a) 1972 to 1984 (508 patents) and (b) 1988 to 1990 (550 patents).

more than one-third of that group was dedicated to ultrasonic motor applications (11%), while the servo displacement and pulse motor applications constituted the rest in equal portions (8%).

The first stage in the development of piezoelectric actuators in the 1980s focused primarily on the inexpensive mass production of devices such as computer hardware, imaging devices, and sensors, and was pursued largely by electronic manufacturing companies. Some examples are the dot-matrix printers produced by NEC, the swing CCD image devices produced by Toshiba, the VCR tracking heads produced by Sony and Matsushita Electric, and the piezoelectric relays produced by OMRON. In the second stage of development in the 1990s, chemical companies, including organic/petrochemical industries such as TOTO Corporation, Tokin, Hitachi Metal, Murata Manufacturing Co., Ube Industry, Tosoh, NTK, Sumitomo Special Metal, Toshiba Ceramics, Taiheiyo Cement, Mitsui Chemical, and NGK, became involved. Their aim was to extend their domain by producing specialized ceramic powders in collaborative ventures with optical and mechanical industries. In these industries, where precision cutting and polishing machines are used extensively, the quality and reliability of the actuator are more important than low cost, and investment in high-quality materials is essential.

You may understand how the patent disclosure can be analyzed. Though the data created in Figure 8.2 are secondary, you can use your knowledge to create a final result equivalent to the primary data.

8.1.3.2 Primary Data

8.1.3.2.1 Private Interview—Piezoelectric Actuator Market

I met or called the corporate executives of individual companies, and collected the following sales forecast. The major markets for ceramic actuators in Japan are summarized in Table 8.1. Mr. Sekimoto, former president of NEC, made a prediction for the piezoelectric actuator market in his New Year's speech in 1987, stating that the market for the devices would reach $10 billion in the near future. An overview of recent activity in the Japanese market related to ceramic actuators is briefly presented here to provide some perspective on the growth that has occurred lately in this area.

Table 8.1 Major Markets for Piezoelectric Actuators in Japan (1994) and the United States (1996)

Market	*Production (pieces/year)*
Ink-jet printers (Epson)	5×10^5
Dot-matrix printers (NEC)	3×10^5
Camera shutters (Minolta)	1×10^5
Camera autofocus systems (Canon)	3×10^4
Parts feeders (Sanki)	1×10^4
Piezoelectric transformers (NEC, Tokin, Mitsui Chemical)	5×10^5

Regarding the production of piezo-actuator elements in Japan: NEC-Tokin and other companies are producing multilayer actuators at a rate of roughly 2 million pieces per year. They are sold at an average rate of $50 per piece, bringing the total market value to about $100 million. Predictions for the next 5 years indicate that the production rate will increase by a factor of 10 while production costs will decrease by a quarter, leading to a total market growth of up to $250 million.

If multilayer actuators are incorporated into the design of certain electronic devices, the anticipated market growth for these products could increase by a factor of 10. For example, the original dot-matrix printer produced by NEC cost on average about $3000, while the later design incorporating multilayer actuators cost only about $100. NEC produced about 100,000 dot-matrix printers in 1986, making them the leader with total sales of $300 million. Epson has started to produce piezoelectric ink-jet printers (unit price about $300) at a rate of 1 million units per year, bringing total sales for this product to about $5 billion.

To sum up the estimated annual sales from these prospective markets alone, approximately $500 million is from the sale of individual ceramic actuator elements, about $300 million is from camera-related applications, and roughly $150 million is related to the sale of ultrasonic motors; a total of nearly $1 billion market volume is anticipated. Mr. Sekimoto's prediction of ceramic actuator sales amounting to nearly $10 billion in the near future may in fact be realized! At the very least, the current trends suggest a bright financial future for the ceramic actuator industry.

8.2 Portfolio Model

Portfolio models are useful tools for corporate managers to develop effective marketing plans.

8.2.1 Portfolio Theory

Boston Consulting Group (BCG) developed the portfolio theory based on the concept of experience curves. Experience curves are similar to learning curves. The number of labor hours it takes to produce one unit of a particular product decreases in a predictable manner as the number of units produced increases. Using this experience curve concept, profit impact of marketing strategies was studied. The firm's return on investment (ROI) and cash flow are influenced by the following seven variables:

1. Competitive position
2. Industry/market environment

3. Budget allocation
4. Capital structure
5. Production processes
6. Company characteristics
7. "Change action" factors

The experience curve includes all costs associated with a product and implies that the per-unit costs of a product should fall, due to cumulative experience, as production volume increases. In a given industry, therefore, the producer with the largest volume and corresponding market share should have the lowest marginal cost. This leader in market share should be able to under-price competitors, discourage entry into the market by potential competitors, and, as a result, achieve an acceptable ROI. The linkage of experience to cost to price to market share to ROI is exhibited in Figure 8.3.

8.2.2 Boston Consulting Group Model

The Boston Consulting Group (BCG) portfolio model is based on the assumption that profitability and cash flow will be closely related to sales volume. Thus, strategic business units (SBUs) are classified according to their relative market share and the growth rate of the market that the SBU (such as each product line) is in. The BCG model is presented in Figure 8.4. Using these dimensions, products are either classified as stars, cash cows, dogs, or question marks, defined as follows:

- *Stars* are SBUs with a high share of a high-growth market. Because high-growth markets attract competition, such SBUs are usually cash users because they are growing and because the firm needs to protect their market share position.
- *Cash cows* are often market leaders, but the market they are in is not growing rapidly. Because these SBUs have a high share of a low-growth market, they are cash generators for the firm.

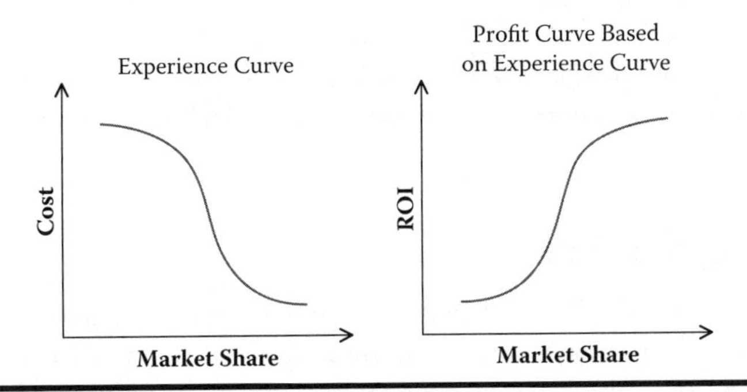

Figure 8.3 The Boston Consulting Group portfolio model: Cost and ROI as a function of market share.

Market Growth Rate	Relative Market Share: High	Relative Market Share: Low
High	Stars	Question Marks
Low	Cash Cows	Dogs

Figure 8.4 The Boston Consulting Group portfolio model: Classification of a strategic business unit.

- *Dogs* are SBUs that have a low share of a low-growth market. If the SBU has a very loyal group of customers, it may be a source of profits and cash. Usually, dogs are not large sources of cash.
- *Question marks* are SBUs with a low share of a high-growth market. They have great potential but require many resources if the firm is to successfully build market share.

A large firm with multiple SBUs will usually have a portfolio that includes some of each of the above four categories. Using this analysis, management must determine what role each SBU should assume, such as whether to expand or diminish each product line.

Regarding a start-up high-tech entrepreneur, we usually start from only one SBU (one product, such as the piezo-motor in the MMI example). Though the market growth rate is most likely very high, we will just start production; that is, the relative market share is almost zero. Most high-tech ventures start from a question mark. A typical product life cycle follows the path below:

1. *Build share.* The first objective should be to build market share, forfeiting immediate earnings. This is appropriate for promising question marks whose share has to grow if they are ever to become a star.
2. *Hold share.* Once your firm creates a star, seek to preserve the SBU's market share. With time the market growth rate saturates and the manufacturing cost decreases. It is expected that the star will transform smoothly into cash cow.
3. *Harvest.* This is expected for the strong cash cow to ensure that it can continue to yield a large cash flow. Harvest may allow market share to decline in order to maximize earnings and cash flow, after spending some time in the product life cycle. Now it is time to create a new question mark in your firm.
4. *Divest.* Sooner or later, the cash cow transforms into a dog with a diminishing market, or with reducing cash generation. It is time to sell or divest this SBU because better investment opportunities exist elsewhere.

You have now learned the product life cycle, which proceeds in a counter-clockwise direction as shown in Figure 8.4, starting from the question mark, turning into a star, cash cow, and finally a dog.

8.3 Marketing Mix Four Ps

The marketing mix is the set of controllable variables that must be managed to satisfy the target market and achieve organizational objectives. These controllable variables are usually classified according to four major decision areas—four Ps: product, price, place (or channels of distribution), and promotion [2].

8.3.1 Product

8.3.1.1 Product Differentiation

Product differentiation is the most essential factor to marketing promoters. You must differentiate your product from your competitors' products in the following ways:

- *Quality:* The functionality of a product from a high-tech entrepreneur is usually higher than the existing products. However, the product also requires reliability and lifetime.

- *Quantity:* Is your start-up firm capable of mass production? If you have a mass-production facility, excellent! If not, do you have a mass-production partner (for example, Cheng Kung Corporation for MMI)? This capability is sometimes required to enter certain industries, such as mobile phone cameras.
- *Intellectual property protection:* Is your product protected through patents (or other intellectual property laws) from imitation manufacturing by other companies?

8.3.1.2 Product Life Cycle

Every product has a life cycle. As shown in Figure 8.5, the product life cycle is segmented into introduction, growth, maturity, saturation, decline, and finally abandonment.

1. *Introduction.* This stage is characterized by research and development (R&D). Sales and profit are usually very low, although costs may be substantial. This stage also corresponds to question marks in BCG's portfolio model.
2. *Growth.* This stage is characterized by increased sales and initial profits, and corresponds to stars. Heavy promotional costs are often incurred, which hinder gross profit.
3. *Maturity.* This stage is characterized by the peak and attempted maintenance of sales levels. It is possible to increase sales, but the marginal cost for unit sales increase is quite substantial.
4. *Saturation.* The maximum profit is usually obtained sometime after the maximum production quantity (saturation) occurs. Notice the cash cow period in Figure 8.5.
5. *Decline.* This stage is characterized by perceived futility in an attempt to maintain market share. Typically, this is accompanied by cost cutting. This stage corresponds to dogs.
6. *Abandonment.* At this stage the product's performance no longer merits inclusion in the firm's product line.

8.3.2 Price

Price is what the customer pays for the product. Four pricing schemes are introduced here: cost-plus, fair/parity, skimming, and penetration pricing.

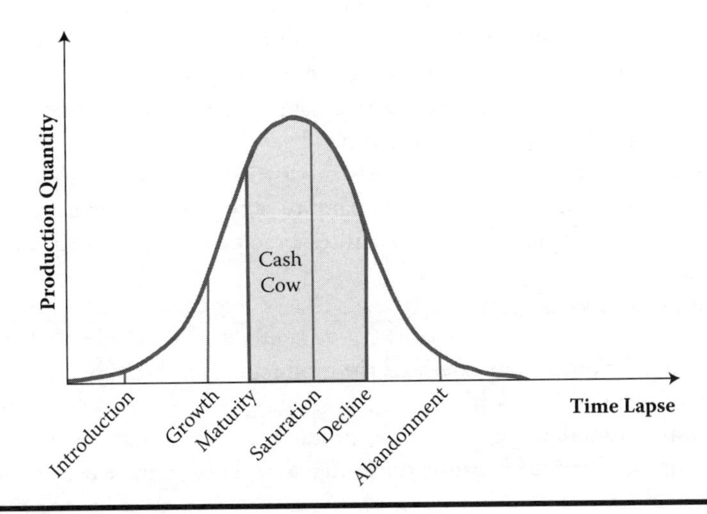

Figure 8.5 Product life cycle.

8.3.2.1 Cost-Plus Pricing

The product should not be sold for less than the manufacturing cost. The concept of cost-plus pricing involves setting a price that factors into a given profit margin (e.g., cost plus 25% profit). This approach may indicate a preoccupation with investment return, and can be particularly unfortunate in the absence of a proper customer-focused orientation. Chapter 6, Section 6.1 introduced the detailed process of how to generate this cost-plus price.

8.3.2.2 Fair/Parity Pricing

Fair pricing is set based on customer-oriented market research. On the other hand, parity pricing involves setting a price that roughly matches that of competing brands within the product class. This approach may suggest that the marketer is not attuned to the importance of differentiation.

8.3.2.3 Skimming Price

This option involves charging a high price relative to other brands within the product class. This can be an effective approach to collect back the initial development investment, if it succeeds. The success depends on the high product quality and differentiated performance. High-tech firms, such as Sony and IBM, usually use this scheme. Because of a strong brand image (e.g., extremely high quality), Sony's products sell well even when the price is 20% higher than other brand products.

8.3.2.4 Penetration Price

This scheme involves charging a low price on the assumption of selling the brand in enormous quantities. If the firm has confidence in immediate share expansion of market, the volume creates profit to the firm.

Most high-tech entrepreneurs will accept a compromise between the skimming price and cost-plus price.

8.3.3 Place

Place, in this context, means the product's channels of distribution or how it is conveyed from the producer to the end user. Its functions include manufacturing, transportation, warehousing, wholesaling, and retailing. The more the intermediate functions or channels are involved, the higher the percentage of selling price it can command.

If an organization controls all the channels of distribution for its product, it is *vertically integrated.* If the manufacturer acquires a company to access the raw materials, it is *backward integrated.* To the contrary, if the wholesaler acquires a retailer to expand distribution, it is called *forward integrated.*

There are three distribution options:

1. *Intensive*—aims for maximum exposure; the product is sold through any responsible wholesaler or retailer who will stock it.
2. *Elective*—aims for moderate exposure; the product is sold through "better" retailers.
3. *Selective*—aims for limited exposure; the product is sold by a single dealer within each trading region.

MMI Sales Division (Retailer) Example

- *FEM computer software.* MMI is an exclusive worldwide distributor of a European provider. The exclusivity is good for MMI, since it forbids any other competitors from selling this product. However, usually this condition is accompanied by a certain norm for the selling quantity. MMI has three subdistributors to cover Japan, Korea, and Taiwan.
- *Multilayer piezo-actuator.* MMI is a nonexclusive distributor of a Japanese manufacturer in the North American territory. There is no selling quantity norm, but MMI has a competitive distributor in the West Coast area.

8.3.4 Promotion

Promotion involves communication of the product attributes and the corporate image in the most favorable light possible to intermediary sellers (i.e., trade advertising and trade promotion) and to end users (i.e., consumer advertising and consumer promotion).

8.3.4.1 Advertising Effectiveness

Advertising media selection is no easy task. Major advertising media include the following:

- Newspaper
- Radio
- Outdoor (billboards)
- Television
- Magazines
- Direct mail
- Internet

The effectiveness of advertising media is measured by cost-per-thousand, or CPM. How much does it cost to reach 1000 people or households with a single exposure? The formula is as follows:

$$\text{CPM} = \frac{\text{Media cost}}{\text{Audience measured in thousands}} \tag{8.1}$$

Refer to Example Problem 8.1 to learn how to calculate CPM.

8.3.4.2 Promotion Mix

Promotion mix relevant to a start-up high-tech firm includes collateral material, public relations, direct mail, trade shows, trade journal advertising, and personal selling.

1. Collateral material (essential)
   a. Tools—Sales literature, brochures, data sheets of the company and products
   b. Distribution—direct mail, trade shows, sales presentation
   c. Cost estimation—$27,500 for 5000 pieces for MMI

2. Public relations (highly recommended)
   a. News release of new products in newspapers or academic journals
   b. Uncertainty of the publication (editor's choice)
   c. Cost: $500 (a basically free publication!)
3. Direct mail (recommended)
   a. Purchasing a mailing list ($5000)
   b. Mailing fee estimation: 1500 at $5 ($7500)
4. Trade show (upon budget allowance)
   a. Booth, travel, and incidental costs
   b. Cost: $100,000 per event
   c. Budget analysis suggests a small event ($25,000 per event)
5. Journal advertising (upon budget allowance)
   a. A series of ads
   b. Cost differs depending on the circulation (refer to Example Problem 8.1)
6. Personal selling (essential, but gradual)
   a. Telemarketing—inbound/inside sales (calls that potential customers make)
      i. Hiring a technical salesperson is required
      ii. Salary and benefits: $50,000 in College Park for MMI
   b. Field sales—outside sales (selling at the customer's site)
      i. Hiring outside salespersons is required (multiple geographically)
      ii. Salary and benefits: $80,000 per person

A Web site is among the collateral materials, but this promotion is much more effective, including functions of public relations, direct mail, trade show, and so forth. Prepare your company's Web site with the highest priority!

Example Problem 8.1

Referring to the data in Table 8.2, calculate the CPM for each trade publication. Then decide which journal is most effective from the CPM's viewpoint for MMI's advertisement.

Solution

CPMs (cost per 1000 circulation) are calculated as $116, $56, $203, and $48 for *Electrical Manufacturing, Electronic Component News, Electronic Manufacturing News,* and *Design News,* respectively. Cheaper two trade publications, *Electronic Component News* and *Design News,* are recommended for advertisement purposes. One-time advertisement costs roughly $7000.

Table 8.2 Trade Publications

Trade Publication	*Editorial*	*Cost per Color Insertion (1 page)*	*Circulation (volumes)*
Electrical Manufacturing	For purchasers and users of power supplies, transformers, and other electrical products	$4100	35,000
Electronic Component News	For electronics original equipment manufacturers (OEMs); products addressed include workstations, power sources, chips, etc.	$6400	110,000

(Continued)

Table 8.2 Trade Publications *(Continued)*

Trade Publication	*Editorial*	*Cost per Color Insertion (1 page)*	*Circulation (volumes)*
Electronic Manufacturing News	For OEMs in the industry of providing manufacturing and contracting for components, circuits, and systems	$5100	25,000
Design News	For design OEMs covering components, systems, and materials	$8100	170,000

Chapter Summary

8.1 Marketing research is the process by which information about the environment is generated, analyzed, and interpreted for use as an aid to marketing decision making.

8.2 Five Ps of marketing research: purpose of the research, plan of the research, performance of the research, processing of research data, and preparation of the research report.

8.3 Boston Consulting Group portfolio model:

Market Growth Rate	**Relative Market Share** High	**Relative Market Share** Low
High	Stars	Question Marks
Low	Cash Cows	Dogs

8.4 Marketing mix four Ps: product, price, place (or channels of distribution), and promotion.

8.5 Product life cycle: introduction, growth, maturity, saturation, decline, and finally abandonment, which correspond to the shifting trend from question marks, to stars, then cash cows, and finally dogs in the portfolio model.

8.6 Four pricing schemes: cost-plus, fair/parity, skimming, and penetration pricing.

8.7 Promotion mix relevant to a start-up high-tech firm: collateral material, public relations, direct mail, trade shows, trade journal advertising, and personal selling. A Web site is one of the most efficient promotion techniques.

8.8 The effectiveness of advertisement media is measured by cost-per-thousand (CPM):

$$\text{CPM} = \frac{\text{Media cost}}{\text{Audience measured in thousands}}$$

Practical Exercise Problems

P8.1 Market Research

What application do you have in mind for your invented device? Research this application market growth from publications such as journal articles, newspapers, websites, and company annual reports, including practical revenue amount (secondary data collection). Estimate the market size for the future 3 years from the previous 3 years' data. Assuming that your start-up firm will take 3% of the total market, how much revenue will your firm receive in the next 3 years?

P8.2 Cost per Thousand

What journals or trade publications do you usually read, and what trade publications do you think are most relevant to your product advertisement? List a minimum of five journals or trade publications, and research the cost for a half-page advertisement and the circulation quantity. Then, calculate the CPMs for these five publications, and choose the two most suitable journals for your product promotion.

References

1. Paul Peter, J., and J. H. Donnelly Jr. *Marketing management.* New York: McGraw-Hill Irwin, 2007.
2. Sobel, M. *The 12-hour MBA program.* Paramus, NJ: Prentice Hall, 1994.
3. Uchino, K., and Jayne R. Giniewicz, *Micromechatronics.* New York: CRC/Dekker, 2003.
4. Reiter's 2006 Camera Phone Report, http://www.cameraphonereport.com/

Chapter 9

Intellectual Properties—How to Protect the Company's Technology

As a high-tech entrepreneur, you may already have a patent or two, or at least may know roughly what a patent is. You will learn more information on intellectual property (IP) in this chapter. IP is essential to a start-up company in order to protect their technological privileges, even against big competitive firms. For what can we apply for a patent? What if we face a patent infringement lawsuit?

9.1 Intellectual Properties

What is IP? Wikipedia describes it as follows [1]:

IP is a legal field that refers to creations of the mind such as musical, literary, and artistic works; inventions; and symbols, names, images, and designs used in commerce, including copyrights, trademarks, patents, and related rights. Under intellectual property law, the holder of one of these abstract "properties" has certain exclusive rights to the creative work, commercial symbol, or invention which it covers.

The related rights include *utility models, industrial design rights,* and *trade secrets.* The utility model is a lower-level right than the patent, applied to a simpler invention. This right is popular in Japan, but not in the United States. The industrial design right is related to a product design. The trademark logo of Micro Motor Inc. (MMI), designed by Barb Shay (Figure 9.1, left side), started as a trademark (™, center) after filing it. After being registered completely, it was changed to a registered trademark (®, Figure 9.1, right side). During the trademark registration process, MMI received notice that the original logo (on the left) may bear similarity to another company's logo (Michigan Molecular Institute, or Mid-Manhattan Institute). Thus, Barb put the full company name in small letters below the original logo.

Figure 9.1 The original logo design (left), trademark (™), and registered trademark (®) of Micro Motor Inc.

Table 9.1 Comparison among Various Intellectual Properties

	Patent	*Utility Model*	*Industrial Design Right*	*Trademark*	*Copyright*	*Trade Secret*
Objective	Invention	Low-level invention	Product design	Product trademark	Production (software, book, etc.)	Know-how, etc.
Protection items	Novelty Register	Novelty Register	Novelty Creativity Register	No confusion Register	Creation	Confidentiality
Protection period	20 years	6 years	15 years	10 years+	50 years after death	No limit
Disclosure	O	O	O after 3 years	Publication	Publication	X
Protecting content	Exclusive operation	Exclusive operation	Exclusive operation	Exclusive usage	Exclusive copyright	Confidentiality

Note that computer algorithms are patents, but most products of computer software are copyrights.

A trade secret is a formula, practice, process, design, instrument, pattern, or compilation of information that is not generally known or reasonably ascertainable, by which a business can obtain an economic advantage over competitors or customers. In some jurisdictions, such secrets are referred to as *confidential information* or *classified information*. Trade secret protection can extend indefinitely and as such may offer an advantage over patent protection, which lasts only for a specifically limited period of time, for example, 20 years in the United States. Coca Cola, the most famous trade secret example, has no patent for its formula and has been very effective in protecting it for many more years than a patent would have. In fact, Coca Cola refused to reveal its trade secret under at least two judges' orders [2].

Table 9.1 summarizes comparisons among various IP rights. In particular, four IP rights—patent, utility model, industrial design right, and trademark—are called *industrial properties*. These IP rights have termination periods of 20, 6, 15, and 10 years, respectively.

9.2 Why Is Intellectual Property Important?

9.2.1 When Your Company Manufactures the Patented Product

The development of a new product generally takes a long time and is expensive. On the other hand, chasing and imitating a product is rather easy. The patent is a kind of protection from other companies imitating or stealing the technology developed by your firm's invested money and manpower. Your firm can initiate a lawsuit to prevent the production of an imitation product by another company.

9.2.2 When Your Company Does Not Manufacture the Patented Product

Your firm can obtain profit or at least expended development cost by transferring or licensing the IP to other companies via *patent royalty*. There are two types of patent royalty in licensing: exclusive and nonexclusive licenses.

9.2.3 Trade Secrets

One of the most important things for the high-tech entrepreneur to keep in mind is trade secret maintenance and protection, which is related to the termination of employees and other job changes. Unlike Japanese employees, American engineers tend to change jobs every several years, resulting in the inevitable transfer of trade secrets, even among competitive firms. Accordingly, we sometimes face a serious conflict with the company at which our former employee found a new position. Typical general conflicts include the following:

- Market research data, research and development (R&D), and marketing strategies
- Similar product lines in the new company
- Know-how in the product-manufacturing processes
- Research proposal ideas
- Customer list

In order to prevent these sorts of problems, the firm needs to legally regulate disloyal behavior through the use of an Employment Agreement. The following is an example of an Employment Agreement for employee researchers, for your reference.

MMI Example

Part of the Employment Agreement:

Confidential Information. Employee recognizes that Employer's business and continued success depend upon the use and protection of confidential and proprietary business information to which Employee has access (the "Confidential Information").

Confidential Information includes Employer and its current or future subsidiaries and affiliates, without limitation, and whether or *not specifically designated as confidential or proprietary:* inventions; all business plans and marketing strategies; information concerning existing and prospective markets and customers; financial information; information concerning the development of new products and services; and technical and nontechnical data related to designs, specifications, compilations, inventions, improvements, methods, processes, materials,

> procedures and techniques; provided, however, that the phrase does not include information that (a) was lawfully in Employee's possession prior to disclosure of such information by Employer; (b) was, or at any time becomes, available in the public domain other than through a violation of this Agreement; (c) is documented by Employee as having been developed by Employee outside the scope of Employee's employment and independently; or (d) is furnished to Employee by a third party not under an obligation of confidentiality to Employer. Confidential Information includes trade secret information. Employee agrees that *during Employee's employment and after termination of employment irrespective of cause,* Employee will use Confidential Information only for the benefit of Employer and *will not directly or indirectly use or divulge, or permit others to use or divulge, any Confidential Information for any reason,* except as authorized by Employer. Employee's obligation under this Agreement is in addition to any obligations Employee has under state or federal law. Employee agrees to deliver to Employer immediately upon termination of Employee's employment, or at any time Employer so requests, all tangible items containing any Confidential Information (including, without limitation, all memoranda, photographs, records, reports, manuals, drawings, blueprints, prototypes, notes taken by or provided to Employee, and any other documents or items of a confidential nature belonging to Employer), together with all copies of such material in Employee's possession or control. Employee agrees that in the course of Employee's employment with Employer, Employee will not violate in any way the rights that any entity has with regard to trade secrets or proprietary or confidential information. Employee's obligations under this Section are indefinite in term and shall survive the termination of this Agreement.

The wording "after termination of employment," in addition to "during employment," is essential to trade secret protection.

9.3 Patent Preparation

You may have a patent or two already, or will have one soon, typically obtained during your MS or PhD thesis period. How did you come up with this invention? Accidentally, during an experiment, or expectedly, associated with your detailed plan?

9.3.1 Patent Idea Search

9.3.1.1 Serendipity

The development of Ivory soap by Procter & Gamble is a good example of serendipity, as already introduced in Chapter 2 [3].

William Procter and James Gamble started their candle manufacturing company in the middle of the nineteenth century (see Figure 9.2). However, after the invention of the light bulb by Thomas Edison, candle sales decreased dramatically, and they needed to change their business strategy. An employee in the Cincinnati factory forgot to turn off his machine when he went to lunch. When he returned, he found a frothing mass of lather filled with air bubbles. He almost threw the stuff away, but instead decided to make it into floating soap, with lots of bubbles. Why was the floating soap such a hot item at that time? Because workers used soap in rivers or ponds during that period, and they occasionally lost soap that did not float.

This is an example of patent creation: a novel idea with a new function or principle and/or a new structure or configuration, completely different from the existing products. However, it is also a mistake that actually worked. I do not recommend that you rely on this sort of lucky mistake too often. We will consider more reliable R&D approaches.

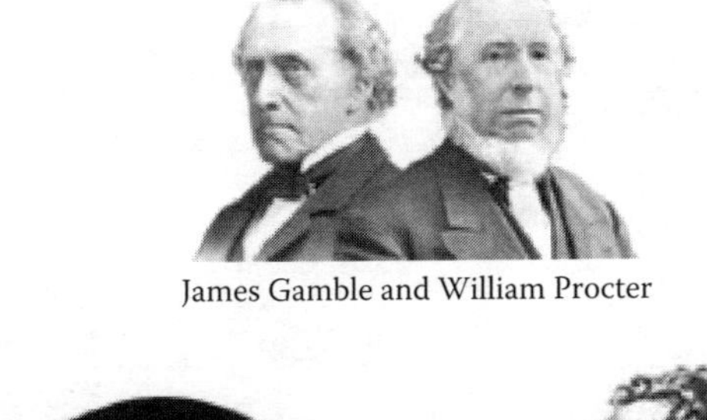

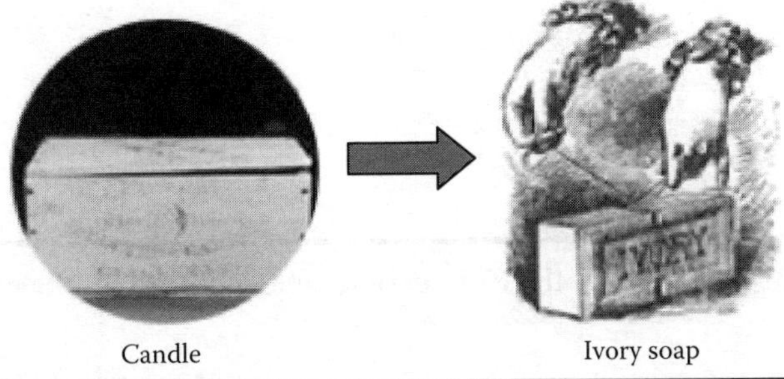

Figure 9.2 Development of Ivory soap by Proctor & Gamble, 1879. (From http://www.ideafinder.com/history/inventions/ivory.htm. With permission.)

9.3.1.2 Systematic Approach

Another approach is to modify the structure or configuration to escape from the existing patent, in order to change the function or performance. Note that the patent content may not need to include the performance improvement or manufacturing cost reduction. The key is to escape from the existing patents.

The lead zirconate titanate (PZT) ceramic tube motor was introduced in the late 1990s [4]. Its structure is illustrated in Figure 9.3a. However, there are two serious problems in this design for actual commercialization: the PZT tube is expensive (more than 40 cents per piece, which is already beyond the targeted motor cost for cellular phone applications), and it is fragile under high compressive stress application for high torque or under a mechanical shock (drop test). Thus, the manufacturer decided to couple simple PZT rectangular plates with an elastically tough metal tube, as shown in Figure 9.3b [5]. Because the PZT plate costs only 2 cents per piece, and the metal tube is less expensive, the total raw material cost was decreased to only 6 cents. Because of this new structure, they could file a patent, though the motor performance (torque and power) was slightly degraded in comparison with the characteristics of the original PZT tube motor.

Now, three different patents are introduced, which are combinations of the metal tube motors with sliders and/or screw mechanisms. Figure 9.4 shows three different designs of zoom/focus mechanisms for cellular phone camera applications, using our metal tube micromotors as key actuators, developed by (a) DEF, Korea, (b) GHJ, United States, and (c) KLM, Taiwan. DEF uses the center of the metal tube coupled with an external screw, while GHJ uses the two end parts of the metal tube coupled with an internal screw. KLM uses the metal tube in a horizontal way coupled with a linear slider, so that its low profile is a significant benefit. These three designs are *extended patents*, which require that a royalty be paid to the basic patent holder once production begins. Another tactic is setting a cross-license with the original patent holder (MMI); this circumvents the royalty, but releases manufacturing rights for the zoom mechanism to MMI.

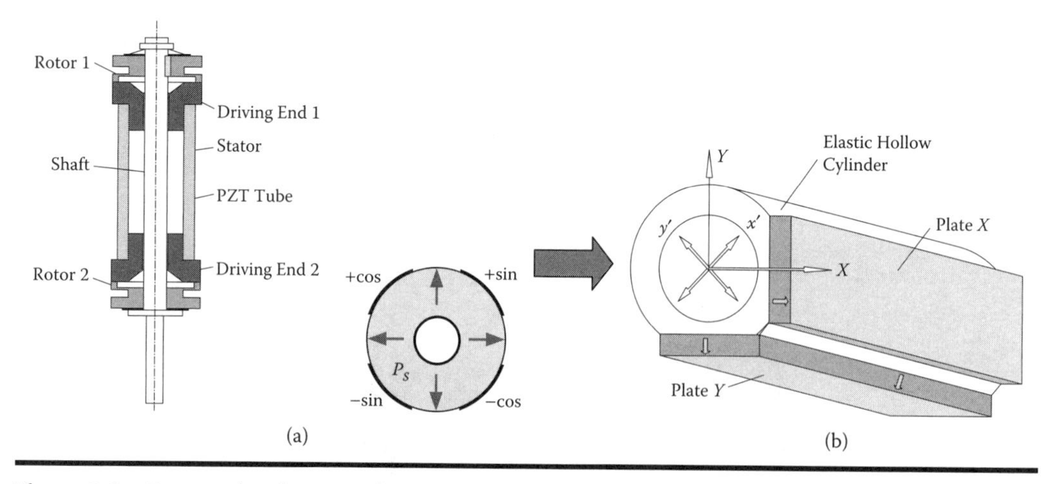

Figure 9.3 Progression from (a) the PZT ceramic tube motor to (b) the metal tube motor [4,5].

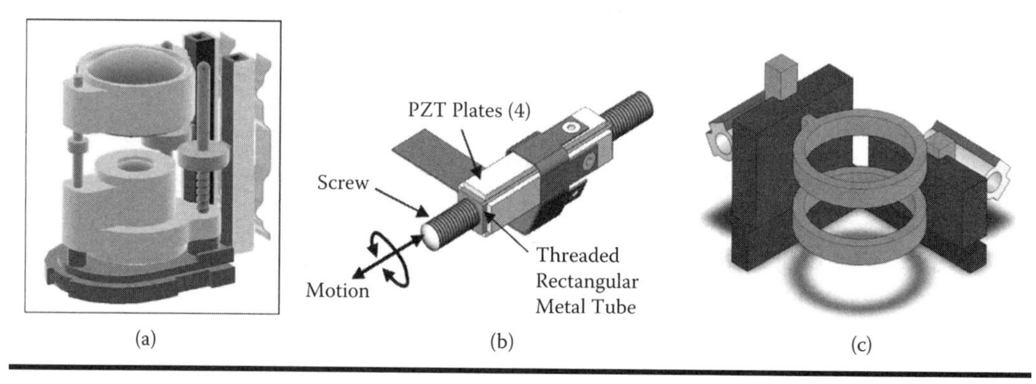

Figure 9.4 Zoom/focus mechanisms for cellular phone cameras using metal tube micromotors: (a) DEF, Korea, (b) GHJ, United States, and (c) KLM, Taiwan.

9.3.1.3 Patentability

Not all patent applications will be accepted. I will introduce an example of possible patent denial here. A blood clot remover that uses the smallest metal tube motor seems to be a good application patent, as schematically illustrated in Figure 9.5. The major problems with this filing are summarized in the next two paragraphs.

Application possibilities were addressed in a previous published paper. That paper has been published on the PZT tube motor, in which a possible intravascular application with the PZT tube motor was reported. Figure 9.6 shows evidence of this, which was exhibited by the patent evaluator. Though a metal tube motor is used in practice in a proposed patent, the intravascular application seems to be obvious, according to the patent evaluator.

A mere guess should not be included in a publication. We find occasionally in journal papers authored by graduate students a sort of "future" application, i.e., "We plan to use this for such-and-such applications." This is a dangerous situation from the patent application viewpoint. After disclosing this interesting idea, if you do not file the application patent with practical designs within a certain time allowance, you may lose the application patent.

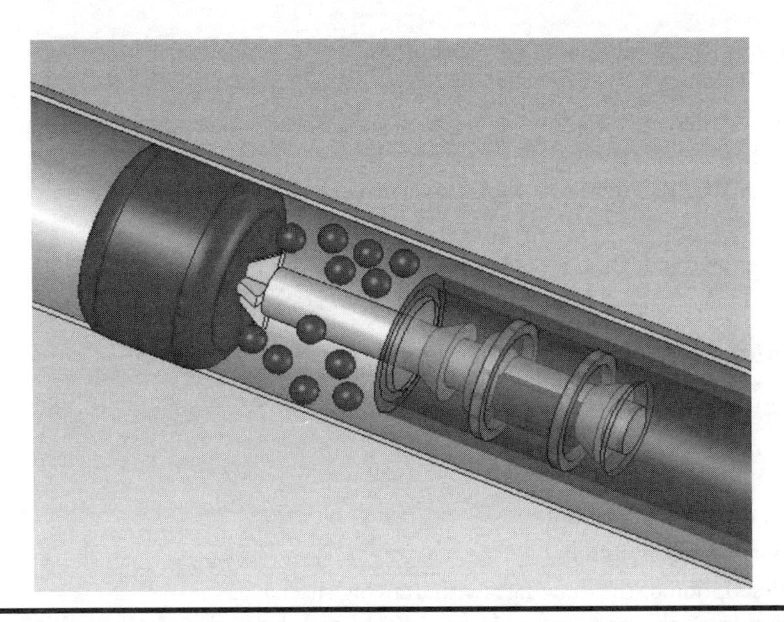

Figure 9.5 Basic design of the blood clot remover.

Piezoelectric Ultrasonic Micromotor with 1.5 mm Diameter

Shuxiang Dong, Siak P. Lim, Kwork H. Lee, Jingdong Zhang, Leong C. Lim, and Kenji Uchino, *Member, IEEE*

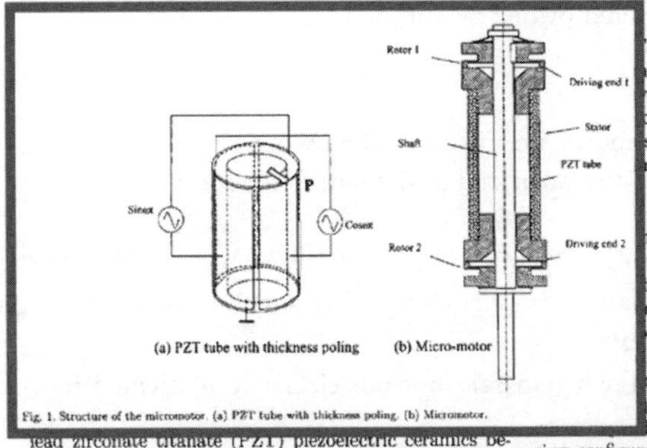

Fig. 1. Structure of the micromotor. (a) PZT tube with thickness poling. (b) Micromotor.

lead zirconate titanate (PZT) piezoelectric ceramics be-

Muralt *et al.* [2] developed hybrid ultrasonic micromotors with a diameter of 4 mm. A standing flexural wave was excited by a PZT thin-film vibrator and converted to an

he stator was a titanium tube coated with a deposited by a hydrothermal technique. The olution speed was 880 rpm at 15 $V_{p\text{-}p}$ diving output torque was calculated to be 7.0 μN-m sient response of the rotor (but was not mea- entally). An improved version of this type later reported by Morita [6], which utilized its stator. Although the torque was greatly pared to the above-mentioned micromotors, was complex and its overall size was much s stator. Thus, it is not suitable for micro- ical system (MEMS) devices. A "Windmill" reviously was developed with a diameter of [7]. A metal ring with inward-pointing fin- dwiched between two uniformly poled PZT ue of 10 μN-m to 10 mN-m was reported, ed upon the stator diameter. Because of the ring configuration, this miniature-type motor is suitable

in order to obtain optimum driving characteristics. This type of stator with a thin diameter and long configuration is well-suited for intravascular applications. The resonant

with a diameter of 1.5 mm. The term micromotor is intentionally used to designate the opening of new and smaller length scales to high-power ultrasonic drives. A PZT ce-

Figure 9.6 A paper published on the PZT tube motor, in which they reported a possible intravascular application with the PZT tube motor. (From Dong, S., S. P. Lim, K. H. Lee, J. Zhang, L. C. Lim, and K. Uchino. *IEEE Transactions on Ultrasonics, Ferroelectrics, and Frequency Control* 50(4):361–367. With permission.)

If we defend this patent, there are two possibilities:

1. Use of a different vibration mode, such as a shear mode rather than the same bending mode
2. Claims on a drill design, etc., excluding a general intravascular application

We may apply for an industrial model rather than a patent in Japan.

9.3.2 Patent Format

A patent consists of the following:

- Abstract
- Background of the invention
- Summary of the invention
- Brief description of the drawings (with drawing sheets)
- Description of the *preferred embodiment(s)* (detailed description, including the principle, experimental results, and the claims)

Figure 9.7 shows an example of a patent for a piezoelectric motor. The cover page with an abstract is shown in Figure 9.7a, and the final page with claims is shown in Figure 9.7b. Claim 1 expresses the widest coverage of the motor design contents, and with increasing claim number the coverage becomes more specific. After claim 11, the drive method is covered. Multiple ideas, such as motor design and its drive method, can be covered by one patent.

The claims are as follows:

1. A rotary ultrasonic piezoelectric motor comprising the following:
    a. Stator having a piezo-ceramic disc polarized in the radial direction
    b. Power source
    c. Rotor operatively connected to said stator
2. The rotary motor as set forth in claim 1 further comprising a metal ring having a plurality of teeth disposed on said ceramic disc
3. The motor as set forth in claim 1 wherein said bottom electrode is divided into an even number of segments
4. The rotary motor as set forth in claim 1 wherein said stator and said rotor are each supported on a shaft extending through a centrally located aperture in said rotor and a centrally located aperture in said stator
5. Other portions have been omitted.

This patent can cover three motor designs: Claims 1 and 2, 1 and 3, and 2 and 3. If another person files a patent with a combination 1 and "said bottom electrode is divided into an odd number of segments," it is out of this patent coverage. This patent cannot cover all rotary motors with a piezo-ceramic disc polarized in the radial direction.

US007095160B2

(12) **United States Patent**
Uchino et al.

(10) **Patent No.: US 7,095,160 B2**
(45) **Date of Patent: Aug. 22, 2006**

(54) **PIEZOELECTRIC MOTOR AND METHOD OF EXCITING AN ULTRASONIC TRAVELING WAVE TO DRIVE THE MOTOR**

(75) Inventors: **Kenji Uchino**, State College, PA (US); **Shuxiang Dong**, Blacksburg, VA (US); **Michael T. Strauss**, Newburyport, MA (US)

(73) Assignee: **The Penn State Research Foundation**, University Park, PA (US)

(*) Notice: Subject to any disclaimer, the term of this patent is extended or adjusted under 35 U.S.C. 154(b) by 0 days.

(21) Appl. No.: **10/855,260**

(22) Filed: **May 26, 2004**

(65) **Prior Publication Data**

US 2005/0001516 A1 Jan. 6, 2005

Related U.S. Application Data

(60) Provisional application No. 60/473,430, filed on May 27, 2003.

(51) **Int. Cl.**
H01L 41/04 (2006.01)
H02N 2/10 (2006.01)
H02N 2/12 (2006.01)

(52) **U.S. Cl.** **310/333**; 310/323.03; 310/323.04; 310/364; 310/365

(58) **Field of Classification Search** ....
310/323.03–323.04, 333, 365, 323.02
See application file for complete search history.

(56) **References Cited**

U.S. PATENT DOCUMENTS

4,622,483 A 11/1986 Staufenberg, Jr. et al. .. 310/328
4,912,351 A * 3/1990 Takata et al. 310/323.16
4,945,275 A * 7/1990 Honda 310/323.02
6,242,849 B1 * 6/2001 Burov et al. 310/328
6,288,475 B1 * 9/2001 Ito et al. 310/323.01
6,518,689 B1 * 2/2003 Yerganian 310/323.06

OTHER PUBLICATIONS

Shuxiang*, H. W. Kim, M. T. Strauss #, K. Uchino and D. Viehland, "A Piezoelectric Shear-Shear Mode Ultrasonic Motor", ICAT/Materials Research Institute, The Pennsylvania State University, University Park, PA 16801 USA, Apr. 2004; *Materials Science & Engineering, Virginia Tech, Blacksburg, VA 24061, USA; #HME, 56 Federal Street, Newbury port, MA 01950, USA.

* cited by examiner

Primary Examiner—Darren Schuberg
Assistant Examiner—J. Aguirrechea
(74) *Attorney, Agent, or Firm*—Gifford, Krass, Groh, Sprinkle, Anderson & Citkowski, PC

(57) **ABSTRACT**

A rotary ultrasonic piezoelectric motor is provided and a method of exciting a flexure traveling wave to drive the motor. The motor includes a stator having a piezoelectric ceramic disc polarized in the radial direction and bounded by a top electrode and a segmented bottom electrode. The motor also includes a power source for applying two pairs of alternating voltages to the bottom electrode segments to excite a shear-shear mode vibration in the stator, resulting in a shear-shear mode flexure traveling wave in the stator. The motor further includes a rotor operatively connected to the stator, and the stator is driven to rotate through a frictional force between the rotor and the stator due to the traveling wave deformation of the stator. A linear ultrasonic piezoelectric motor and method of exciting a flexure traveling wave to linearly drive the motor is provided. The motor includes a stator having a rectangular piezoelectric ceramic plate that is polarized in the longitudinal direction. The motor also includes a power source for applying two pairs of alternating voltages to the bottom electrode segments to excite a shear-shear mode vibration in the stator, resulting in a shear-shear mode flexure traveling wave in the stator. The motor further includes a slider operatively connected to the stator, and the stator is driven to move linearly through a frictional force between the slider and the stator due to the traveling wave deformation of the stator.

18 Claims, 8 Drawing Sheets

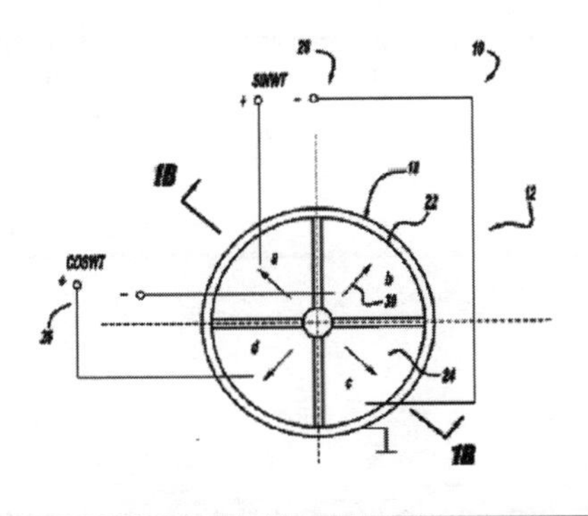

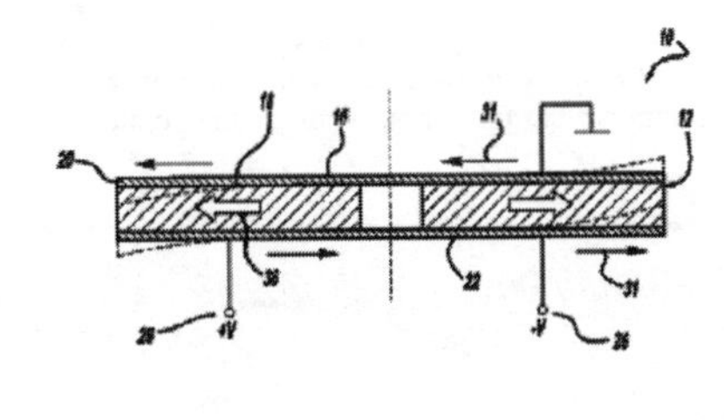

(a)

Figure 9.7 A patent example on a piezoelectric motor. (a) Cover page; and (b) last page (includes the patent claims).

US 7,095,160 B2

9

What is claimed is:

1. A rotary ultrasonic piezoelectric motor comprising:

a stator having a piezoelectric ceramic disc polarized in the radial direction, and bounded by a top electrode positioned on an upper side of said ceramic disc and a bottom electrode positioned on a lower side of said ceramic disc, and the bottom electrode is divided into segments;

a power source, wherein said power source applies two pairs of alternating voltages to said bottom electrode segments to excite a shear mode vibration in said stator, resulting in a shear-shear mode flexure traveling wave in said stator; and

a rotor operatively connected to said stator, wherein a portion of said rotor in contact with said stator is driven to rotate through a frictional force between said rotor and said stator due to said traveling wave deformation of said stator.

2. The rotary motor as set forth in claim **1** further comprising a metal ring having a plurality of teeth disposed on said ceramic disc.

3. The motor as set forth in claim **1** wherein said bottom electrode is divided into an even number of segments.

4. The rotary motor as set forth in claim **1** wherein said stator and said rotor are each supported on a shaft extending through a centrally located aperture in said rotor and a centrally located aperture in said stator.

5. The rotary motor as set forth in claim **4** further comprising a holding means for supporting said stator.

6. The rotary motor as set forth in claim **5** further comprising a pressing means for preloading said stator against said rotor.

7. The rotary motor as set forth in claim **6** wherein said pressing means is a spring.

8. The rotary motor as set forth in claim **1** wherein said piezoelectric disc is operated at a flexure resonance mode of B_{02}.

9. The rotary motor as set forth in claim **1** wherein said piezoelectric disc is operated at a flexure resonance mode of B_{12}.

10. The rotary motor as set forth in claim **1** wherein a phase difference of said applied voltage is changed from 90 degrees to −90 degrees to change the rotational direction of said rotor.

11. A method of exciting a shear-shear mode vibration in a piezoelectric ceramic disc for a rotary ultrasonic piezoelectric motor having a rotor and a stator, said method comprising the steps of:

polarizing the piezoelectric disc for the stator in a radial direction, wherein the piezoelectric ceramic disc is bounded by a top electrode positioned on an upper side of the ceramic disc, and a bottom electrode positioned on a lower side of the ceramic disc, and the bottom electrode is divided into segments;

applying a pair of alternating voltages with a phase shift of 90 degrees to the bottom electrode segments from a power source;

exciting a shear-shear mode vibration in the stator; and

producing a shear-shear mode flexure traveling wave in the stator causing a portion of the rotor in contact with the stator to rotate through a frictional force between the rotor and the stator due to the traveling wave deformation of the stator.

12. A method as set forth in claim **11** further comprising the step of changing a phase difference of the applied voltage from 90 degrees to −90 degrees to change the rotary motion of the rotor.

13. A rotary ultrasonic piezoelectric motor comprising:

a stator having a piezoelectric ceramic disc polarized in the radial direction, and bounded by a top electrode positioned on an upper side of said ceramic disc and a bottom electrode positioned on a lower side of said ceramic disc, and the bottom electrode is divided into segments;

a metal ring having a plurality of teeth disposed on said ceramic disc;

a power source, wherein said power source applies two pairs of alternating voltages to said bottom electrode segments to excite a shear mode vibration in said stator, resulting in a shear-shear mode flexure traveling wave in said stator; and

a rotor operatively connected to said stator, wherein a portion of said rotor in contact with said stator is driven to rotate through a frictional force between said rotor and said stator due to said traveling wave deformation of said stator, and said stator and said rotor are each supported on a shaft extending through a centrally located aperture in said rotor and a centrally located aperture in said stator.

14. The rotary motor as set forth in claim **13** further comprising a holding means for supporting said stator.

15. The rotary motor as set forth in claim **14** further comprising a pressing means for preloading said stator against said rotor.

16. The rotary motor as set forth in claim **15** wherein said pressing means is a spring.

17. The rotary motor as set forth in claim **13** wherein said piezoelectric disc is operated at a flexure resonance mode of B_{02}.

18. The rotary motor as set forth in claim **13** wherein said piezoelectric disc is operated at a flexure resonance mode of B_{12}.

* * * * *

10

(b)

Figure 9.7 *(Continued)*

9.4 Patent Infringement (Lawsuits)

Patent infringement is the act of utilizing a patented invention without the permission or license of the patent proprietor. The scope of patented invention is defined in the claim section of each granted patent of each country. The patent is examined under the concerned law and granted by each country separately, and is enforceable only within the countries where the patent is granted. In general, patents are not worldwide permission or protection.

Since the 1840s, the expression "patent pirate" has been used as a pejorative term to describe those who infringe on a patent and refuse to acknowledge the propriety of the inventor. Samuel F. B. Morse, inventor of the telegraph, for example, complained in a letter to a friend in 1848 [6]:

> I have been so constantly under the necessity of watching the movements of the most unprincipled set of pirates I have ever known, that all my time has been occupied in defense, in putting evidence into something like legal shape that I am the inventor of the Electro-Magnetic Telegraph! Would you have believed 10 years ago that a question could be raised on that subject?

Those who accuse others of being patent pirates say that they take advantage of the high cost of enforcing a patent to willfully infringe valid patents with impunity, knowing that the average small inventor does not have the financial resources required to enforce their patent rights. In the United States, for example, an inventor must budget $1 million or more in order to initiate patent litigation. Patent pirates also take advantage of countries where patent rights are difficult to enforce and willfully infringe in those countries. The U.S. government identified some countries explicitly with this patent-pirate conspiracy, such as China and Thailand.

Ironically, the term "pirate" has also been used to describe patent owners that vigorously enforce their patents, also known as "patent trolls." Thus, whether one deliberately infringes a patent or vigorously enforces a patent, he may be referred to as a pirate by those that feel he is overstepping his bounds. Finding an interesting patent in a foreign country, and submitting a similar patent in the United States seems to be against business ethics, but we should be aware that many of these incidents occur merely to collect the royalty (patent mafia).

9.4.1 Patent Infringement Example Problem

Shinsei Industry in Japan developed and commercialized the world's first surface wave-type ultrasonic motor (USM). Its original patent was applied in Japan in 1981, and then expanded to the United States. It has been selling the motors worldwide (including the United States) for several years. On the other hand, XYZ Company in the United States developed a slightly modified USM and applied for its patent in 1994 only in the United States.

XYZ started to sell its motor products to NASA in the United States. Does this constitute a patent infringement against Shinsei? After reading the following background information on the related patents, provide your comments on this issue.

9.4.2 Background of the Related Patents

Figure 9.8 summarizes patent designs related to the piezoelectric USMs: (a) a basic patent design of the USM created by Shinsei Industry, which covers any traveling wave-type motors; (b) one example design by Shinsei, which is commercialized in practice; and (c) a modified design by XYZ.

Shinsei's patent (Japan and worldwide) was submitted in 1981, which covers any traveling wave-type motors in patent claim 1: a practical example design is shown in Figure 9.8b. XYZ's patent (United States) was submitted in 1994 and uses two PZT plates with different electrode patterns, which is illustrated in Figure 9.8c. Thus, the patent expert can say that the patent design of Shinsei with one PZT plate is different from that of XYZ with two PZT plates.

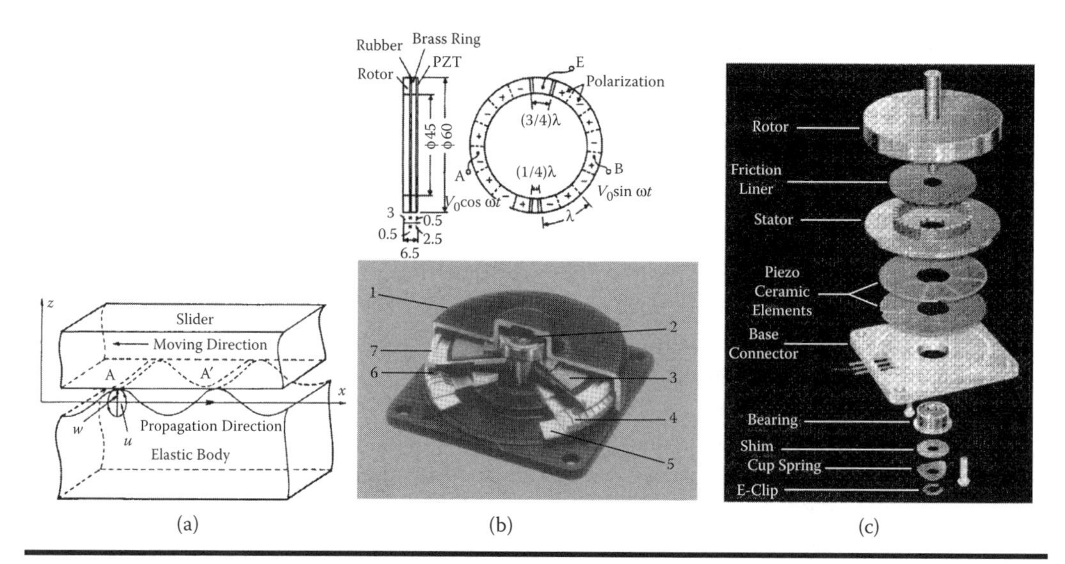

Figure 9.8 Patent designs related to the piezoelectric ultrasonic motors: (a) basic patent design of the ultrasonic motor by Shinsei Industry, covering traveling wave-type motors; (b) one example design by Shinsei, which is commercialized, and (c) modified design by XYZ, USA.

9.4.3 Comments by the Author

1. Shinsei's patent claim 1 describes first the motor idea of the traveling wave usage and exhibits a wide coverage, which covers any traveling wave-type motors (this is a basic patent).
2. Though XYZ's design with two PZT plates (Figure 9.8c) can be recognized as different from Shinsei's practical product with one PZT ring plate (Figure 9.8b), its patent is a sort of extended patent. The evaluation at the U.S. Patent Office had less insight on XYZ's application at that time; or, they should have approved the patent coverage only on the modified part. In conclusion, if XYZ started to commercialize the motors without paying the royalty to Shinsei, it would be considered a patent infringement.
3. Canon in Japan developed a slightly modified USM for its camera's automatic focusing mechanism, as shown in Figure 9.9. They decided to pay the royalty to Shinsei until the patent expiration (2001 = 1981 + 20 years), because the patent covers any traveling wave-type motors.
4. However, XYZ sold its motors to NASA. If this product delivery was made through a research contract (i.e., deliverable items), this sale is not considered regular commercialization. Thus, Shinsei cannot request the royalty payment.
5. If XYZ sells its motors to unrestricted markets or customers, Shinsei could bring a lawsuit against XYZ. Another important comment relates to the cost of this patent infringement lawsuit. Supposing that Shinsei would win and settle on this lawsuit by negotiating to receive a royalty (5% of the sales price), Shinsei would need to take into account the minimum cash budget of $300,000 for starting the lawsuit to pay for the court and lawyers, regardless of winning or losing. Should XYZ sell 500 motor units per year in the United States with an average sales price of $1000, Shinsei would receive only $25,000 per year as the royalty. There is no profit for Shinsei, compared to the initial $300,000 investment.
6. Finally, Shinsei's wide-coverage patent had already expired, and no lawsuit had been initiated.

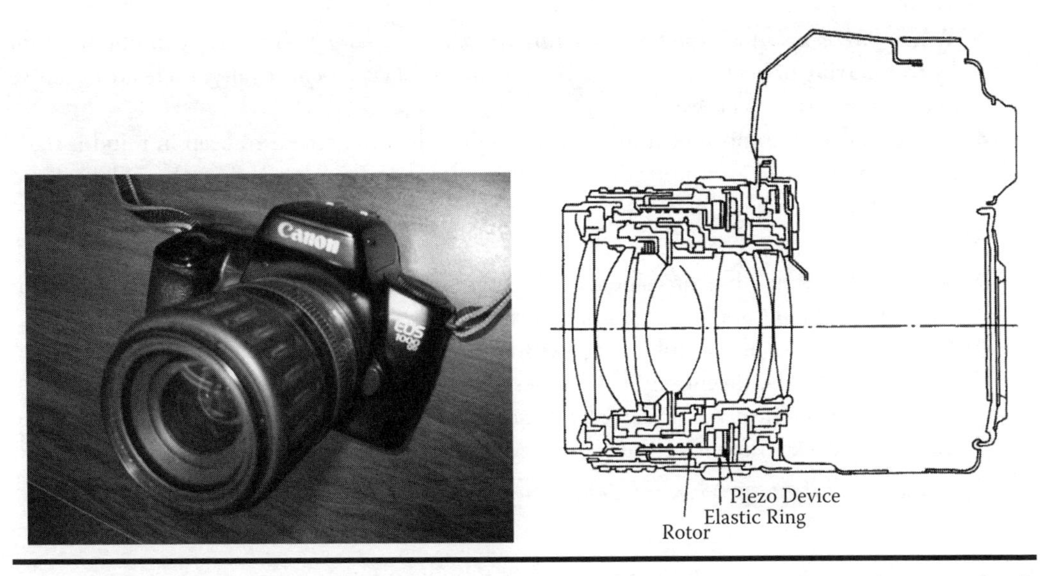

Figure 9.9 Camera auto-focus mechanism with a traveling wave-type ultrasonic motor developed by Canon, Japan.

Chapter Summary

9.1 Comparison among various intellectual properties.

	Patent	*Utility Model*	*Industrial Design Right*	*Trademark*	*Copyright*	*Trade Secret*
Objective	Invention	Low-level invention	Product design	Product trademark	Production (software, book etc.)	Know-how, etc.
Protection items	Novelty	Novelty	Novelty	No confusion	Creation	Confidentiality
	Register	Register	Creativity	Register		
			Register			
Protection period	20 years	6 years	15 years	10 years+	50 years after death	No limit
Disclosure	0	0	0 after 3 years	Publication	Publication	X
Protecting content	Exclusive operation	Exclusive operation	Exclusive operation	Exclusive usage	Exclusive copyright	Confidentiality

9.2 The importance of IP

- *When your company manufactures the patented product*: The patent is a form of protection against other companies imitating or stealing the technology developed by your firm's invested money and manpower.

- *When your company does not manufacture the patented product*: The firm can obtain profit by transferring or licensing the IP to other companies via patent royalty; there are exclusive and non-exclusive licenses.
- One of the most important things for the high-tech entrepreneur to keep in mind is trade secret maintenance and protection, which is related to the termination of employees and other job changes.

9.3 Patent preparation

- Major problems in filing for a patent are 1) guesses should not be included, and 2) application possibilities should not have been previously discussed.
- A patent consists of the following items: 1) abstract; 2) background of the invention; 3) summary of the invention; 4) brief description of the drawings (with drawing sheets); and 5) description of the preferred embodiment(s), i.e., a detailed description, including the principle, experimental results, and the claims.

9.4 Be cautious of *patent infringement* and *patent pirates.*

Practical Exercise Problems

P9.1 Patent Evaluation

Figure 9.10 exemplifies four piezoelectric ceramic actuator-related ideas, which may be patentable. They are related to (a) component design, (b) drive scheme program software, (c) manufacturing process, and (d) composition of the material. Evaluate the patentability of these new ideas. Note that we have excluded from Figure 9.10 the situations in which the idea was already well known or reported elsewhere previously.

P9.1.1 Model Answers

1. *Multilayer designs.* A basic patent of mine submitted in 1978. Unfortunately, I have not profited much from this patent, because worldwide production started after the patent expired. Do not invent too early.

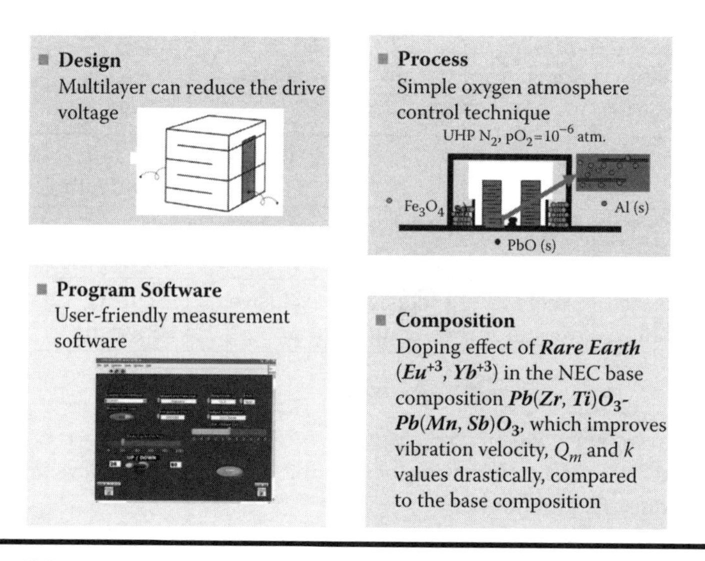

Figure 9.10 Possible patentable ideas related to piezoelectric ceramic actuators.

2. *Program software*. This should not be a patent, but a copyright. Note that computer software belongs to a copyright.
3. *Manufacturing process*. Filing a patent on the manufacturing process is *not* recommended! This falls under trade secrets. Because the process cannot be verified technically, patent piracy could occur. This will merely create a disadvantage by disclosing the know-how in the firm. Remember the attitude of Coca Cola. NEC filed the patent of the multilayer actuator manufacturing process 5 years after the commercialization (i.e., after establishing their prestigious status in the multilayer actuator industry). NEC's application is merely to prevent a patent pirate from applying for a patent.
4. *Modified composition*. This is a sort of subsidiary patent, which MMI took in order to cross-license it with NEC's basic composition patent.

P9.2 Invention in the Company

When a company employee creates an invention, who is the owner of this intellectual property?

Global consensus: An invention by a company employee in his or her company task belongs to this employee, but the company will obtain a nonexclusive operation license to this invention. Also, the company can request the employee to transfer the IP ownership by paying a reasonable amount of cash to this person. An example of a calculation made by Tokyo Local Court in Japan against one of the major Japanese electronic companies is indicative. The total revenue during the patent effective period until now was estimated at $6 billion. A reasonable patent royalty was suggested at the rate of 20% of the sales, which was equal to $1.2 billion ($6 billion × 0.20). Assuming that the contributions by the company and the employee to this patent are equal, the employee is eligible to receive 50% of the total royalty, leading to $600 million. However, the actual settlement was reached via reconciliation, by paying only $6 million to this employee.

Example practice in a U.S. university: Employees at U.S. universities (professors, postdoctoral fellows, graduate students) are requested by the employer to sign patent right waiver agreements (to the university) simultaneous with the employment agreements. The employee is merely an inventor, and the university is the patent submitter and its owner. The university controls licensing of the patent to the client companies, and collects the royalty. The royalty is returned in part to the patent inventor, but in practice the rate is much less than 50% of the total royalty.

When a patent is created during contract research, who will retain the patent rights, research consignor or consignee?

Global consensus: The patent created during the contract research usually belongs to the research consignor, if no particular agreement is exchanged on the patent ownership.

Example practice in a U.S. university: Unlike the global consensus, U.S. practice dictates that the patent created during the contract research usually belongs to the research consignee, which is usually written in the research contract agreement. This is reasonable for research contracts from federal institutes such as the Defense Advanced Research Projects Agency (DARPA), Office of Naval Research, Army Research Office, and National Institutes of Health, because federal institutes cannot be manufacturers of the invention. However, this is sometimes applied also to a commercial company. Because most international companies try to follow the global consensus, American corporations sometimes experience a serious conflict with the contract consignor when exchanging research contract forms.

An engineer P invented a new piezo-motor on June 1, and its patent was filed on August 1. Another engineer Q invented the same motor on June 15, and its patent was filed on July 1. Which engineer will receive the patent award in practice?

Global consensus: Most countries respect the first filing date of the patent. From this viewpoint, engineer Q will receive the patent, and engineer P's request will be rejected.

Practice in the United States: Interestingly, the United States respects the first invention date of the patent, as long as this invention date can be verified. From this viewpoint, engineer P will receive the patent, and engineer Q will be rejected. Remember how important it is to keep a written note about your invention in your lab notebook with the date and a signature by a witness.

You can easily imagine a possible serious international lawsuit, when an attractive and competitive patent, e.g. in the medical or pharmaceutical area, is submitted to be covered worldwide. Depending on the country, the patent winner may change.

References

1. http://en.wikipedia.org/wiki/Intellectual_property.
2. Pendergrast, M. *For God, Country & Coca-Cola*, 2nd ed., New York: Basic Books, 2000.
3. http://www.ideafinder.com/history/inventions/ivory.htm.
4. Dong, S., S. P. Lim, K. H. Lee, J. Zhang, L. C. Lim, and K. Uchino. Piezoelectric ultrasonic micromotors with 1.5 mm diameter. *IEEE Transactions on Ultrasonics, Ferroelectrics, and Frequency Control* 2003;50(4):361–367.
5. Koc, B., S. Cagatay, and K. Uchino. A piezoelectric motor using two orthogonal bending modes of a hollow cylinder. *IEEE Transactions on Ultrasonics, Ferroelectrics, and Frequency Control* 2002;49(4):495–500.
6. Morse, S. F. B. "Samuel F. B. Morse, his letters and journals" 1914, Part 5 out of 9. www.fullbooks.com (accessed June 10, 2006).

Chapter 10

Human Resources—Who Should We Hire?

A typical high-tech start-up has less than 10 employees. It requires only basic knowledge about general human resources (HR) management. However, because many PhD students in U.S. universities are international students, the probability of hiring non-U.S. citizens is quite high. Therefore, you will need to learn the SBIR restrictions for non-U.S. citizens, the visa application process, etc. A comparison of HR management between the United States and Japan is also discussed at the end of this chapter.

I am deeply indebted to Bohlander and Snell's *Managing Human Resources* [1] for the material in this chapter.

10.1 Legal Issue Essentials in Human Resources Management

10.1.1 Key Laws

- *Equal Pay Act of 1963*: This act outlaws discrimination in pay, employee benefits, and pensions based on the worker's gender.
- *Civil Rights Act of 1964*: Discrimination is prohibited on the basis of race, color, religion, sex (gender), or national origin.
- *Employment Act of 1967*: This prohibits private and public employers from discriminating against people age 40 or older.
- *Equal Employment Opportunity Act of 1972*: Item 10.1.2 was amended by broadening the coverage of the act, including state and local governments and public and private educational institutions.
- *Civil Rights Act of 1991*: This provides for compensatory and punitive damages and jury trials in cases involving intentional discrimination; requires employers to demonstrate that job practices are job related and consistent with business necessity; extends coverage to U.S. citizens working for American companies overseas.

10.1.2 Civil Rights Act Coverage

The Civil Rights Act and the Equal Employment Opportunity Act cover a broad range of organizations:

- All private employers in interstate commerce who employ 15 or more employees for 20 or more weeks per year
- State and local governments
- Private and public employment agencies, including the U.S. Employment Service
- Joint labor-management committees that govern apprenticeship or training programs
- Labor unions having 15 or more members or employees
- Public and private educational institutions
- Foreign subsidiaries of U.S. organizations employing U.S. citizens

Certain employers are excluded from the coverage of the Civil Rights Act:

- U.S. government-owned corporations
- Bona fide, tax-exempt private clubs
- Religious organizations employing people of a specific religion
- Organizations hiring Native Americans on or near a reservation

10.1.2.1 Bona Fide Occupational Qualification

Under the Civil Rights Act, employers are permitted limited exemptions from antidiscrimination regulations if employment preferences are based on a bona fide occupational qualification (BFOQ). A BFOQ permits discrimination when employer hiring preferences are a reasonable necessity for the normal operation of the business. Business necessity is defined as a work-related practice that is necessary to the safe and efficient operation of an organization. However, a BFOQ is a suitable defense against a discrimination charge only when age, religion, gender, or national origin is an actual qualification for performing the job (discrimination is never allowed due to race or color).

10.2 Employee Collection

When you start a high-tech firm, will you be the president or vice president of this firm? The organization chart of Micro Motor Inc. (MMI) is re-cited from Chapter 4 in Figure 10.1. If you accept the vice president position, as Barb Shay did, you need to find a partner to be the president. Furthermore, recruiting a research and development (R&D) division director and research engineers is key.

10.2.1 Corporate Executives

10.2.1.1 Qualifications of Corporate Executives

Core skills include experience, decision making, resourcefulness, adaptability, team building, and maturity. Augmented skills include technical proficiency, negotiation, strategic thinking, delegation, and cultural sensitivity, in particular when the firm is expanding internationally.

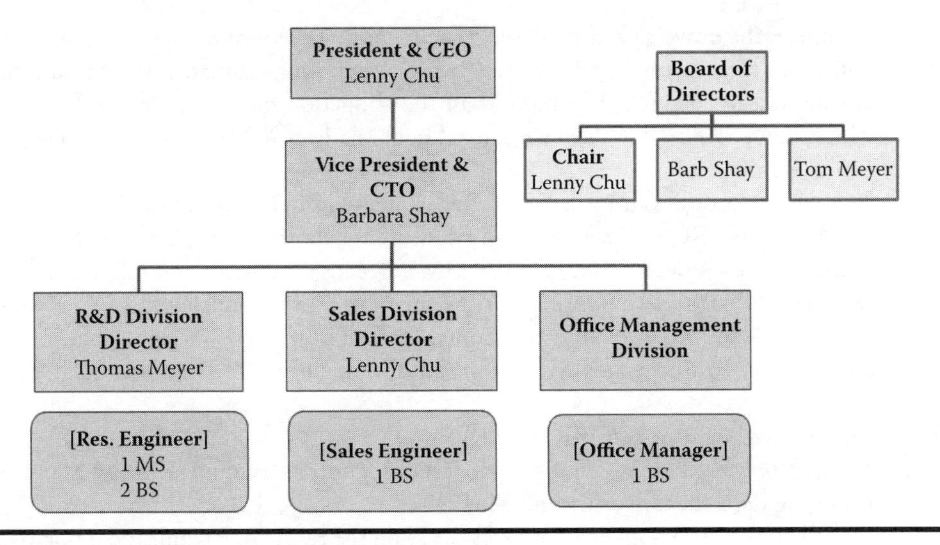

Figure 10.1 MMI organization chart (second year).

10.2.1.2 Searching Methods

If you are a university faculty member like Barb Shay at MMI, some possible scenarios when looking for president and research director candidates are as follows:

- Consulting companies: Barb hired Mr. Lenny Chu, President of Cheng Kung Corporation, as president and financial officer, as well as a financial sponsor.
- Venture capitalists: They may help you find a suitable financial officer.
- University or state government outreach offices.
- University alumni connection: Barb hired former graduate student Tom Meyer as the R&D division director.

10.2.1.3 Conflict of Interest with Yourself

If you are still university faculty when you start the company, how can you legally escape from the so-called conflict of interest? I strongly recommend that you exchange an Employment Agreement Appendix, which is added to the regular University Employment Agreement, with the applicable department in terms of conflict of interest disclosure with your university, which includes defining your working time in your company.

This is most important to avoid illegal accusations. If you are hired by the university 100%, you cannot be a principal investigator (PI) of the federal contract research in your company. Even if you can work on Saturday or Sunday, the federal contract does not allow assignment of your working time out of the regular 40-hour work week. Also if you are hired by the university 100%, your intellectual properties, based on your professional expertise, belong to the university if you signed an employment agreement with the university.

10.2.1.4 Agreement Example

1. Employment Conditions
   a. *Working time.* Barb Shay is currently receiving 9 months salary from the State University of Pennsylvania (SUP), and 25% salary from MMI. In practice, she is working for

6 hours in the university through one year (including the summer semester), teaching an obligatory three courses and instructing more than four graduate assistants and several visiting researchers, which is more than the obligation for a full-time professor in the department. She is allowed to work for 2 hours daily at MMI, located in College Park, Pennsylvania.

   b. *Role.* Shay is a professor in the XXX Department, and Director of the Research Center for Actuators (RCA), located in the Research Laboratory Building at SUP. She is founder, vice president, and CTO at MMI. Because President and CEO Lenny Chu, who has an MBA, is responsible for the financial and accounting tasks, and we hired a capable R&D division director, Tom Meyer (PhD), who is responsible for R&D, Shay's 2-hour workday at MMI is sufficient for operating MMI, in particular, the sales division.

2. Federal Research Contracts (SBIR, STTR)
   a. *Research involvement.* When a federal research contract is applied for by a joint team including both the university and MMI:
      i. When the RCA is involved, Shay is the PI on the RCA/SUP team, and is not involved with the MMI team.
      ii. When the other research group is representing SUP, Barb Shay is involved as co-PI of the MMI team.
   b. *Salary paid from the research contract.* Shay's salary will not be paid from both sides (SUP and MMI) for each research program.
3. Intellectual Properties
   a. *Patent at RCA/SUP.* The intellectual property generated through the RCA research with the students is applied through the university's Intellectual Property Office.
   b. *Patent at MMI.* The intellectual property generated at MMI during Shay's MMI working time with MMI employees or the partner company employees is filed by MMI without including SUP as a submitter.
   c. *Intellectual property belonging.* In order to clarify the intellectual property belonging, Shay and her employees use separate lab notebooks to clarify the demarcation of work at SUP and MMI.

10.2.1.5 Enterprise Incentive Plans

Common enterprise incentive plans include profit sharing, stock options, and employee stock ownership plans.

Profit sharing. After setting the base salary minimum, the bonus will depend on the actual profit at year-end. Many Japanese firms offer a significant bonus in the salary (equivalent to 3 to 8 months salary).

Stock options. Stock option programs are sometimes implemented as part of an employee benefit plan. However, for a start-up firm, these are occasionally used for executives' incentive. Stock options are also a popular method to reduce the actual cash expenses for the corporate officers. There are two types of stock options: *compensation* and *incentive*. The employee who receives compensation stock options must pay federal income tax (the income on the W-2 form is increased by this amount), while the person who receives incentive stock options does not need to pay income tax. However, the maximum percentage of the incentive stock option over the total stock is limited (typically 10%). In the MMI scenario, Barb received her salaries (annual \$60,000) by the compensation stock option (total \$60,000) for the first year, then the incentive stock option ($30,000) and the compensation stock option ($30,000) for the second year. She received 33% (incentive stock option, 10 shares; compensation stock option, 30 shares) of MMI's stocks in total. However, notice that she did not receive any cash (corresponding to the $90,000 of the compensation stock options) from MMI, although she paid the income tax for this nominal additional income. She decided to take this personal financial risk for starting up her own company by asking for the financial compensation from her husband.

Employee stock ownership plans. These are stock plans in which an organization contributes shares of its stock to an established trust for the purpose of stock purchases by its employees. This plan is adopted by relatively large firms. The established trust qualifies as a tax-exempt employee trust under section 401(a) of the Internal Revenue Code.

10.2.2 Subordinates Collection

10.2.2.1 Job Description/Interview

The job description for engineers can include the required knowledge, such as ceramic manufacturing skills or knowledge of piezoelectric characterization techniques.

There are no regulations about the interview questions on his or her engineering and technological background. However, the HR Employment Act restricts the following questions (you must rephrase the questions):

- Bad: Are you married? Do you have family?
  Recommended: Don't ask about marital status.
- Bad: Do you have a car?
  Recommended: Will you have any problem getting to work on time?

10.2.2.2 Hiring Students—Conflict of Interest

If you are a university faculty member operating the company in parallel, do not hire your students directly; this is another legal issue to be remembered.

10.2.2.3 Agreement Example

MMI Student Involvement Example

Graduate and undergraduate students who are involved in a research contract relating to MMI (MMI direct research contracts and/or SBIR/STTR subcontracts) are hired at the RCA/SUP via a graduate assistantship or undergraduate student wage payroll. They are not hired or paid directly by MMI.

10.2.2.4 Employee Turnover

The U.S. Department of Labor suggests the following formula for computing turnover rates:

$$\frac{\text{Number of separations during the month}}{\text{Total number of employees at midmonth}} \times 100(\%) \tag{10.1}$$

The turnover rate in the United States is 10 times higher than the rate in Japan, which may be attributed to cultural differences (refer to Section 10.4).

One of the most important things for the high-tech entrepreneur to keep in mind is trade secret maintenance and protection, which is related to termination of employees and other job changes. Unlike Japanese employees, American engineers tend to change jobs every several years, resulting in the inevitable transfer of trade secrets, even among competitive firms. Accordingly, we sometimes face a serious conflict with the company at which our previous employee has found a new position. Typical general conflicts include: (1) market research data and R&D and marketing strategies, (2) similar product lines in the new company to which he or she moved, (3) know-how in product-manufacturing processes, (4) research proposal ideas, and (5) the customer list.

In order to prevent these sorts of problems, the firm needs to legally regulate disloyal behavior through use of the Employment Agreement. Refer to Chapter 9, Section 9.2.3 for an example agreement.

10.2.2.5 Employee Benefits

Employee benefits legally require employer contributions, which include social security insurance, unemployment insurance, and worker's compensation insurance.

Bonus-type benefits include healthcare benefits, payment for time not worked, and pension plans. These are not mandatory for a small start-up, but they add incentive for the employees to work longer. They are summarized below:

- Healthcare benefits include health insurance, vision care, and dental insurance.
- Payment for time not worked includes vacations with pay, paid holidays, sick leave, and severance pay. Severance pay is a one-time payment given to employees who are being terminated.
- There are multiple pension plans, among which 401(k) saving plans are most relevant to a small firm.

A significant change in pension coverage has been the tremendous growth of tax-deferred 401(k) saving plans, which are named after section 401(k) of the Internal Revenue Code. The popularity of 401(k) plans is driven primarily by (a) the ability of the employer to transfer plan funding to employees, (b) the ability to transfer responsibility for investment choices to employees, and (c) the fact that employee contributions to 401(k) plans represent tax-free investing. 401(k) savings plans are particularly popular with smaller employers who find these pension plans less costly than defined-benefit programs.

10.2.2.6 Safety and Health

According to Occupational Safety and Health Administration (OSHA) statistics, in 2002 there were 5.5 million injuries or illnesses in private-sector firms. These occupational safety and health accidents are both numerous and costly to employers. Thus, managers are expected to know and enforce safety and health standards throughout the organization.

The employer must

1. Be familiar with mandatory OSHA standards
2. Post the OSHA poster

3. Make sure employees have and use safe, properly maintained tools and equipment
4. Report within 8 hours any accident that results in a fatality or the hospitalization of three or more employees

10.2.3 Outsourcing, Offshoring, and Employee Leasing

Over the past 25 years, the relationship between companies and employees has shifted from person-based to transaction-based in the United States. Unlike in Japan and Europe, U.S. employees do not work for one employer over the course of their lifetimes. More people are choosing to work on a freelance or contract basis, or to work part-time. Outsourcing is evidence of this trend.

Outsourcing is hiring someone outside the company to perform tasks that could be done internally. Companies often hire the services of accounting firms, for example, to take care of financial services. MMI has outsourcing companies that help with electronic circuits designing and materials machining. *Offshoring* is the business practice of sending jobs to other countries. MMI uses Cheng Kung Corporation as an offshore manufacturing facility for ceramic devices.

Small companies tend to sign *employee leasing* agreements with professional employer organizations (PEOs). Though the wages for the employee are equivalent to a fully employed worker, additional employment costs for the employee benefits, such as 401(k) pension plans and health insurance, can be shifted to the PEO. Thus, total costs can be reduced. MMI hired their office manager using this practice.

10.3 International Employees

10.3.1 SBIR/STTR Restrictions

A firm's eligibility for applying for SBIR/STTR programs is again cited here from Chapter 3:

- 500 or fewer employees
- Annual revenue under $5 million
- A company that is at least 51% owned and controlled by one or more individuals who are citizens of the United States, or permanent resident aliens in the United States, or
- A company that is at least 51% owned and controlled by another business that is itself at least 51% owned and controlled by individuals who are citizens of, or permanent resident aliens in the United States

10.3.1.1 Workforce

The firm should also be aware of the restrictions and regulations on the workforce. Depending on the security level of the program, there are various restrictions on the workforce:

- In most cases, the PI should be a U.S. citizen or a permanent resident alien.
- Lowest level: Workers can be aliens with an eligible working visa (H-1, J-1, etc.), in addition to U.S. citizens and permanent residents.
- Middle level: Workers should be U.S. citizens and permanent residents.
- High level: Workers should be only U.S. citizens.
- Highest level: Workers should be only U.S. citizens with a minimum of one person with security clearance.

10.3.1.2 Clearances

Individuals who need to have confidential/secret/top secret (C/S/TS) clearances because of their job or access to federal government assets will be required to sign the Security Clearance Form (TBS/SCT 330-60e) [2]. The clearance levels are as follows:

- Confidential (Level I)
  - In addition to the ERS checks, foreign employments, immediate relatives, and marriages/common-law relationships must be declared and screened.
  - This level of clearance will grant the right to access protected and classified information up to the confidential level on a need-to-know basis. Department heads have the discretion to allow for an individual to access secret-level information without higher level clearance on a case-to-case basis.
- Secret (Level II)
  - Same as confidential.
  - This level of clearance will grant the right to access protected and classified information up to the secret level on a need-to-know basis. Department heads have the discretion to allow for an individual to access top secret–level information without higher level clearance on a case-to-case basis.
- Top secret (Level III)
  - In addition to the checks at the secret level, foreign travels, assets, and character references must be given. A field check will also be conducted prior to granting clearance.
  - This level of clearance will grant the right to access all protected and classified information on a need-to-know basis.

10.3.2 Visa Application

Permanent resident visa holders do not have any restrictions on working in the United States. However, your company needs to help different visa holders with updating their eligible working visa.

Let us consider two typical scenarios for hiring aliens: (1) hiring an engineer immediately after graduation from the university, and (2) hiring an engineer from another company (including a post-doc, research associate, or faculty at a university).

10.3.2.1 Hiring an Engineer Immediately after Graduation from the University

Usually an international student or graduate student in a university has an F-1 visa (student visa). The F-1 visa holder has eligibility to work in a company with a special working permit, optional practical training (OPT). This allows for one year to stabilize his or her actual working visa (H-1).

10.3.2.2 Hiring an Engineer from Another Company

This applicant usually has an H-1 working visa. Because this working visa is issued based on his previous company, your company needs to apply the same category H-1 visa through your company.

Table 10.1 summarizes visa classifications that allow an alien to work in the United States.

Table 10.1 Visa Classifications That Permit Work in the United States [3]

Visa Classification	*Definition*
E-1, E-2	Treaty trader or treaty investor
F-1	Foreign academic student, when certain conditions are met
H-1B, H-1C, H-2A, H-2B, H-3	Temporary worker
I	Foreign information media representative
J-1	Exchange visitor, when certain conditions are met
K-1	Fiancé of a U.S. citizen
L-1	Intracompany transferee
M-1	Foreign vocational student
O-1, O-2	Temporary worker in the sciences, arts, education, business, or athletics
P-1, P-2, P-3	Temporary worker in the arts, athletics in an exchange or cultural program
Q-1, Q-2	Cultural exchange visitor
R-1	Temporary religious worker with a nonprofit organization
TC	Professional business worker admitted under U.S.-Canada Free Trade Act (NAFTA)
TN	Professional business worker admitted under NAFTA

If your company submits all necessary visa application documents directly, the application costs just less than $1000. However, if you use an immigration attorney for this task, it usually costs $8000 to $12,000. As an executive of a start-up company, you should carefully calculate the additional cost required in hiring international engineers. Most companies discuss this issue carefully with the candidates , including the compensation and salary of the candidates in the first and second years of their employment.

10.4 Human Resources Management in the United States and Japan

With increasing global, transnational, and multinational business opportunities (refer to Chapter 13), we need information regarding the host country's business culture and management styles. The core skills required for expatriate managers include cultural sensitivity and team building, in addition to resourcefulness and decision making. For example, there are significant differences in human resources management styles between the United States and Japan. I will analyze these differences, particularly in terms of organizational structures, management styles, leadership, hiring, compensation, performance appraisal, and training and education systems. The management styles can be symbolized by a regatta in the United States and *mikoshi* in Japan. Employee productivity is evaluated by a differential method in the United States, while an integral method is used in Japan. I will discuss how these human resources management differences originate from differences in culture and lifestyle. The United States is an individual-based society while Japan is a group-based society.

10.4.1 Introduction to Human Resources Management

For 19 years starting in 1975, I occasionally had joint appointments as a university professor and a company executive (standing auditor and deputy director of R&D Center) in Japan. In 1991, I was brought to Pennsylvania State University to be a founding director of the International Center for Actuators and Transducers (ICAT) to transfer technologies I had developed in Japan. This was because I was known worldwide as one of the pioneers in the field of piezoelectric actuators—using piezoelectric materials to move something mechanically directly from an electrical signal.

One of my most memorable experiences happened at the opening ceremony of the new research center, ICAT. I prepared the ceremonial speech by myself, based on my long Japanese executive career. I said, "It is my honor to be nominated as the founding director of this new research center. This center is dedicated to you 14 faculty members. I do not have any particular plan for the center operation at present. I would like to ask all of you to provide me your ideas on how to manage this center. I would like to compromise as much as I can to meet your desires."

However, during the reception after the opening ceremony, I was chastised by a higher ranking director of Pennsylvania State University. He said, "Your speech was not good for a university research center director. You should mention your ideas on how to manage the center explicitly, like 'I want to do this, first, then that, etc. Do you folks have any objections to my plan?' You need to show your strong mind first." I was really shocked by this criticism and felt a deep culture division between the United States and Japan, in particular, in the leadership style.

In Japan, the director is usually chosen from the original research center members. Without having a formal election, the director is selected through an underwater negotiation in the institute. Bringing in a new director from a different country rarely happens in Japan. In Japan, the center's staff would respect a quiet and moderate director, and would not welcome an address expressing a new direction. A director who will not aggressively alter the center's status is desirable.

H. Mintzberg [4] categorized the manager's work roles into the following 10 categories:

The interpersonal roles:

1. Figurehead
2. Leader
3 Liaison

The informational roles:

4. Nerve center
5. Disseminator
6. Spokesman

The decision-making roles:

7. Entrepreneur
8. Disturbance handler
9. Resource allocator
10. Negotiator

It is worth noting that Americans emphasize interpersonal roles, such as leadership, for a manager, while Japanese society welcomes strong decision-making roles, such as negotiator, disturbance handler, and resource allocator.

Table 10.2 summarizes keywords at a glance for understanding the HR management differences between the United States and Japan. We will consider the details in the following sections. This article is based on a chapter from a book I wrote, "The Difference between Japan and the United States in Research and Development Policy," published in 1987 [5].

10.4.2 Individual versus Group

10.4.2.1 Living Philosophy

Japanese industries still use the basic concept of *permanent employment*, provided the employee is loyal to the company. "Industrial warriors" are still highly respected in Japanese industrial society. Their lifestyles are arranged around the company schedule (group decision). Even though Japanese employees have more than two weeks of paid holidays per year, in practice, it is difficult to use more than a couple of days continuously because of work environment pressure. For example, many Japanese friends of mine have unfortunately passed away due to stress-related illnesses caused by executive management jobs.

In contrast, the American lifestyle centers on the individual and their family. Even directors in American companies can easily take more than a week off during the summer and at Christmas. In an extreme case, one sales engineer did not go to a tradeshow that was the most important to the company product's promotion, because that day coincided with her daughter's birthday.

In Japan, the only time employees are free to take vacation is during the Golden Week from April 30 to May 5. This simultaneous mandatory vacation causes major traffic, rail, and air

Table 10.2 Comparison between the United States and Japan in HR Management

	United States	*Japan*
Living philosophy working style	Individual	Group
Industry type performance appraisal	Differential	Integral
Management	Regatta	*Mikoshi* (portable shrine)

overcrowding throughout Japan, because more than 20 million Japanese people travel during this week. Of course, there is traffic congestion in the United States around Thanksgiving, Memorial Day, and Labor Day, but the situation for taking vacations is still diffused in the United States, in comparison with Japan. Note that the motivation of "big nation travel" originates from individual intention to go home for the holidays in the United States, but it originates from the company's operation schedule in Japan. The Japanese enjoy this one-week vacation at resorts or in an entertainment place such as Disneyland.

In the United States, employees seem to change companies frequently, often within 5 to 7 years of being hired. This mobility is driven by salary increases when accepting new employment. It reduces loyalty to the current employer. Figure 10.2 compares how salaries increase with time in the United States and Japan, respectively. Americans expect a jump in salary when changing companies, while Japanese expect to stay 10 years in one company with small raises, and thereafter receive exponential increases in his or her salary. This exponential compensation system encourages the employee to stay in one company permanently. A worker who changes employers roughly every 8 years in Japan would not get much increase in his or her compensation, as illustrated in Figure 10.2b.

This difference in employment and salary traditions affects the career paths of employees. In the United States, research engineers tend to stay within their area of expertise. For example, American colleagues of mine have spent their entire careers studying piezoelectric devices. When their employers have changed strategic directions, they changed employers to continue their work on the same or similar topics.

Japanese companies, in contrast, move research engineers along with products, giving them experience in different functions. For example, at Murata Manufacturing Co., a research engineer who develops a promising electronic device in a research center is strongly encouraged to become a factory director to start mass producing this new device, away from research. Once manufacturing development is finished, this engineer is again encouraged to become the sales division manager, promoting the product's sales. These successive position and task transfers are analogous to parenting a child, the invention being his or her child. The engineer is responsible for the child from birth to adulthood (sales). Without experiencing all roles, the employee will not be promoted to higher management positions in the company.

Similarly, company presidents in Japan will not be recruited from other companies, except rarely in emergency situations, such as when Nissan and Sony recruited new presidents from the other companies. On the contrary, this is very common in the United States; for example, the current president of Pennsylvania State University came from the University of Nebraska-Lincoln,

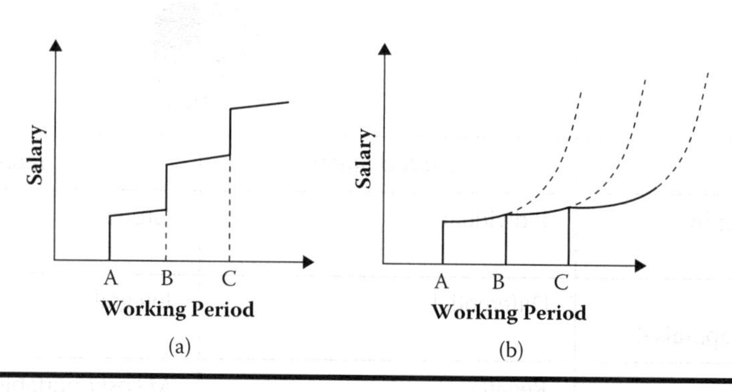

Figure 10.2 Compensation increment system: (a) United States and (b) Japan.

not from the Penn State faculty. Mr. Lee Iacocca was recruited by Chrysler from a competitive company, Ford.

Labor union structures also follow similar individual and group philosophies. Figure 10.3 compares the structural difference of labor union in the United States and Japan. In the United States, labor unions are based on each individual's specialty; electricians join the Electrician Union, mechanics join the Mechanics Union, and so on. In Japan, most companies have their own labor union; all employees join the same labor union, which allows the employees to transfer to any position or section in the company (from a research laboratory to a manufacturing facility, or to a sales division). As long as an employee does not insist on a particular area, permanent employment is guaranteed by the company.

F. W. Taylor proposed the Four Principles of Scientific Management in the late nineteenth century [6]:

1. The deliberate gathering of the great mass of traditional knowledge by the means of time and motion study
2. The scientific selection of workers and then their progressive development
3. The bringing together of science and the trained worker, by offering some incentive to the worker
4. The complete re-division of the work establishment, to bring about democracy and cooperation between the management and the workers

Japanese management closely follows Taylor's original principles. Japanese managers guarantee employment to workers under the supposition of their loyalty to the company. Productivity increases are in the common interest of both managers and workers because profits are more equally divided between the parties. In Japan, typically the salary ratio between the president and the lowest-ranked worker is less than 20 to 1. For example, the president of NEC, one of the largest Japanese companies, earns $500,000, while a McDonald counterperson earns $25,000. This creates a more trusting relationship between the employee and employer in Japan. In the United States the ratio is more than 300 to 1. Rarely do Japanese employees read their employment agreements because they trust the company. Often, the company does not have an actual

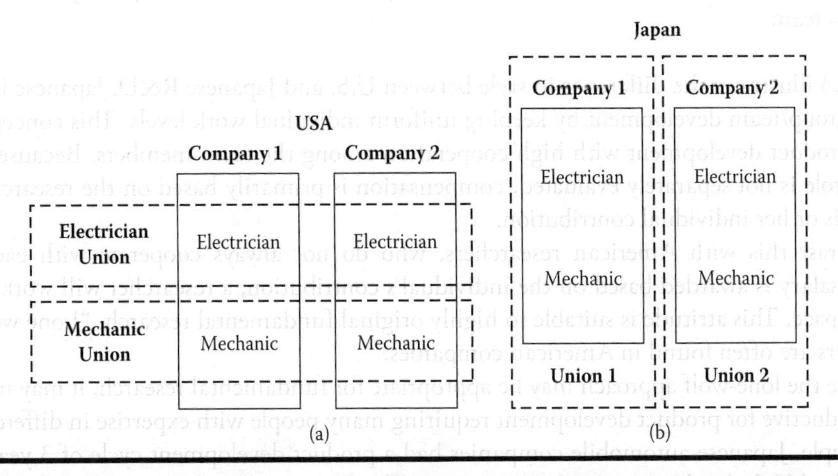

Figure 10.3 Structural differences of labor unions in (a) the United States and (b) Japan, which are based on individual specialty and on company/group, respectively.

employment agreement signed by the employee. I do not recall whether I signed the employment agreement when I was hired by Sophia University in Japan—there was no official agreement form! The employee does not negotiate salary individually because it is primarily determined by his or her age, with a large exponential salary increase after more than 10 years of service.

In the United States, salary is totally different. American employees normally negotiate for higher salaries when applying for a position, without regard for the company's financial situation at that time. These sorts of negotiations reinforce the concept of the employee working for his or her best interests, with the company a secondary consideration. Managers interpret this as lack of loyalty, knowing the employee can change companies at any time. F. Taylor's Theory of Scientific Management was meant to address this issue. In Japan, Taylor's concept is accepted as common practice by both managers and workers.

Ford Motors recently laid off 20,000 workers. Of course, Japanese industries make inevitable minimum layoffs sometimes, but the layoff is not very popular. My friend, a Taiheiyo Cement Corporation executive, told me about his company policy: they reduced the salaries of the managers (higher than the assistant manager level) by 20% uniformly, and distributed this amount to the lower-lever workers without executing layoffs in the beginning of twenty-first century, when Japan's economy faced a serious recession. This is another example of scientific management, which American managers do not want to follow.

10.4.2.2 Working Style

One of my former Japanese graduate students is currently a general manager at the Mitsubishi Electric Research & Development Center. He described the working policy of Mitsubishi, using traditional Japanese proverbs:

- "A tall tree catches much wind." All researchers need to keep a uniform development speed among groups and teams (see Table 10.3).
- "Do not do a task today which can be postponed till tomorrow." This was modified from the original proverb "Do a task today which you can do today," which seems to be similar to the "go slow" policy of the regular worker, as pointed out by F. Taylor [6]. However, note that the motivation in Japan is not to actually slow down, but to keep the pace all together as a team.

Table 10.3 illustrates the difference in style between U.S. and Japanese R&D. Japanese industries respect group/team development by keeping uniform individual work levels. This concept is suitable to product development with high cooperation among the team members. Because an individual's role is not separately evaluated, compensation is primarily based on the researcher's age, not on his or her individual contribution.

Contrast this with American researchers, who do not always cooperate with each other. Because salary is awarded based on the individual's contribution, a researcher will work at his or her own pace. This attitude is suitable to highly original fundamental research. "Lone wolf"-type researchers are often found in American companies.

While the lone-wolf approach may be appropriate for fundamental research, it may not be the most productive for product development requiring many people with expertise in different areas. For example, Japanese automobile companies had a product development cycle of 3 years, while the standard U.S. development cycle was 5 years. The faster Japanese system came about because the team was able to put working environmental pressure on the slowest employees. Chrysler was

Table 10.3 Working Style Differences between the United States and Japan

US	*Japan*
Individual pace	Group/team uniform pace
Research-oriented originality	Development-oriented cooperation
Salary based on individual ability	Salary based on age

(a)

(b)

Figure 10.4 (a) Cubicle style office (U.S. style), and (b) all-in-one style office (Japanese style).

only able to improve after the former president Lee Iacocca publicly mandated an effort to reduce the cycle time from 5 to 2 years. Ford Motor Company instituted a similar effort to reduce the cycle time to 3 years with the successful introduction of the Taurus.

Office structure also reflects the differences between the U.S. individual and Japanese group emphases. Figure 10.4 shows typical offices in the United States and Japan. Individual cubicles are popular in U.S. industries for privacy of employees, while Japanese industries do not allow private cubicles. Instead, everyone is in one room. A manager can observe employee behavior all the time. This system encourages the slowest workers to keep up with the group.

10.4.3 Differential versus Integral

10.4.3.1 Industry Type

The percentage of manufacturing industries to total industries is 24% in the United States and 34% in Japan, and the percentage of science and technology students to total university students is 5% in the United States and 20% in Japan. In other words, Japan has four times more engineers (the university student numbers are almost the same) and more manufacturing industries, while the United States has fewer engineers and many more financial, insurance, and legal corporations.

American industries are a differential type of industry, where the profit is created from the change in the price. The profit from the stock market does not depend on the absolute stock price, but on the time derivative of the stock price. In contrast, the manufacturer in Japan is an integral type, where the total investment or accumulated price is valued.

10.4.3.2 Employment and Performance Appraisal Criteria

Most Japanese employees put a priority on employment stability, aiming for permanent employment. Next, they are interested in the pension system after retirement. These issues are discussed when they interview for their first job. In contrast, Americans seek better pay and tend to change jobs and companies often. These differences can be explained using the example of "hardware-like" and "software-like" employment. Software can be immediately used, but is usually used only for a short period. It is in the category of an "expenditure." To the contrary, hardware requires training or experience, such as education programs, in order to use it, but it remains in the company in the category of "equipment" forever.

These employees' attitude differences originate from the employer's performance appraisal criterion. As described in Figure 10.2, the Japanese salary system invests in the future capability of the worker based on permanent employment (hardware-like). Thus, "fresh employee education programs" are very popular in Japanese companies. We have multiple Japanese graduate students in our research center at Pennsylvania State University, who are completely supported financially by their companies, under the supposition that these employees will return to their home companies after finishing their degrees and contribute more than this initial investment. On the other hand, even if they are very capable, a young employee cannot expect a high salary. It will take 10 years to get a significant salary increase. In other words, the accumulation of work done over a long period in a company (integral work, or accumulation of output in terms of the time worked) determines the salary. This salary system guarantees employment through old age, but suppresses the working motivation of young engineers. This situation is depicted in Figure 10.5.

In contrast, the American salary system is based on a way to purchase the present capability of the worker based on the employee's transfer in the future (software-like). Independent of age or time worked for the company, the salary is determined by the profit delivered by this worker for 1 year (or an equivalently short period), that is, the derivative/differential of the capability in terms

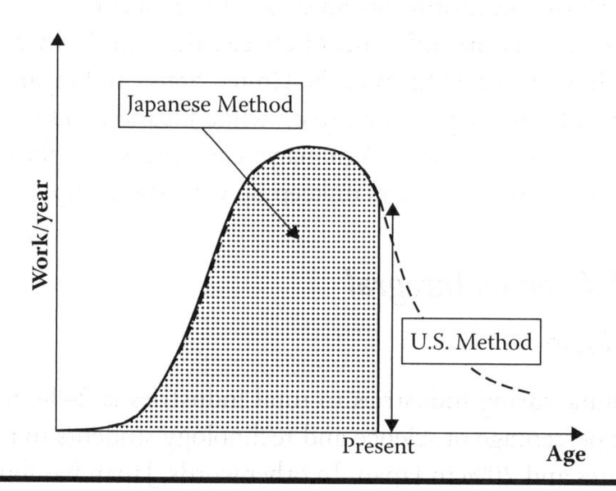

Figure 10.5 Performance appraisal criteria: (a) derivative method in the United States, and (b) integral method in Japan.

of a year (see Figure 10.5). This salary system motivates young engineers to work harder in the short term. However, with age and a slowing of fresh ideas, there is a big risk of salary reduction or unemployment. As a result, this compensation system will not create long-term loyalty to the company.

It is interesting to note that these employment and performance appraisal criteria affect the university teaching curriculum. Because there are no regular fresh-employee education programs in most American industries, a candidate's abilities and knowledge are checked thoroughly at the job interview, in particular, his university training. Therefore, students tend to take courses that will be helpful for obtaining a job. Consequently, courses that are not popular will be automatically eliminated from the university curriculum, particularly in graduate curricula.

10.4.4 Regatta versus Mikoshi

10.4.4.1 Management Structure

Figure 10.6 illustrates the management structures of American and Japanese industries. The American structure resembles a regatta, where only the coxswain (company president) sitting in the back knows the destination, and other oarsmen (employees) just row synchronously to the cox's command. The oarsmen don't communicate with each other in general, which is a suitable management structure for lone wolf-type engineers. If the president does not command correctly, the regatta will quickly pile up. Similarly, the American economy can change significantly, depending on the U.S. president's control. Further, American managers seem to prefer more power in a bigger company, which is illustrated in Figure 10.7a. This may be called a "whale" or "brontosaurus" type.

In contrast, Japanese structure resembles *mikoshi* (portable shrine, or the company), where all of the carrying men (employees) know the destination of *mikoshi*. Walking about, staggering left and right, the *mikoshi* does not go straight to the destination, but will reach the destination finally, even if the flagman (president) standing on the *mikoshi* falls down during travel. The employees behave as a team with a patriotic loyalty to the *mikoshi,* not to a person or a president. In a similar fashion, even if the Japanese prime minister were to change, the Japanese economy would not change much. In contrast to the whale type for the American managers, the Japanese managers prefer a "sardine" structure (Figure 10.7b). Loose coupling by medium- and small-sized companies creates a power similar to a big whale. But, unlike a whale, a group of sardines can change shape adaptively according to an enemy's presence. Based on each sardine's patriotic and synchronized intention, these companies can make a *keiretsu* (industry family tree).

A related discussion is found in an article titled "The Fourth Economic Crisis" by N. Makino [7], in which he uses an analogy to *kabuki* and musical theaters. *Kabuki* attracts the audience with one or two key actors (there are no actresses in *kabuki*, only male actors), which resembles American industries, such as Mr. Iacocca when at Chrysler. In contrast, the musical is an assembly of many minor actors and actresses, which more closely resembles Japanese industry's situations.

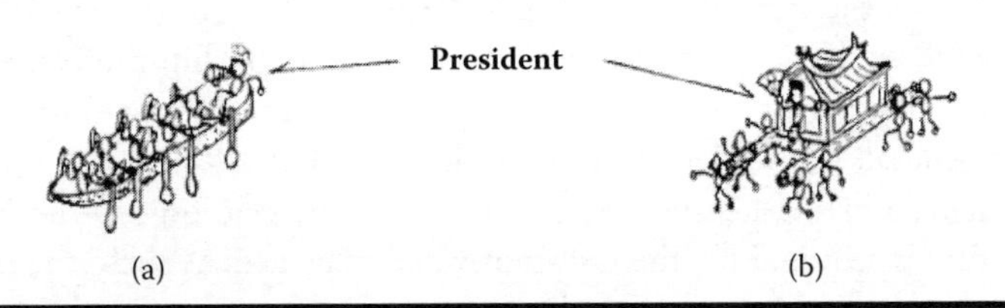

Figure 10.6 Management structure differences between (a) the United States and (b) Japan.

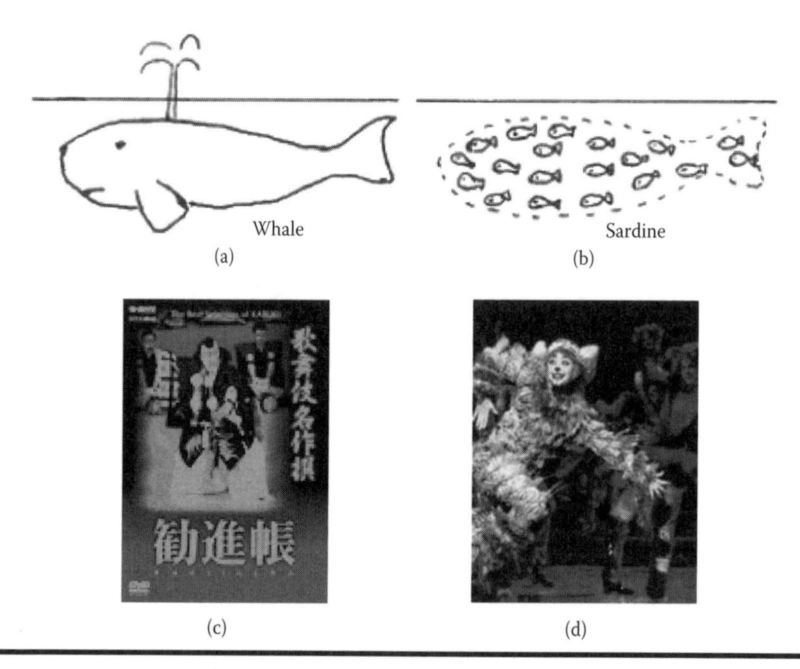

Figure 10.7 Difference between the United States (a and c) and Japan (b and d) in terms of the company structure that managers desire.

Exemplified by the long-run musical *Cats*, the teamwork is brilliant, but not many audiences remember the names of the actors and actresses (see Figure 10.7c, d).

It is interesting to compare my above description with the concepts of Theory X and Theory Y by D. McGregor [8]. Theory X is a set of propositions of the conventional view of management's task in harnessing human energy to fill organizational requirements, indicative of an autocratic management style, while Theory Y is based on more adequate assumptions about human nature and human motivation, and therefore has broader dimensions indicative of an egalitarian management style. Figure 10.8 is a summary description provided by M. Sobel [9], where the autocratic and custodial management styles reflect the Theory X mindset, and the participative and collegial management styles reflect the Theory Y mindset. We can say that the American management style is basically Theory X based, while the Japanese management style is basically Theory Y based. Again, this seems to have originated from the differences between hunting and farming races.

The Contingency Model, proposed by F. E. Fiedler for management leadership, is another interesting analysis [10]. Figure 10.9 shows the performance of relationship- and task-motivated leaders in different situational-favorability conditions. The situational-favorability dimension is indicated on the horizontal axis, extending from the most favorable situation on the left to the relatively least favorable situation on the right, in terms of leader–member relations, task structure, and leader position power. Here the vertical axis indicates the group or organizational performance; the solid line on the graph is the schematic performance curve of relationship-motivated leaders, while the dashed line indicates the performance of task-motivated leaders. In Japan, manager–employee relations are good in general, but leader power is weak (second and fourth columns). In order to obtain high performance, a low task structure is required for the relationship-motivated manager, and a high task structure is required for the task-motivated manager. As such, the manager education program for task-motivation became popular in Japan.

In contrast, when we consider an American manager with strong leadership but poor leader–member relations in general (fifth and seventh columns), high performance will be obtained only

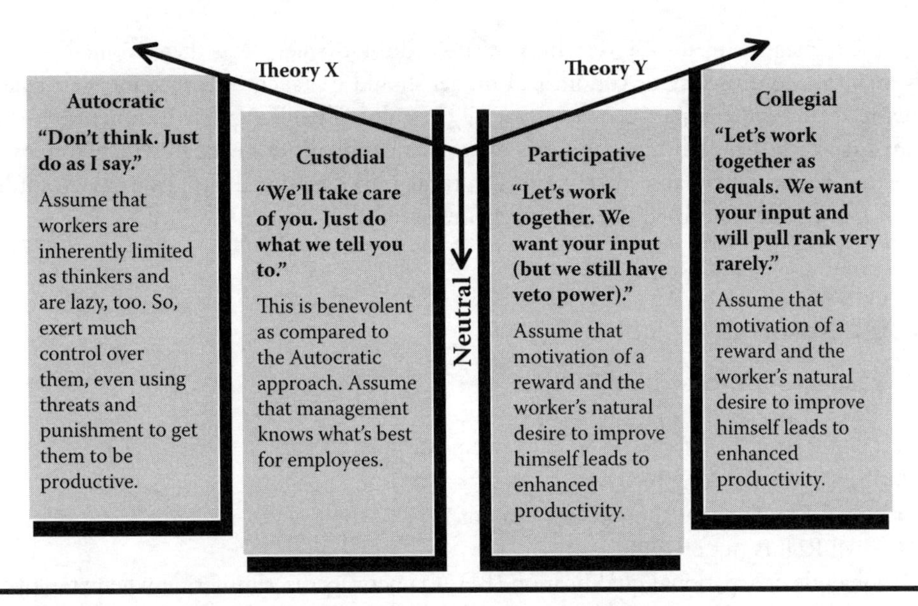

Figure 10.8 Management styles: Theory X and Theory Y. (Based on Sobel, M. 1994. *The 12-hour MBA program*. Upper Saddle River, NJ: Prentice Hall.)

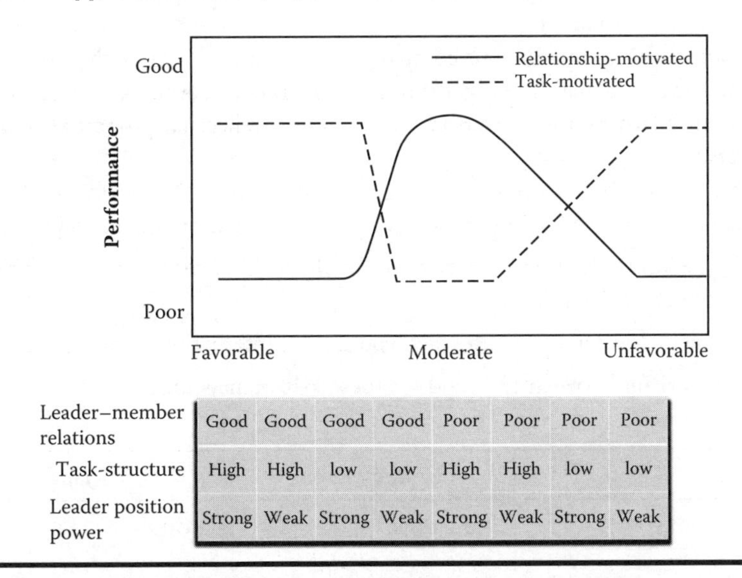

Leader–member relations	Good	Good	Good	Good	Poor	Poor	Poor	Poor
Task-structure	High	High	low	low	High	High	low	low
Leader position power	Strong	Weak	Strong	Weak	Strong	Weak	Strong	Weak

Figure 10.9 The performance of relationship- and task-motivated leaders in different situational-favorableness conditions. (Based on Fiedler, F. E. *Journal of Contemporary Business* 1974;3(4):65–80.)

when the task structure is high for the relationship-motivated manager. I believe that this is why the current MBA courses for American managers emphasize the human relationship.

10.4.5 Concluding Remarks

I took two MBA courses—one in Japan and the other in the United States. The American MBA class emphasized the humanistic (Theory Y) approach, starting from the mechanistic (Theory X) approach. In contrast, the Japanese MBA class tried to teach a more mechanistic approach, because the managers are already very humanistic. I do not believe that there is one best way for management. A contingency or conditional model should be taken into account, particularly to consider human resources management in racially or culturally different environments.

Your high-tech firm will hire engineers of various nationalities, through which you, as a corporate executive, will inevitably learn various cultural and educational backgrounds. Accordingly, your HR management skills will be adaptively improved.

Chapter Summary

10.1 Key human resources-related laws:
- Equal Pay Act of 1963
- Civil Rights Act of 1964
- Employment Act of 1967
- Equal Employment Opportunity Act of 1972
- Civil Rights Act of 1991

10.2 A bona fide occupational qualification (BFOQ) permits discrimination when employer hiring preferences are a reasonable necessity for the normal operation of the business.

10.3 If you are a university faculty member, when you start a company, do not create a conflict of interest. Consider the following:
- Define your working time in your company: (1) If you are hired by the university 100%, you cannot be a principal investigator of the federal contract research in your company. (2) If you are hired by the university 100%, your intellectual properties will belong to the university.
- Do not hire your students directly through your company. This is a serious legal issue!

10.4 Outsourcing and employee leasing are popular styles adopted by small businesses.

10.5 SBIR/STTR programs often have restrictions on aliens; U.S. citizenship or permanent resident status is sometimes required.

10.6 Carefully consider additional visa cost when you hire an international engineer in your firm.

10.7 Below is a comparison between U.S. and Japanese HR management:

<table>
<tr><th></th><th>United States</th><th>Japan</th></tr>
<tr><td>Living philosophy</td><td rowspan="2">Individual</td><td rowspan="2">Group</td></tr>
<tr><td>Working style</td></tr>
<tr><td>Industry type</td><td rowspan="2">Differential</td><td rowspan="2">Integral</td></tr>
<tr><td>Performance appraisal</td></tr>
<tr><td>Management</td><td>Regatta</td><td>Mikoshi (portable shrine)</td></tr>
</table>

Practical Exercise Problems

P10.1 Organization Chart

Generate an organizational chart (president, vice president, director, etc.) for your start-up company by identifying actual persons.

P10.2 Conflict of Interest

Supposing that you started your own company by keeping your job in the present university, prepare an agreement with your university without conflict of interest in terms of your working time, salary, and advising students.

References

1. Bohlander, G., and S. Sness. *Managing human resources,* 14th ed. Mason, OH: Thomson, 2007.
2. Wikipedia, "Security Clearance" http://en.wikipedia.org/wiki/Security_clearance.
3. Social Security Online, "Visa Classifications that Allow You to Work in the U.S." http://www.ssa.gov/immigration/visa.htm.
4. Mintzberg, H. Managerial work: analysis from observation. *Management Science* 1971;18(2):B97–B110.
5. Uchino, K. Difference between Japan and the US in research and development policy. *Sophia* 1992;41(2):213–224.
6. Taylor, F. W. The principles of scientific management. *Bulletin of the Taylor Society* December 1916:13–23.
7. Makino, N. *The fourth economic crisis*, p. 202–203. Tokyo: Hiraku Publication Company, 1997.
8. McGregor, D. M. *The human side of enterprise*. Management Review (November) American Management Association, 1957.
9. Sobel, M. *The 12-hour MBA program*. Upper Saddle River, NJ: Prentice Hall, 1994.
10. Fiedler, F. E. The contingency model—new directions for leadership utilization. *Journal of Contemporary Business* 1974;3(4):65–80.

Chapter 11

Business Strategy—Why It Is Important, and How to Set It Up

A business strategy is a company's game plan. Just as a football team needs a good game plan to have a chance for success, a company must have a good strategic plan to compete successfully against business competitors. Because the profit margin of a normal firm is not large in most industries, even a slight error in the strategic plan becomes fatal. A business strategy includes the following:

- Developing a vision and mission
- Identifying the organization's external opportunities and threats
- Determining internal strengths and weaknesses
- Establishing long-term objectives
- Generating alternative strategies
- Choosing particular strategies to pursue

In this chapter, I will focus on (1) SWOT (strengths–weaknesses–opportunities–threats) analysis for analyzing overall internal and external environments, (2) STEP (social/cultural, technological, economic, and political) four-force analysis for external factors, and (3) using Porter's five forces to analyze proximate internal environment. Product portfolio management (PPM) developed by Boston Consulting Group (BCG), which was introduced in Chapter 8, is another important analytical tool for business strategy. I am indebted to *Strategic Management* by F. David [1] for the material used to describe general issues in this chapter.

11.1 SWOT Matrix Analysis

Figure 11.1 shows the basic concept of the SWOT matrix. The horizontal axis is a measure of the strength–weakness of the firm (your company), while the vertical axis is a measure of the opportunity–threat of the industry (in our example, the piezo-actuator area). The former is called *internal environment analysis*, and the latter is called *external environment analysis*. All companies

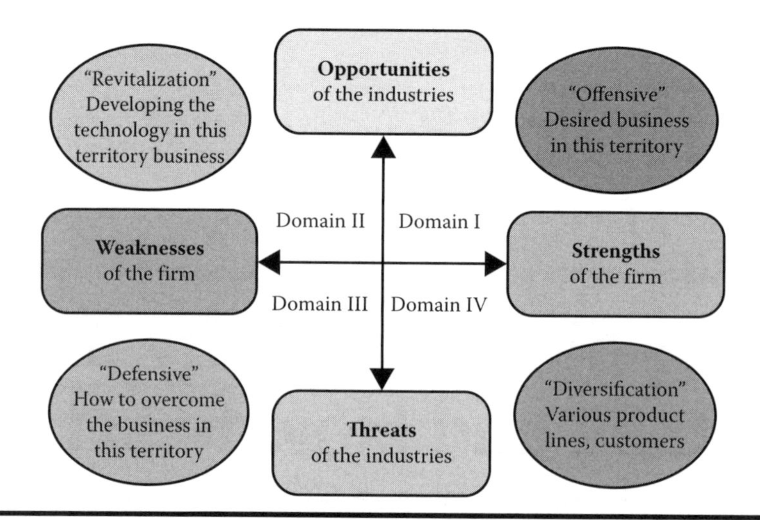

Figure 11.1 General concept of the SWOT grid.

are positioned in one of these four domains in Figure 11.1, where you will also find the suggested strategy that each type of company should take.

The SWOT matrix is an important tool that helps managers develop four types of strategies: SO (strengths–opportunities), WO (weaknesses–opportunities), ST (strengths–threats), and WT (weaknesses–threats). Identifying key external and internal factors is the most difficult part of developing a SWOT matrix and requires good judgment. There is not one best set of matches. Key external and internal environmental factors are discussed in Sections 11.2 and 11.3, respectively. The following four strategies can be taken, once your firm's grid position is determined.

SO strategies use a firm's internal strengths to take advantage of external opportunities. All managers would like their organizations to be in an "offensive" position in which internal strengths can be used to take advantage of external trends and events. Organizations generally will pursue WO, ST, or WT strategies to get into a situation in which they can apply SO strategies. When a firm has major weaknesses, it will strive to overcome them and make them strengths. When an organization faces major threats, it will seek to avoid them to concentrate on opportunities.

WO strategies aim to improve internal weaknesses by taking advantage of external opportunities (this is called *revitalization*). Sometimes key external opportunities exist, but a firm may have internal weaknesses that prevent it from exploiting those opportunities. For example, there may be a high demand for piezoelectric multilayer actuators (MLAs) to control the amount and timing of fuel injection in diesel automobile engines (the opportunity), but certain piezo-device manufacturers may lack the technology required for producing these devices (the weakness). One possible WO strategy would be to acquire this technology by forming a joint venture with a firm that has competency in this area. An alternative WO strategy would be to hire and train people who have the required technical capabilities. Because the piezoelectric actuator industry generally has better opportunities in external environments (domain I or II in the grid position) than the electromagnetic conventional motor industry, the small firms in the piezo-device industry usually follow these WO strategies.

ST strategies use a firm's strengths to avoid or reduce the impact of external threats. Diversification in the product line is the most popular strategy to overcome the external threats. An example of ST strategy can be found in the case of Texas Instruments using an excellent legal department (its strength) to collect nearly $700 million in damages and royalties from nine Japanese and Korean firms that infringed on patents for semiconductor memory chips (the threat). This is an example of how the firm used its strength against a competitors' invasion.

WT strategies are "defensive" tactics directed at reducing internal weakness and avoiding external threats. An organization facing numerous external threats and internal weaknesses may indeed be in a precarious position. In fact, such a firm may have to fight for its survival, merge, retrench, declare bankruptcy, or choose liquidation.

11.2 STEP Four-Force Analysis for External Environments

Though the piezo-actuator industries generally have a better opportunity in external environments (domain I or II), if the competitive electromagnetic motor takes a larger share in a particular device, the external environment becomes a threat (domain III or IV). The STEP four forces are considered external factors.

11.2.1 Social/Cultural Forces

Though piezoelectric transformers (PT) were used on a trial basis in color TVs in the 1970s, serious problems were found in the mechanical strength (collapse occurred in the device) and with heat generation in color TVs, leading to the termination of production for two decades (refer to Chapter 2). However, the development of laptop computers with liquid crystal displays requiring very thin and electromagnetic noise-free transformers to start the glow of a fluorescent back-lamp accelerated after the 1990s. This is a good example of how social demand creates the technology development (need-push).

The best technological device is not necessarily the best-selling device. A best seller is related to consumer attitudes and social trends. Table 11.1 shows how the Japanese consumer attitude has

Table 11.1 Social/Consumer Attitude Trends in Japan with Time in Several Decades

1960s	Heavier	Ship manufacturing
	Thicker	Steel industry
	Longer	Building construction
	Larger	Power plants (dam)
1980s	Lighter	Printers, cameras
	Thinner	TVs, computers
	Shorter	Printing communications period
	Smaller	Walkman, air conditioners
2000s	Beautiful	Well-known brand apparel
	Amusing	TV games
	Tasteful	Cellular phones (private communication)
	Creativity	"Culture" center made-to-order shoes

changed over time. Compared to the technological trends in the 1980s, when the keywords were "lighter," "thinner," "shorter," and "smaller," we can say that "beautiful," "amusing," "tasteful," and "creative" are the keywords in the 2000s [2]. The present best-selling cellular phones rely on their sophisticated function and artistic design, rather than the technology. Artistic sense should be added during device development in the twenty-first century "Engineering Renaissance" era [3]. Harmony between science and art (Sci-Art) will be the key in the future.

11.2.2 Technological Forces

11.2.2.1 Specifications

COPAL developed a piezo-bimorph shutter for a Minolta dual-zoom camera in the late 1980s [4]. The original design required a drive voltage of 120 V, which was rejected by Minolta, because an additional voltage supply was required in the camera. The solution was to increase the drive voltage by thickening the lead zirconate titanate (PZT) plate, rather than to decrease the drive voltage. The final bimorph was designed to be driven by 240 V, which was the strobe voltage supply originally installed in the camera. The drive voltage of an actual electronic component is decided by the battery specifications: 1.5, 3, 9, 12, or 240 V.

11.2.2.2 Cost Minimization

Some researchers are working to reduce the drive voltage by reducing the layer thickness of piezoelectric MLAs. However, it is not recommended for large MLAs used for diesel injection valve control applications in diesel automobiles (Figure 11.2). Figure 11.3 graphs the piezo-stack price and its electronic driver cost as a function of drive voltage [5]. The MLA price increases with reduced drive voltage (i.e., reducing each layer's thickness), while the driver cost increases with increasing voltage. The minimum total cost is obtained around 160 V, leading to a layer thickness of 80 μm. This sort of "standard" thickness can be derived from the cost minimization principle rather than the performance. Note that a 20-μm layer thickness is not technologically difficult in today's MLAs.

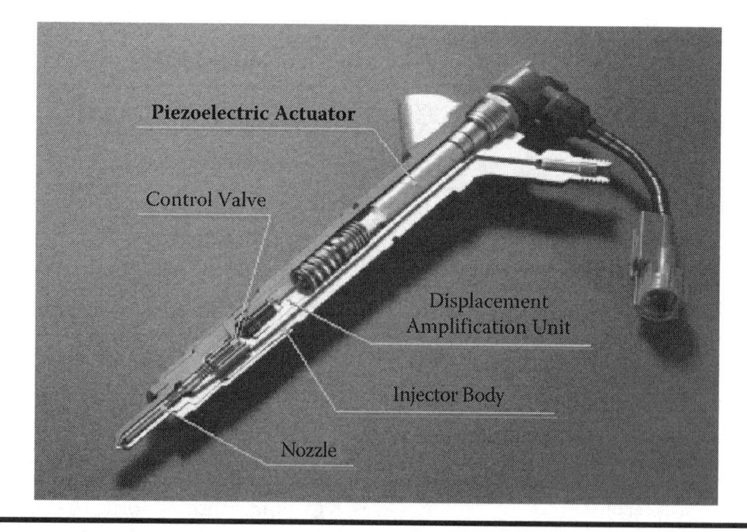

Figure 11.2 Common rail-type diesel injection valve with a piezoelectric multilayer actuator. (From Fujii, A. *Proceedings of JTTAS*, 2005. With permission.)

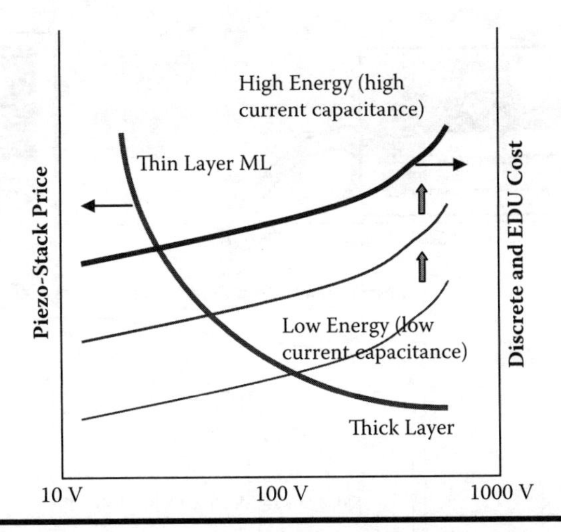

Figure 11.3 Drive voltage dependence of piezo-stack price and the electronic driver cost. (From Fujii, A. *Proceedings of JTTAS*, 2005. With permission.)

11.2.3 Economic Forces

Figure 11.4 illustrates the manufacturing cost calculation processes for MLAs with tape-cast automatic equipment and with a cut-and-bond manual production process, already discussed in Chapter 6. For the tape-casting equipment, the initial investment (fixed cost of equipment is $300,000) is expensive, with a low slope of variable cost (raw materials cost) as a function of production quantity. On the other hand, the cut-and-bond process requires a steep slope of variable cost provided by the labor fee in addition to the raw materials cost. We can find an intersection between these two lines (point T in Figure 11.4). This product quantity is the threshold above which the equipment installation starts to provide a better profit (a sort of break-even point).

Because labor costs are expensive in developed countries such as the United States and Japan, the introduction of a tape-casting facility is usually recommended, when the production quantity exceeds only 0.1 million pieces per year. Due to the lower labor cost in developing countries such as Thailand and China, the production threshold quantity is dramatically high, i.e., about 2 million pieces in MLA production. If the production quantity is 1 million pieces, the manual production process in these countries with line-workers' labor is actually cheaper than tape-casting facilities. This is the motivation for a corporation to locate its factory or find an original equipment manufacturer (OEM) partner in a foreign country.

11.2.4 Political/Legal Forces

The twenty-first century has been called the "century of environmental management." We are facing serious global problems such as the accumulation of toxic wastes, the greenhouse effect, contamination of rivers and seas, and lack of energy sources, such as oil, natural gas, etc. In 2006, the European community implemented restrictions on the use of certain hazardous substances, called the Restriction of Hazardous Substance (RoHS) Directive, which explicitly limits the usage of lead (Pb) in electronic equipment. Therefore, in the future we may need to regulate the usage of PZT, the most widely used piezoelectric ceramic material. Japanese and European communities may experience governmental regulation on PZT usage within 10 years. Lead-free piezo-ceramics started to be developed after 1999. Figure 11.5 shows statistics for various lead-free piezoelectric ceramics.

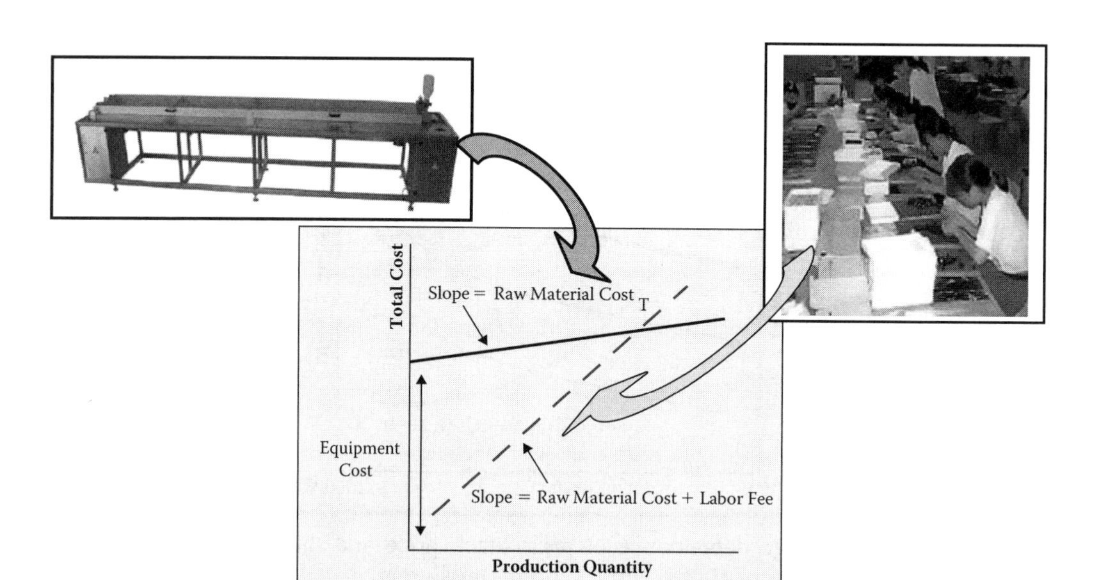

Figure 11.4 Total cost calculation comparison for a multilayer product between a tape-cast equipment automatic production and a cut-and-paste manual production.

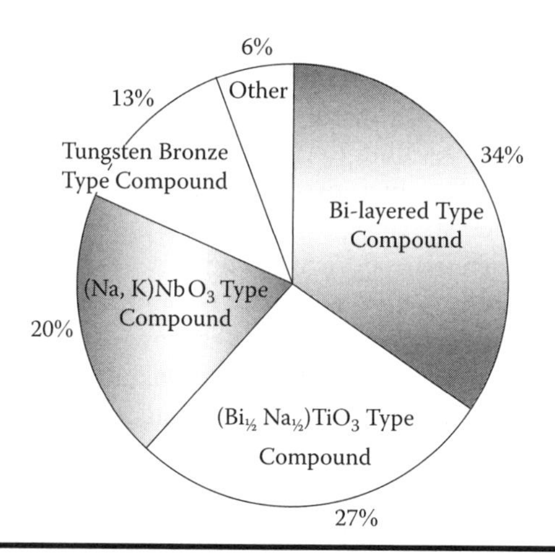

Figure 11.5 Patent disclosure statistics for lead-free piezoelectric ceramics as of 2001. (Total number of patents and papers is 102.)

The share of papers and patents for bismuth compounds (bismuth layered type and (Bi,Na) TiO_3 type) exceeds 61%. This is because bismuth compounds are easily fabricated in comparison with other compounds. Figure 11.6 shows strain curves for oriented and unoriented (K,Na,Li) (Nb,Ta,Sb)O_3 ceramics, based on the best current data reported by the Toyota Central Research Lab [6]. Note that the maximum strain reaches up to 1500×10^{-6}, which is equivalent to the PZT strain. RoHS seems to be a significant threat to the piezoelectric industry, which has primarily only PZT piezo-ceramics thus far. However, this is an opportunity for a company such as Toyota, which is preparing alternative piezo-ceramics to gain more of the piezoelectric device market share.

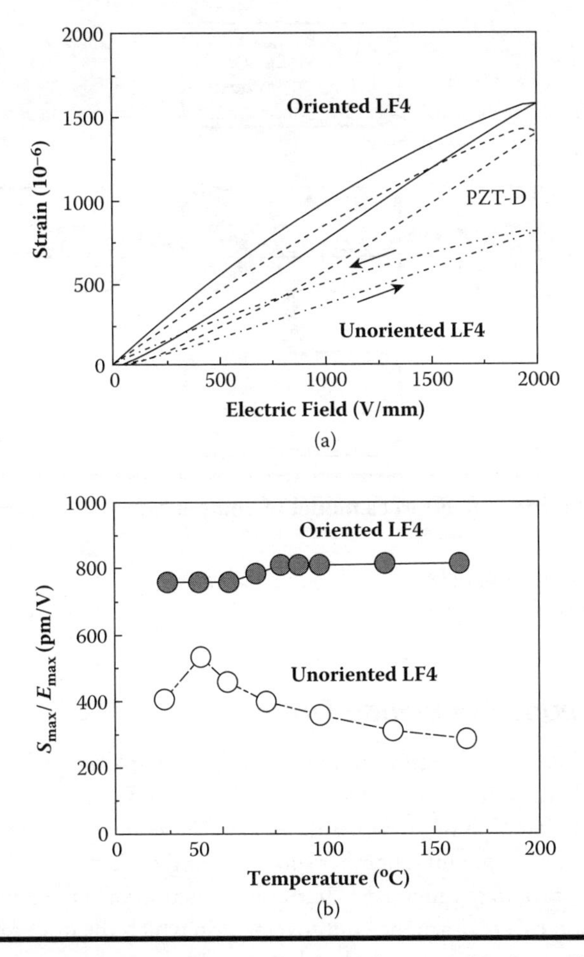

Figure 11.6 Strain curves for oriented and unoriented (K,Na,Li)(Nb,Ta,Sb)O_3 ceramics. (From Saito, Y. *Jpn J Appl Phys* 1996;35:5168–5173. With permission.)

Another issue is related to automobiles. Diesel engines are recommended because they consume less purification energy, which contributes to the reduction of global warming. However, the conventional diesel engine generates toxic exhaust gases such as SO_x and NO_x. In order to solve this problem, new diesel injection valves were developed by Siemens with piezoelectric MLAs. Figure 11.2 shows a common rail-type diesel injection valve [5].

11.3 Five-Force Analysis for Proximate Environments

Figure 11.7 shows the concept of Porter's five-forces model for proximate environment analysis [7]. The nature of competitiveness in a given industry can be viewed as a composite of five forces:

1. Rivalry among competing firms
2. Potential entry of new competitors
3. Potential development of substitute products

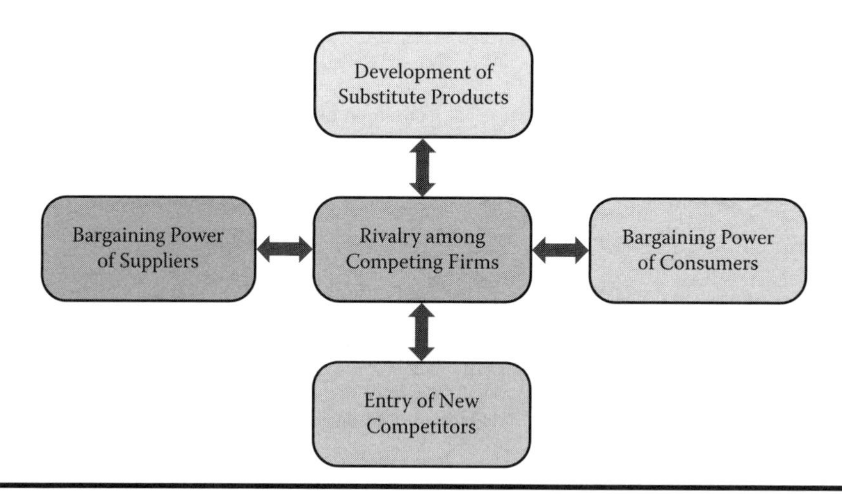

Figure 11.7 Michael Porter's five-forces model of competition.

4. Bargaining power of suppliers
5. Bargaining power of consumers

11.3.1 Rivalry among Competing Firms

Figure 11.8a shows the head structure of the Epson Mach ink-jet printer; here the so-called cut-and-bond method was initially used by the supplier, Philips, to manufacture the PZT unimorph [8]. Three years later, NGK in Japan developed the cofiring technology to manufacture the PZT unimorph on a zirconia ink-chamber substrate, and took over the supplier position from Philips. The cofiring technology provided a dramatic reduction of the manufacturing cost of the printer head, in addition to a performance improvement in which the unnecessary vibration of the ink chamber was suppressed. As you may already know, Philips' piezo-device division was closed soon after losing Epson, a large customer. This episode teaches us the importance of making a continuous effort to develop new technology, because every product has a life cycle.

11.3.2 Development of Substitute Products

Though the responsivity of a piezo-actuator in the Epson printer head (Figure 11.8a) is much faster than the competitive Canon bubble-jet printer, the resolution of the piezo-unimorph printer head is lower than the bubble-jet type, because the unimorph-covering area is rather large. In order to realize a much finer nozzle arrangement, Epson developed the piezo-multilayer type as shown in Figure 11.8b [8]. This arrangement provides finer resolution and quicker speed. Although it costs more than the unimorph type, this development is strategically important to compete with Canon's printers.

The most competitive substitute for the piezoelectric actuator is the electromagnetic (EM) motor. Figure 11.9 compares the metal tube–type microultrasonic motor (USM, 1.5 mm $\phi \times 4$ mm long) to a cellular phone EM motor with a similar output power level (50 mW). Note the significant differences in volume and weight, by a factor of 20. Also, the torque is 10 times higher and the efficiency of the USM is superior to the conventional EM motor. Although the USM is presently a threat to the EM motor in terms of performance, the historically lower price of the EM motor is a big threat to the piezo-actuator.

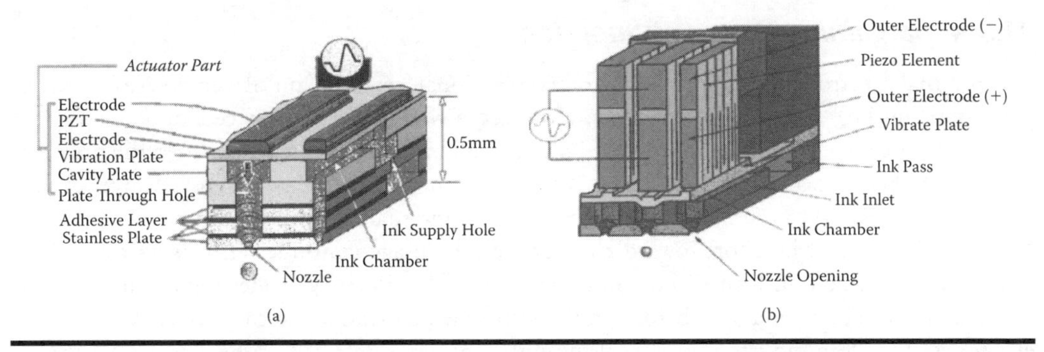

Figure 11.8 Piezoelectric ink-jet printer heads: (a) conventional unimorph-type and (b) new multilayer-type. (From Kurashima, N. *Proceedings of Machine Technical Institute Seminar, MITI,* Tsukuba, Japan, 1999. With permission.)

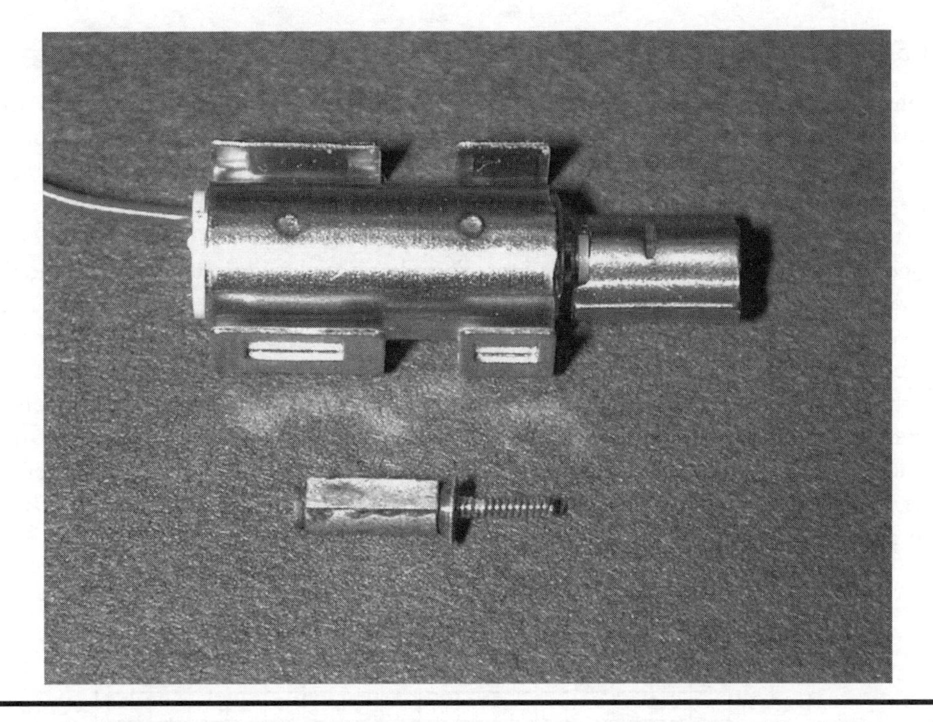

Figure 11.9 Comparison between the piezo-ultrasonic motor (bottom) and the EM motor (top), with similar power levels.

11.3.3 Entry of New Competitors

With expanding applications and markets for piezo-actuators, raw-material suppliers (upstream industries) such as cement, chemical, and even steel companies moved into the piezo-actuator business. Original customers (downstream industries) such as computer, home, and office electronics companies also started to manufacture the actuators themselves. Prosthetic arm manufacturers and measuring equipment companies had already started to make USMs, which was a threat to the present piezo-actuator industry. However, this could also be a good opportunity to find collaborative business partners.

11.3.4 Bargaining Power of Suppliers

The internal electrode material choice for conventional ceramic multilayer devices is silver/palladium (Ag/Pd), as it was in capacitors. However, owing to the Russian economy crisis in the late 1990s, the price of palladium increased dramatically with a peak in 2000, when the price was 10 times higher than in 1990 (Figure 11.10). Because of this raw-material cost change, capacitor industries started to use nickel (Ni) (a base metal) internal electrodes. Although the price of Pd is stabilized now, the effort toward cheaper material usage continues. This is an example of how a threat can be transformed into an opportunity. The necessity creates the technology. The piezo-actuator industry is a little behind in the shift to widespread use of the copper (Cu) internal electrodes. The piezo-industries are still using Ag/Pd or platinum (Pt) for the internal electrodes. EPCOS recently started to commercialize Cu-embedded MLAs for a diesel injection valve application, in which the actuator cost is currently the major bottleneck [9].

11.3.5 Bargaining Power of Consumers

The consumer's power depends on the application area. In IT/office equipment applications, the customers of piezo-actuators are giant electronic companies (IBM, Dell, Motorola, Samsung, Nokia, etc.) that have strong bargaining power against the component companies, which are "price takers." Thus, the current development target is to reduce the production cost to meet the customer's desire. MLAs and USMs currently need to be manufactured for much less than US$3 per piece for cellular phone applications. For robotic applications, how nanopositioning is demanded will be the key to estimating the future market size, which will determine the actuator price because it is performance/cost oriented. In the biomedical area, price is not a very important factor, but the specs and performance, such as very low drive voltage and confined size, are critical to the design. Fortunately the piezo-actuator industry is the price fixer in this area. Finally, the environmental businesses are expanding these days. Accordingly, the demand for actuators/transducers in this area is quickly increasing. Because it is still in the prototype production stage, there is no threat from customers ("skimming price" period, which was introduced in Chapter 8), but the actuator manufacturer needs to obtain the interest of the customers.

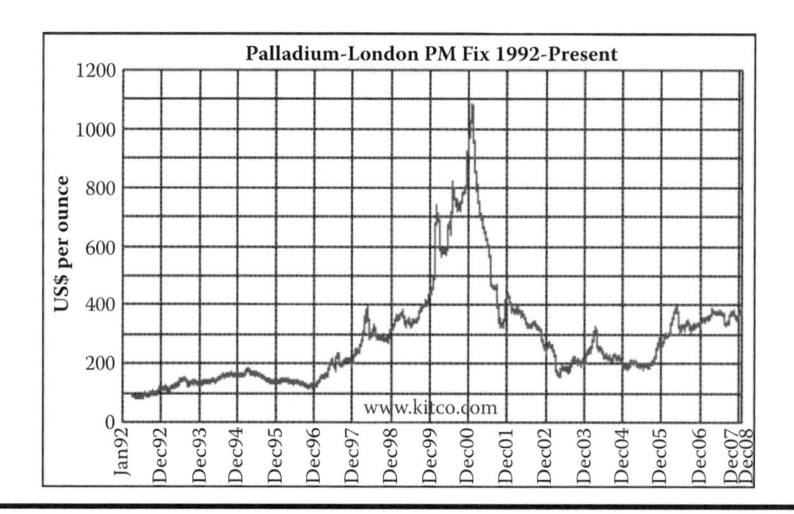

Figure 11.10 Palladium price change over years.

11.4 Business Strategy Format

We will consider two business strategy case studies in this section. First, Micro Motors Inc.'s (MMI's) expansion scenario: MMI received an inquiry from Saito Industries to supply multimillion pieces of motors. Should MMI shift to a mass-production company? Second, MMI's restructuring scenario: small high-tech entrepreneurs do not always face lucky situations. If we face a serious financial crisis, how can we solve it? This fictitious story may encourage you to deeply consider unlucky situations. The business strategy should be submitted to the corporate board of directors and needs to obtain permission from the board.

An example format for the business strategy is discussed in Section 11.5.

List of Contents

Executive Summary
Section 1: Background of the Strategic Planning
- 1.1 Necessity of SWOT analysis
- 1.2 Corporation description and the researcher's standpoint

Section 2: External Environment Analysis
- 2.1 Remote environment
  - 2.1.1 Technology maturity
  - 2.1.2 Component market
  - 2.1.3 Application device market
  - 2.1.4 Environmental regulations
- 2.2 Proximate environment
  - 2.2.1 Rivalry among competing firms
  - 2.2.2 Potential entry of new competitors
  - 2.2.3 Potential development of substitute products
  - 2.2.4 Bargaining power of suppliers
  - 2.2.5 Bargaining power of consumers

Section 3: Internal Environment Analysis
- 3.1 Managerial orthodoxy
  - 3.1.1 Mission and mission statement
  - 3.1.2 Formal organization
  - 3.1.3 Informal organization
  - 3.1.4 Authority system
  - 3.1.5 Managerial skills
- 3.2 Operations
  - 3.2.1 General corporation operations
  - 3.2.2 Mass-production division operations proposal
- 3.3 Financial situation
  - 3.3.1 Present financial situation
  - 3.3.2 Production division financial estimation
- 3.4 Members
- 3.5 Marketing
  - 3.5.1 R&D division
  - 3.5.2 Sales division
  - 3.5.3 New production division

Section 4: Strategic Position

11.5 Business Strategy Case Study I: MMI's Expansion

11.5.1 Shall We Shift to a Mass-Production Company? Executive Summary

MMI was founded by Barb Shay as a spin-off company from the State University of Pennsylvania (SUP), financially sponsored by Cheng Kung Corporation, Taiwan. Barb, who is the founder, vice president, and CTO of MMI, is well known as one of the active researchers of piezoelectric actuators. She has been developing various piezoelectric actuators and USMs over the past 10 years. Recently, her group developed a micromotor called metal tube type consisting of a metal hollow cylinder and two PZT rectangular plates. This development was the trigger for founding MMI. In collaboration with Saito Industries in Japan, MMI further developed the world's smallest camera module with both optical zooming and auto focusing mechanisms for a mobile phone application, by utilizing micromotors slightly modified from the original design.

MMI received an inquiry from Saito Industries to manufacture tens of millions of zoom/focus mechanisms for their cellular phone application. Because MMI is basically a research and development (R&D) company, that is, receiving research funds from a client company and creating prototype products for that company's sake, MMI hesitates to accept 10-million-piece manufacturing requests.

This SWOT analysis-based strategic plan provides sufficient information regarding the decision of whether MMI should shift to a mass-production company. Although the request from Saito is an attractive opportunity to expand MMI, because MMI has strength in its manufacturing technology (strength because this is actually their product developed at SUP and MMI), MMI may exhibit weaknesses in the manpower and financial resources required to set up a factory by itself. The monthly operating cost of more than 10 times the present expenditures may initiate a serious cash flow problem, leading to the bankruptcy of MMI (a threat). Even if MMI started production successfully, other large competitors may chase them with similar or alternative products in a couple of years, which is an external threat.

Supposing that you are a consultant to MMI, consider the best advice for them. MMI is facing at present a decision among three choices: (1) to start mass production in collaboration with their partner company, Cheng Kung; (2) to decline this manufacturing request; or (3) to transfer this manufacturing task to a competitive company with some back margin (1% of the sales cost). The SWOT method was adopted to analyze MMI's external and internal environments in terms of this zoom/focus mechanism production, including financial analysis (investment, cash flows) and production planning (how many products at which factory) with the linear programming tools.

In order to meet Saito's request, MMI will use Cheng Kung as an OEM. The tape-cast automation manufacturing systems will be installed in Cheng Kung's Taiwan factory, and low-cost labor in Chiang Mai's Thailand factory will be used for the manual cut-and-bond manufacturing process.

A detailed monthly production plan and cash flow were forecast, taking into account cash receipts from Saito (2 months after the product delivery), equipment cost payments (1 month after the installation), variable manufacturing cost, net cash flow for each month, payments to a production partner Cheng Kung (in the last month of each purchase order [PO] period), and cumulative cash balance at MMI. As noted in the details provided in the main part of the plan, the largest negative cumulative cash balances, –$218,000 13 months after the contract, may be rather critical to the MMI financial situation. However, the following financial backups may cover this deficit: (1) MMI has a cash balance of $100,000 at present, and (2) the partner, Cheng Kung, can help MMI financially by waiting for the payment.

Including the above financial evaluation, a complete SWOT analysis can be made for both external and internal environments, in particular for the mass production of camera modules for Saito. In order to numerically evaluate the strengths–weaknesses and opportunities–threats, new parameters—S/W score and O/T score—are as follows:

$$\text{S/W score} = [(\text{number of strengths}) - (\text{number of weaknesses})] \div [(\text{number of strengths}) + (\text{number of weaknesses})] \quad (11.1)$$

$$\text{O/T score} = [(\text{number of opportunities}) - (\text{number of threats})] \div [(\text{number of opportunities}) + (\text{number of threats})] \quad (11.2)$$

Under the supposition of a SWOT analysis category weight, the total S/W and O/T scores are calculated as total S/W score = 0.615 and total O/T score = 0.065.

With these numbers, MMI's position can be assigned as strategic position I (offensive), although rather close to strategic position II (diversification). Regarding the mass production of camera modules for Saito, MMI has a good position (score = 0.615) in its internal environment, but is almost neutral (score = 0.065) in its external environment. The two key reasons for threats (negative numbers) are (1) Saito's strong price-setting power, and (2) lead compound regulation in the future.

In conclusion, moving forward to mass production of camera modules for Saito Industries is highly recommended without hesitation. Even if the PO termination were to happen in the middle of the 2-year agreement period, the exit plan evaluates that there would not be any deficit or risk to MMI.

11.5.2 Background of This Strategic Planning

11.5.2.1 Necessity of SWOT Analysis

The company's introduction and its products—metal tube and multilayer-type USMs—are omitted here because you already learned about these in Chapters 2 and 6.

As mentioned previously, this SWOT analysis-based strategic plan provides sufficient information regarding the decision of whether MMI should shift to a mass-production company. The SWOT method will be adopted to analyze the external and internal environments of MMI in terms of this zoom/focus mechanism production, including financial analysis (investment, cash flows) and production planning (how many products at which factory) with the linear programming tools learned in Chapter 7. A recommendation will be provided for decision making in this report.

11.5.3 External Environment Analysis

MMI's mission is "Your Development Partner," by supporting technology development (R&D division) and by providing commercial devices (sales division) to the client, federal institutes, and private companies. Therefore, the analysis on the external environment should be made prior to considering the internal environment.

11.5.3.1 Remote Environment

After 30 years of intensive research and development of piezoelectric actuators, the focus has drastically shifted to applications. Piezoelectric shutters (Minolta Camera) and automatic focusing mechanisms (Canon) in cameras, dot-matrix printers (NEC), and ink-jet printers (Epson), piezoelectric part-feeders (Sanki), and diesel injection valves (Siemens) have been commercialized and mass produced on a scale of several tens of thousands to millions of pieces per month. Throughout this period of commercialization, especially over the last decade, new actuator designs and methods of drive and control have been developed to meet the requirements of the latest applications. At a somewhat slower pace, advances in device reliability and strength have likewise occurred, although some considerable work is yet to be done in further extending the lifetime of devices and ensuring consistent performance over that period. An overview of these recent trends in development will be presented in this section.

11.5.3.1.1 Technology Maturity

Please refer to Chapter 8, Section 8.1.3, "Marketing Research Example" in order to complete the Business Strategy Report.

In general, taking into account the difficulty in overcoming the reliability requirements such as operating temperature range and lifetime, we engineers recognize the development pecking order in a sequence, such as camera > office equipment > automobile. Because the MLA has already been utilized for the diesel injection valve, we could say that piezoelectric actuator development has already gone into a maturing period. Further, we can say that the industrial competition has started to share the market by reducing the actuator price drastically, which is not only related to technological development, but also to the industrial managerial decision [10]. In order to calculate the O/T score and S/W score, we will use the following scoring table for each category. The following table suggests that in terms of technology maturity, MMI receives 1 point in "Opportunities."

External Environment Analysis	*Opportunities*	*Threats*
Remote environment		
Technology maturity	Technology is maturing already	None

11.5.3.1.2 Piezoelectric Actuator/Ultrasonic Motor Market

Please refer to Chapter 8, Section 8.1.3, "Marketing Research Example" in order to complete the Business Strategy Report. This part was also widely cited from my textbook, *Micromechatronics* [10].

To sum up, of the estimated annual sales from these prospective markets alone, approximately $500 million came from the sale of individual ceramic actuator elements, about $300 million from camera-related applications, and roughly $150 million was related to the sale of USMs. A total of nearly $1 billion is anticipated. Mr. Sekimoto's prediction of ceramic actuator sales amounting to nearly $10 billion in the near future may in fact be realized! At the very least, the current trends suggest a bright financial future for the ceramic actuator industry.

Figure 11.11 illustrates the product life cycle and/or industry growth curve [11]. The piezoelectric actuator industry is still in the preliminary growth stage, expecting much steeper growth and a significant increase in income. In addition to the current applications in information technologies (microactuators for computers, cellular phones), robotics, and precision machinery (positioners), piezoelectric actuators will provide supporting tools for biomedical equipment, nanotechnologies (nanomanipulation tools), and energy areas (micropumps of fuel cell systems). Thus, the market should significantly increase from the current annual $1 billion to $10 billion (or even higher) in 5 to 7 years.

External Environment Analysis	*Opportunities*	*Threats*
Remote environment		
Piezoelectric actuator market	Beginning of growth stage Billion $ revenue expectation	None

11.5.3.1.3 Camera Module Market

The number of mobile phones in the world will reach 1 billion by 2010. Eighty-seven percent of them will feature a compact camera, an amazing percentage. Even if MMI starts camera module production at a rate of 10 million pieces per year for Samsung, taking into account that the required total module number is around 700 million, our contribution is less than 1.4% of the

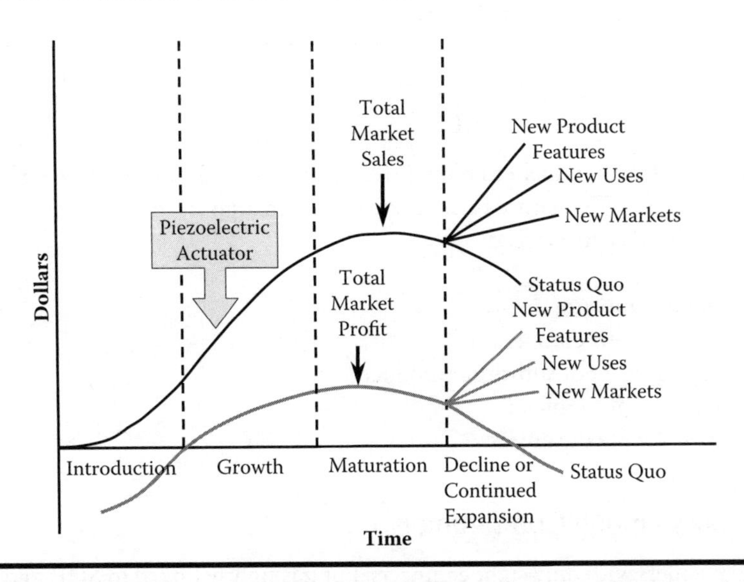

Figure 11.11 The product life cycle and/or industry growth curve. The piezoelectric actuator industry is still in the preliminary growth stage.

demand. After finishing their 2-year contract with Samsung, MMI may find the next customer/client in the mobile phone production area.

External Environment Analysis	*Opportunities*	*Threats*
Remote environment		
Camera module market	Rapid growth in cell phones More significant growth in camera phones	ME motor may be a substitute

11.5.3.1.4 Environmental Regulation

As previously mentioned, the twenty-first century has been called the "century of environmental management." In 2006, the European community implemented restrictions on the use of certain hazardous substances—the RoHS Directive—which explicitly limits the usage of lead in electronic equipment. RoHS could be a significant threat to piezoelectric companies, which have only PZT piezo-ceramics. However, because ICAT/Penn State is already studying (Na,K)(Ta,Nb)O_3-based piezo-ceramics extensively, once strict regulations start, MMI will change the piezo-materials and develop similar piezo-devices in the market.

External Environment Analysis	*Opportunities*	*Threats*
Remote environment		
Environmental regulation	None for module production	PZT may be regulated in future

11.5.3.2 Proximate Environment

Michael Porter's five-forces model of competitive analysis is widely used for developing external environmental strategies. The nature of competitiveness in a given industry can be reviewed as a composite of the following five forces:

1. Rivalry among competing firms
2. Potential entry of new competitors
3. Potential development of substitute products
4. Bargaining power of suppliers
5. Bargaining power of consumers

11.5.3.2.1 Rivalry among Competing Firms

MMI is a small business with only eight engineers, but it is tightly linked to SUP, and Cheng Kung Corporation in Taiwan in terms of technological, financial, and mass-production strengths.

As a small firm, there are several competing firms in the United States and Europe. The key to competition against small firms is differentiated high technology. Because MMI has all sets of technologies—materials, devices, and drive/control systems—they do not have particular weaknesses against these small businesses. KLM Group is their strongest competitor. However, it is very doubtful that KLM Group will move into cellular phone camera module production. DEF is a direct competitor to MMI in USMs. However, because they are also a licensee of MMI's patent (metal tube type), and Cheng Kung is its major supplier of piezo-ceramic plates, MMI may use DEF as an OEM manufacturing partner. If MMI moves into camera module manufacturing, it will need to compete with big firms.

The key to competition against large firms is production cost effectiveness. Because MMI has a strong partner, Cheng Kung, as an OEM facility, mass-production capability already exists. Further, because Cheng Kung has a manufacturing factory in Chiang Mai, Thailand, in addition to a factory in southern Taiwan, the manufacturing cost (especially labor cost) is competitive even with mainland China. We can conclude that rivalry among competing firms is therefore not very important to MMI.

External Environment Analysis	*Opportunities*	*Threats*
Proximate environment		
Rivalry among industry		No particular threat

11.5.3.2.2 Potential Entry of New Competitors

There were many new entrants in the piezoelectric actuator field in Japan from the middle of the 1980s until the middle of the 1990s, when Japanese industries were enjoying a so-called bubble economy. Many materials companies, including metal, steal, polymer, and chemical manufacturers, started to join the fine- or advanced-ceramic industry, and to work on piezoelectric actuators. However, after the economic recession in the middle of the 1990s, more than half of the new entrants abandoned the piezo-actuator division, because commercialization success was not easy for them. The main reason for this high entrance barrier is the necessity of a long accumulation of manufacturing know-how. Actuator applications, unlike sensors, require especially high quality and reliability of piezo-ceramics under high electric field conditions, which the new entrants could not initially imagine. Thus, no potential new entry can be expected at present.

On the contrary, in the United States we can find some new entrants, originally spun off from university research centers. However, most of them are application-oriented or device/system companies, and none of them has ceramic manufacturing capability. Thus, no adversary can be found in the United States in the mass production of the camera modules.

External Environment Analysis	*Opportunities*	*Threats*
Proximate environment		
New competitors		No particular threat

11.5.3.2.3 Potential Development of Substitute Products

Piezoelectric actuators were originally introduced to replace EM motors. In this sense, the strongest competitor or substitute is still an EM motor. The efficiency of EM motors drastically decreases with a decrease in power and size, while piezoelectric motors are not as sensitive to the level of power and size. The key reason for efficiency degradation in EM motors is the inevitable heat generation via Joule heat of the thin copper coil wires.

The advantages of piezo-motors over EM motors are summarized below:

- More suitable to miniaturization
  - Stored energy is larger than that of an EM type
  - 1/10 smaller in volume and weight
- No electromagnetic noise generation
  - Magnetic shielding is not necessary
- Higher efficiency
  - Efficiency insensitive to power size
- Nonflammable
  - The piezo-device is safer for overload or short-circuit at the output terminal

In conclusion, piezo-motors have advantages over EMs in performance. The only disadvantage is their higher cost, because of the generally small manufacturing quantity.

External Environment Analysis	*Opportunities*	*Threats*
Proximate environment		
Substitute products (against EM motors)	Significant advantages in performance Piezo is expanding the share	Possible high manufacturing cost

11.5.3.2.4 Bargaining Power of Suppliers

The raw materials of piezoelectric actuators include the following:

- PZT ceramic raw materials—PbO, ZrO_2, TiO_2, and other oxides
- Electrode materials—Ag/Pd, Pt
- Elastic materials for mechanical amplification and/or protection—brass, steel, etc.
- Electric lead wires—Cu, Ag

Except for Ag/Pd and Pt, the raw materials are generally much cheaper than EM motors. Only the rare-metal electrode materials are expensive (i.e., not negotiable, unfortunately).

External Environment Analysis	*Opportunities*	*Threats*
Proximate environment		
Power of suppliers		No particular threat

11.5.3.2.5 Bargaining Power of Consumers

The bargaining power of consumers is the largest threat to piezoelectric material/device manufacturers. The consumers are usually the biggest electric/electronics companies, such as GE, IBM, Apple, Toshiba, Sony, and Samsung; they decide the purchase price and initiate negotiations. The

Table 11.2 Expected Purchase Order from Samsung EM to MMI in 96 weeks (2 years)

PO No.	*1*	*2*	*3*	*4*	*5*	*6*	*7*	*8*
Week	1–12	13–24	25–36	37–48	49–60	61–72	73–84	85–96
Quantity	10 K	50 K	250 K	1000 K	2000 K	2400 K	2700 K	3000 K
Requested unit price	$10.00	$5.00	$2.80	$1.80	$1.20	$1.10	$1.00	$1.00
Total payment	$100 K	$250 K	$700 K	$1800 K	$2400 K	$2640 K	$2700 K	3000 K

component companies, such as MMI, are always "price takers," and need to improve the manufacturing process and reduce the manufacturing cost to generate profit.

Table 11.2 shows an example of Saito Industries' PO for camera zoom/focus modules requested from MMI. The production quantity increases dramatically, up to 3 million pieces per quarter. To the contrary, the unit cost decreases continuously from $10 to $1 over 2 years, so that the total sales amount starts to saturate in the last half of the contract period.

External Environment Analysis	*Opportunities*	*Threats*
Proximate environment		
Power of consumers		MMI is price taker against Saito Industries

11.5.4 Internal Environment Analysis

11.5.4.1 Managerial Orthodoxy

Managerial orthodoxy is useful to describe how well an organization is managed [12]. We will adopt this classification in this section.

11.5.4.1.1 Mission and Mission Statement

MMI's mission statement is "Your Development Partner," because they support technology development (R&D division) and provide commercial devices (sales division) to their clients, federal institutes, and private companies. MMI will also provide special and customized products and services. The purpose of this strategic plan is to consider a new division—a manufacturing division. The mission statement need not be revised because of this mass-production work task. MMI still maintains the position, "Saito's Development Partner." However, because MMI has been an R&D-oriented, small-scale manufacturer up to this point, once MMI starts 10-million-piece production of the piezo-components it may need to restructure by adding at least a production control manager and test engineers for the products.

11.5.4.1.2 Formal Organization

The MMI organization chart is shown in Figure 11.12. There are three divisions: R&D, sales, and office management. Once MMI starts mass production of camera modules, an additional division—manufacturing—may be created.

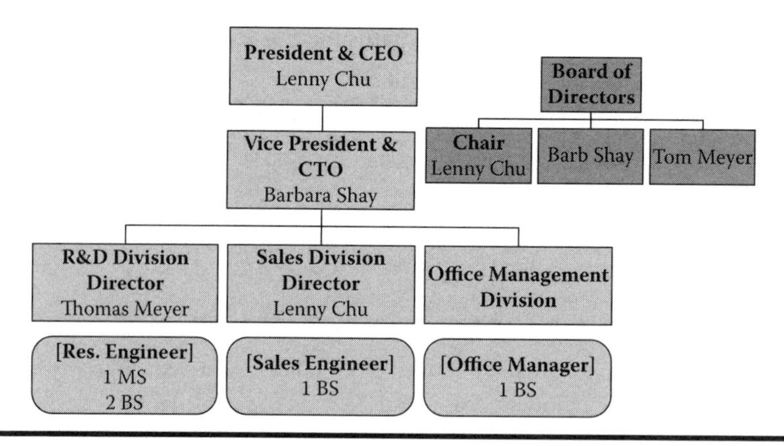

Figure 11.12 MMI organization chart (second year).

11.5.4.1.3 Informal Organization

Because MMI does not want to create a "two bosses dilemma" in the organization, Vice President Shay gives orders to the subordinates directly, with permission from President Chu. Shay's status in the company is special: because she is the founder, and Chu is not always in College Park, she is practically operating MMI.

The informal organization of MMI also includes Cheng Kung Corporation in Taiwan. Cheng Kung is a major stockholder and an OEM to MMI, which makes the mass production of camera module parts possible without significant investments for purchasing the factory and equipment.

11.5.4.1.4 Authority System

Vice President Shay has authority over daily operations, but President Chu represents the firm officially, and advises Shay occasionally. The organization in Figure 11.12 shows a reasonably expanding vertical and horizontal structure.

11.5.4.1.5 Managerial Skills

Vice President Shay has a physical restriction for operating MMI: because she is a professor at SUP 75% of her time, she can spend only 25% of her time at MMI (only 2 hours per day). Although President Chu is not always in College Park, because he spends some of his time at the Taiwan Cheng Kung office, he is always monitoring MMI's corporate account status. Further, because Tom Meyer has small-business experience and is capable as the R&D director, Shay's managerial handicaps can be compensated.

Internal Environment Analysis	*Strengths*	*Weaknesses*
Managerial orthodoxy	Strong mission statement Balanced combination of formal and informal organizations Consistent authority system Excellent managerial skills	None

11.5.4.2 Operations

11.5.4.2.1 General Corporation Operations

MMI has already spent $660,000 over revenue in the initial 2 years, in order to keep eight employees for start-up; $330,000 was the initial capital, and $330,000 was obtained as a loan from the partner company, Cheng Kung Corporation. The current monthly running cost is about $40,000 (in the third year), which will be maintained in this year. Their revenue at the end of the third year reached $514,000 from the R&D programs and $114,000 from the piezo-product sales, dramatically improving the financial situation, and leaving $113,000 in cash at present.

11.5.4.2.2 Mass-Production Division Operations—Proposal

MMI provided the MLA components to Saito Industries, initially with only 1000 pieces for their test purpose, which were manufactured at MMI using SUP facilities. Once this test is finished, much larger production will be requested, finally reaching more than 10 million pieces per year. MMI needs to develop a production plan for supplying actuators to Saito Industries.

The customer's demands include the following: (a) the supply quantity will be exponentially increased within 2 years, and (b) the unit price will be drastically decreased with the quantity. As mentioned previously, MMI has three manufacturing factories: MMI in Pennsylvania (with no tape-cast facility); a partner company, Cheng Kung factory in Taiwan (with one tape-cast facility); and Chiang Mai factory in Thailand (with no tape-cast facility). Labor cost is drastically decreased in a sequence of Pennsylvania, Taiwan, and Thailand. Taking into account the transportation period, which can be counted as a manufacturing delay, MMI will distribute the manufacturing load to these three places to minimize the manufacturing cost. The manufacturing factory locations of these three facilities are shown in Figure 1.14, in Chapter 1.

An OEM agreement was set between MMI and Cheng Kung with the following four conditions: (1) MMI will provide all the raw materials to both Cheng Kung Taiwan and Chiang Mai factories, shipped from College Park, Pennsylvania. (2) All the direct labor wages will be paid by MMI. (3) When required, MMI will purchase tape-cast equipment for Taiwan factory without delay. (4) After each PO period (3 months), MMI will share half of the gross profit with Cheng Kung.

11.5.4.2.2.1 Tape-Casting Facility versus Manual Cut-and-Bond Production

Please refer to Practical Exercise Problem P7.1 in Chapter 7.

11.5.4.2.2.2 Constraints from the Customer (Saito Industries)

Saito asks MMI for a 12-week delivery for each PO. The requested MLA quantity and the MLA product unit price are summarized in Table 11.2. Because MMI is much smaller than Saito, MMI is merely a price-taker with only a slight negotiation possibility.

11.5.4.2.2.3 Manufacturing Capability

Raw materials, such as piezoelectric ceramic powder and Ag/Pd metal electrode paste, are synthesized at MMI's laboratory in Pennsylvania and can be transported to MMI's factory in Pennsylvania with no delay, to Taiwan Cheng Kung with a 1-week delay for one way, and/or to Chiang Mai factory with a 2-week delay for one way. For both a tape-cast machine and a human production line, one production batch can be finished biweekly. So, during the 12-week PO period, MMI in Pennsylvania can run six production batches. However, due to the transportation time loss, Taiwan Cheng Kung and Chiang Mai factory can run five and four production batches over 12 weeks. Only Taiwan Cheng Kung has a tape-cast facility. Thus, the other factories need to purchase the necessary number of tape-cast apparatuses, with increasing production quantity: \$300,000 per set in the United States and Thailand, but \$250,000 in Taiwan, which may provide a privilege to the Taiwan factory. One tape-cast apparatus can manufacture a maximum of 42,000 multilayer pieces per biweekly batch. Regarding the workers, MMI in Pennsylvania, Taiwan Cheng Kung, and Chiang Mai factory can hire a maximum of 60 people (10 lines), 240 people (40 lines), and 1200 people (200 lines), respectively, due to factory space. Six workers will make one manufacturing line, which can manufacture an average of 2100 pieces per biweekly batch. The average biweekly salary for one manufacturing line (six workers) is \$6250, \$1562, or \$312 for the United States, Taiwan, or Thailand, respectively. Table 11.3 summarizes the necessary data for the three manufacturing plants.

MMI should, in general, use the cheapest labor in Thailand as much as possible. However, we need to take into account the freight cost and the manufacturing time loss due to the transportation.

11.5.4.2.2.4 Multilayer Actuator Production Plan

In Chapter 7, Practical Exercise Problem P7.1 describes the detailed linear programming model for the cost minimization production plan for the multilayer USM fabrication. The results are summarized in this section. Figure 11.13 shows a graph of expected PO from Saito to MMI in these 96 weeks (12 weeks for each PO period), visualized from Table 11.2. The MLA unit price drops dramatically, from the initial \$10 skimming price down to \$1, with an increase in purchase quantity from 10,000 to 3 million pieces. Thus, total sales amount (revenue) increases more slowly than the quantity.

Figure 11.14 summarizes the number for new equipment purchase and line-workers for the three locations: MMI in Pennsylvania, Cheng Kung in Taiwan, and Chiang Mai factory in Thailand. Because Taiwan Cheng Kung has one tape-cast apparatus, MMI does not need to purchase additional apparatus until the sixth PO period. One, and two new sets should be purchased in the sixth, seventh, and eighth PO periods, respectively. Note that all purchases should be made

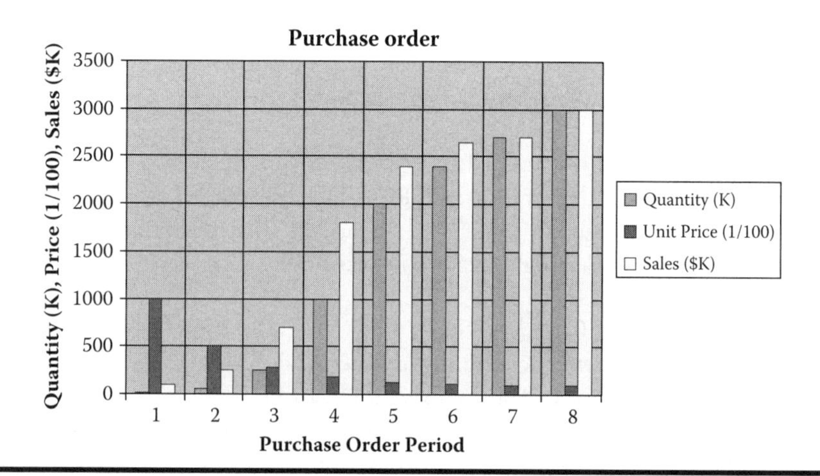

Figure 11.13 Graph of an expected purchase order from Saito to MMI over 96 weeks (2 years). Note that in the unit price scale, the quantity for each purchase order period divided by 100 equals the price per unit.

Table 11.3 Estimation Sheet for the Manufacturing Capability of Three MMI Facilities

Manufacturing Plant		*MMI, Pennsylvania*	*Cheng Kung, Taiwan*	*Chiang Mai, Thailand*
	Maximum number of working rotation	6	5	4
Tape-cast apparatus	Production capability per system (pieces biweekly)	42,000	42,000	42,000
	Total production capability per system (pieces during one purchase order period)	252,000	210,000	168,000
	Price per system	$300,000	$250,000	$300,000
	Number of present apparatuses	0	1	0
	Maintenance fee per system (biweekly)	$8000	$2000	$400
	Total maintenance fee per system (during one purchase order period)	$48,000	$10,000	$1600
Manual process	Biweekly salary for one manufacturing line (6 workers)	$6250	$1562	$312
	Total salary for one manufacturing line during one purchase order period)	$37,500	$7810	$1248
	Production capability per one line (pieces biweekly)	2100	2100	2100
	Total production capability per one line (during one purchase order period)	12,600	10,500	8400
	Maximum workers/ maximum lines	60/10	240/40	1200/200
Transportation	Transportation time required (weeks)	0	2	4

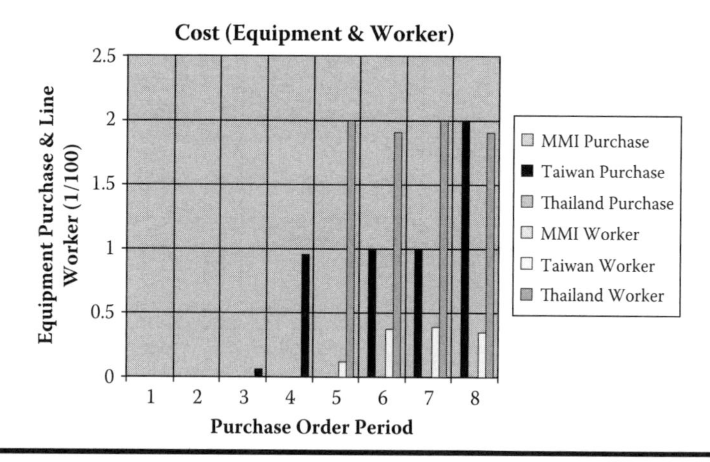

Figure 11.14 Numbers for new equipment purchases and line workers for three locations, MMI at Pennsylvania, Taiwan Cheng Kung, and Chiang Mai Factory.

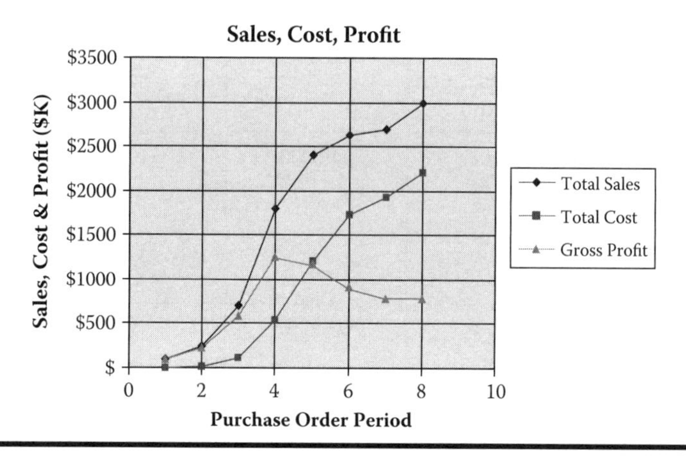

Figure 11.15 Total sales, total manufacturing cost, and gross profit (sales–cost) plotted in PO period sequence.

through the Taiwan factory. As was mentioned already, this is due to the lower purchase cost of the tape-cast system in Taiwan. Regarding the line workers, Chiang Mai is the highest priority. Once the maximum hiring allowance of 200 has been reached, the second choice, Taiwan, will start to hire workers.

Figure 11.15 shows the total sales, total manufacturing cost and gross profit (sales—cost) as a function of the PO period sequence. Because there is a tape-cast apparatus in the Taiwan factory initially, they do not need to incur manufacturing cost until PO period 3. Thus, the maximum gross profit will be reached around PO period 4. Because MMI needs to continuously spend $250,000 for the equipment, due to increasing PO quantities, the gross profit decreases after the maximum PO period 4. The gross profit finally saturates around $800,000 per PO period (PO periods 7 and 8), which corresponds to a profit/sales rate of 26%.

Internal Environment Analysis	*Strengths*	*Weaknesses*
Operations	An OEM partner, Cheng Kung Two factories in Taiwan and Chiang Mai, Thailand In the MMI/Cheng Kung production capability	None

11.5.4.3 *Financial Situation*

11.5.4.3.1 Present Financial Situation

The start-up capital of $330,000 was collected from entrepreneur Barb Shay, her friend Jackie Wang, and the Cheng Kung Corporation. Further, MMI has a loan agreement of $330,000 for the first 2 years with the partner company, Cheng Kung in Taiwan, which should be paid back in 5 years with an annual interest rate of 8%. As mentioned in the previous section, MMI has an ending cash balance of $100,000 at present.

In addition, MMI's primary bank loan (interest rate of 8%) was set for flexible cash flow up to $50,000 at present (increased up to $200,000 next year), for covering emergency financial deficits for 90 days, which may occur from government or client payment delays. This short-term loan should be paid off as soon as MMI receives the contract money.

11.5.4.3.2 Mass-Production Division Financial Estimation

Rapidly expanding sales causes intense pressure for inventory and accounts receivable buildup, draining the cash resources of the firm. Although Figure 11.15 suggests promising gross profits for MMI, the monthly cash flow should be carefully forecasted to estimate the cash to be loaned from our partner company without making a fatal cash deficit. Another important factor is the unexpected termination of the PO during the 2 years.

11.5.4.3.2.1 Monthly Cash Flow

We adopt here the following assumptions:

1. MMI will use Cheng Kung as an OEM. All necessary manufacturing costs (direct materials, direct labor, and tape-casting equipment) will be provided directly by MMI. Net cash flow during each PO period (3 months or 12 weeks) will be shared equally with Cheng Kung.
2. All purchases by Saito Industries will be credit. Payment from Saito will be due 8 weeks after the delivery. For example, for PO period 2, the products should be delivered by the end of 24th week (sixth month), but the cash will come in the 32nd week (eighth month).
3. On the contrary, MMI's payment for new tape-cast equipment should be made 4 weeks after the purchase. For example, for PO period 6, a new tape-casting system should be installed in the beginning of the 61st week, and the payment will be made in the 64th week.
4. The variable manufacturing cost (direct materials, direct labor, and manufacturing overhead costs) should be paid in the month of purchase without delay.

Figure 11.16 and Table 11.4 are monthly summaries of cash receipts from Saito (2 months after the product delivery), equipment cost (1 month after the installation), variable manufacturing cost, net cash flow for each month, payment to Cheng Kung (in the last month of each PO period), and cumulative cash balance at MMI. Notice the largest negative cumulative cash balances, –$74,100 in month 4a and –$218,000 in month 5a, which may be rather critical to MMI's finances.

However, remember MMI has the following financial backups:

1. MMI has an ending cash balance of $100,000 at present.
2. The partner, Cheng Kung Corporation, may wait for the profit share payment in month 4c, so that the cash flow deficit can be avoided by the cash receipt in month 5b.

After completing the 2-year manufacturing contract, MMI can sell the used tape-casting equipment to Cheng Kung at discount prices, which will create additional cash. Of the four newly purchased machines, one may be sent to the MMI College Park facility. However, due to space limitations and the lack of need for larger manufacturing capability, the remaining three will be sold to Cheng Kung with an exponential depreciation rate (1- and 2-year-old equipment by one-half and one-quarter of the original price, respectively).

11.5.4.3.2.2 Exit Plan

There is no risk of PO contract termination by Saito Industries during these 2 years, because the cash flow in the last month of each PO period, such as month 4c or 6c in Figure 11.16 exhibits all positive numbers, or positive gross profit. Purchase of tape-cast equipment is made at the beginning of each PO period and will be compensated by payment at the end of each PO period.

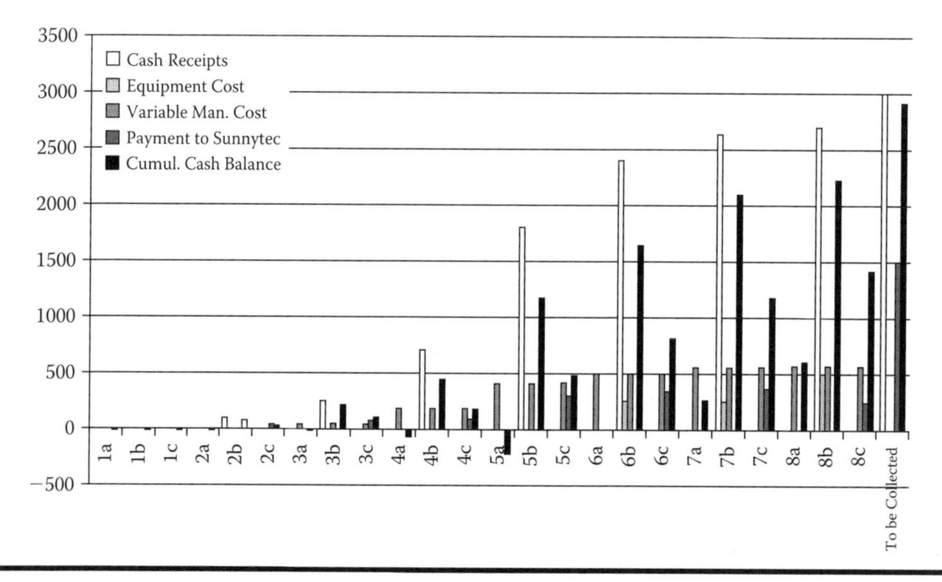

Figure 11.16 Monthly summary of cash receipts, equipment costs, variable manufacturing costs, payment to Cheng Kung, and cumulated cash balance at MMI.

Table 11.4 Monthly Summary of Cash Receipts, Equipment Cost, Variable Manufacturing Cost, Net Cash Flow, Payment to Cheng Kung, and Cumulative Cash Balance at MMI

Month	*1a*	*1b*	*1c*	*2a*	*2b*	*2c*	*3a*	*3b*	*3c*	*4a*	*4b*	*4c*	*5a*	*5b*	*5c*	*6a*	*6b*	*6c*	*7a*	*7b*	*7c*	*8a*	*8b*	*8c*	*To be collected*
Cash receipts					100			250			700			1800			2400			2640			2700		3000
Cash payments																									
Equipment cost																	250			250			500		
Variable man. cost	1.7	1.7	1.7	7	7	7	37	37	37	178	178	178	405	405	405	493	493	493	558	558	558	571	571	571	
Net cash flow	–1.7	–1.7	–1.7	–7	93	–7	–37	213	–37	–178	522	–178	–405	1395	–405	–493	1657	–493	–558	1832	–558	–571	1629	–571	
Payment to Cheng Kung			0			39.5			69.5			83			293			336			358			244	1500
Cumulative cash balance	–1.7	–3.4	–5.1	–12.1	80.9	34.4	–2.6	210.4	104	–74.1	448	187	–218	1177	478	–14	1643	815	257	2089	1173	602	2231	1416	2916

Internal Environment Analysis	*Strengths*	*Weaknesses*
Financial situation	Backup plan: cash in bank ($100,000) and request for payment wait to Cheng Kung	Large deficit in 4th and 5th PO period
	Exit plan seems to be no problem	

11.5.4.4 Members

MMI has eight employees: Barb Shay is a young, active vice president and CTO who supervises all MMI operations; President Lenny Chu is an experienced businessperson with vast accounting and financial knowledge; R&D Director Tom Meyer is a self-motivated and reliable engineer; other research and sales engineers are also highly skilled.

The key to personnel selection is self-motivation. Certain research norms are assigned to research engineers on a weekly basis. They may finish in less than the regular 40-hour work week or need to work for 60 hours or more per week. On the other hand, MMI introduced incentives to sales engineers, in proportion to their sales amount above a certain threshold. MMI will hire a new manager/sales engineer and a research engineer for a new production division, both of whom will work directly with Saito Industries.

Internal Environment Analysis	*Strengths*	*Weaknesses*
Members	Leadership by a young CTO	Lack of total employee number
	High-skilled employees	
	Self-motivated employees	

11.5.4.5 Marketing

11.5.4.5.1 R&D Division

The R&D division targets federal agencies such as the Defense Advanced Research Projects Agency (DARPA), Navy, Army, National Institutes of Health (NIH), and Telemedicine and Advanced Technology Research Center (TATRC) for obtaining research funds (from $3 billion funds in total). Although MMI does not presently have particular competitors in micromotors, piezo-transformers, piezoelectric energy harvesting, and microrobot fields (a strength), because of the high growth potential, multiple competitors will appear in the next 3 to 4 years, which may be a possible risk (a threat).

11.5.4.5.2 Sales Division

The sales division has an all-in-one concept, like e-Bay, which is different from other competitive distributors (a strength). The customers for the piezoelectric products are federal institutes and universities that are working in the IT, biomedical, and energy areas. MMI will advertise in three ways: their Web site, direct e-mail/mail to the customers, and customer site visits.

11.5.4.5.3 New Production Division—Camera Module

This division will be created to work for Saito Industries. Not many efforts are required at present for marketing, because this division already has Saito as a large customer. Once production with Saito is terminated after 2 years, MMI may face a problem.

Internal Environment Analysis	*Strengths*	*Weaknesses*
Marketing	Unique R&D technology	Existence of chasing companies
	Unique sales policy	Termination of Saito contract
	Rapid growth in sales amount	
	MMI has already a large customer, Saito	

11.5.5 Strategic Position

The SWOT analysis for external and internal environments in terms of a new mass-production division of camera modules is summarized in Table 11.5. The SWOT analysis is further quantitatively made by using a strategic position grid. There are several strategic position grids, but the one developed by McIlnay [13] will be used here. McIlnay's strategic position grid uses a horizontal axis to depict the strengths and weaknesses in an organization's internal environment and a vertical axis to illustrate the opportunities and threats in its external environment (see Figure 11.17).

Table 11.5 Summary of the SWOT Analysis for External and Internal Environments in Terms of a New Mass-Production Division of Camera Modules

External Environment Analysis	*Opportunities*	*Threats*
Remote Environment		
Technology maturity	Technology is maturing already	None
Piezo-actuator market	The beginning of growth stage	None
	Billion $ revenue expectation	
Camera module market	Rapid growth in cell phones	ME motor may be a substitute
	More significant growth in camera phones	
Environmental regulation	None for module production	PZT may be regulated in future
Proximate environment		
Rivalry among industry		No particular threat

(Continued)

Table 11.5 Summary of the SWOT Analysis for External and Internal Environments in Terms of a New Mass-Production Division of Camera Modules *(Continued)*

External Environment Analysis	*Opportunities*	*Threats*
New competitors		No particular threat
Substitute products (electromag. motors)	Significant advantages in performance	Possible problem is high manufacturing cost
	Piezo is expanding the share	
Power of suppliers		No particular threat
Power of consumers		MMI is a price taker against Saito
Internal environment analysis	*Strengths*	*Weaknesses*
Managerial orthodoxy	Strong mission statement	None
	Balanced combination of formal and informal organizations	
	Consistent authority system	
	Excellent managerial skills	
Operations	An OEM partner, Cheng Kung	None
	Two factories in Taiwan and Chiang Mai, Thailand	
	In the MMI/Cheng Kung production capability	
Financial situation	Backup plan: cash in bank ($100K) and payment postpone request for Cheng Kung	Large deficit in 4th and 5th PO periods
	Exit plan seems to be fine	
Members	Leadership by a young CTO	Lack of total employee number
	Highly-skilled employees	
	Self-motivated employees	
Marketing	Unique R&D technology	Existence of chasing companies
	Unique sales policy	Termination of Saito contract
	Rapid growth in sales amount	
	MMI already has a large customer, Saito	

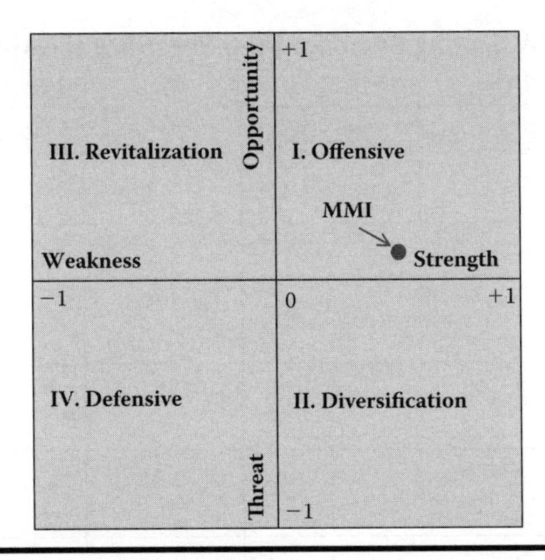

Figure 11.17 The McIlnay strategic position grid. MMI's position is in Region I.

Strategic position I is often called the *offensive* position (SO strategy), while strategic position IV is called the *defensive* position (WT strategy). Strategic position II, where a company has a positive internal environment but a negative external environment, is called the *diversification* position (ST strategy). The appropriate strategy is to spread the risk in the threatening external environment. Strategic position III, where a company has a weak internal environment but an opportunity-filled external environment, is called the *revitalization* position (WO strategy) because the appropriate strategy is a brisk turn-around to take advantage of the promising external environment.

In order to numerically evaluate the strengths–weaknesses and opportunities–threats, new parameters, S/W score and O/T score, were introduced in Section 11.5.1, Equations 11.1 and 11.2, and are repeated here:

S/W score = [(number of strengths) – (number of weaknesses)] ÷ [(number of strengths) + (number of weaknesses)]

O/T = [(number of opportunities) – (number of threats)] ÷ [(number of opportunities) + (number of threats)]

The S/W score is equal to 1 when all the analyses show strengths, while it is −1 when all the analyses show weaknesses. The number suggests the degree of strength/weakness. Similarly the O/T score varies from +1 to −1 continuously according the degree of opportunity/threat.

According to the above definitions, Table 11.5 can be converted to Table 11.6. S/W and O/T scores widely deviate from −1 to 1 depending on the analysis category. Under the supposition of category weight shown in Table 11.6, the total S/W and O/T scores can be calculated as

- Total S/W score = 0.615
- Total O/T score = 0.065

With these numbers, MMI's position can be plotted in Figure 11.17 by a solid circle, which is in strategic position I, though rather close to strategic position II. Regarding the mass production of camera modules for Saito Industries, MMI is in a good position (score = 0.615) in its internal

Table 11.6 O/T and S/W Scoring Board for MMI Regarding the Mass Production of Camera Modules

External Environment Analysis	*Opportunities*	*Threats*	*O/T Score*	*Weight*	*Total O/T Score*
Remote environment					
Technology maturity	1	0	1	0.1	0.1
Piezo-actuator market	2	0	1	0.1	0.1
Camera module market	2	1	0.33	0.2	0.066
Environmental regulation	0	1	–1	0.1	–0.1
Proximate environment					
Rivalry among industry	0	0	0	0.1	0
New competitors	0	0	0	0.05	0
Substitute products (EM motors)	2	1	0.33	0.15	0.0495
Power of suppliers	0	0	0	0.05	0
Power of consumers	0	1	–1	0.15	–0.15
			Total	1	0.0655
Internal Environment Analysis	*Strengths*	*Weaknesses*	*S/W Score*	*Weight*	*Total S/W Score*
Managerial orthodoxy	4	0	1	0.1	0.1
Operations	3	0	1	0.3	0.3
Financial situation	2	1	0.33	0.3	0.099
Members	3	1	0.5	0.1	0.05
Marketing	4	2	0.33	0.2	0.066
			Total	1	0.615

environment, but is almost neutral (score = 0.065) in its external environment. The key two reasons for threats (negative numbers) are (1) Saito Industries' strong price-setting power, and (2) lead compound regulation in the future.

11.5.6 Recommendations, Goals, and Objectives

MMI presently faces three choices: (1) to start mass production in MMI in collaboration with their partner company, Cheng Kung Corporation in Taiwan, (2) to decline this manufacturing, or (3) to transfer this manufacturing task to a competitive company with some back margin (1% of the sales cost). Choice 2 will cause no risk, but no change. Choice 3, to transfer the manufacturing task to a competitive company who is a patent licensee of the metal tube USM from SUP, may create some continuous income (total sales $13,590,000 × 1% = $135,900). However, if this competitor was successful in the manufacturing business, it would be a major competitor against MMI in the future (a big future threat).

My recommendation for MMI to follow choice 1 is made under the supposition that the MMI managers will take the offensive position in this business opportunity.

11.5.6.1 Recommendation 1: SWOT Analysis

As long as the MMI position can be assigned in strategic position I (offensive), with the total S/W and O/T scores 0.615 and 0.065, respectively, choice 1, moving forward to mass production of camera modules for Saito Industries, is highly recommended without hesitation. It is not necessary to pick choice 2 and decline the offer. Choice 3, transferring the manufacturing task with some back margin, may be a second choice when the total S/W score is lower than 0.2.

In MMI's external environment, though, there are threats such as

- Saito's strong price-setting power
- Lead compound regulation in the future

The opportunities seem to be stronger, which keeps the total O/T score positive. Unfortunately these threats cannot be solved solely by MMI, but are the destiny of a small firm. I have no recommendation on these items.

In MMI's internal environment, the situation is rather optimistic, as they have many strengths. The only weaknesses are as follows:

- Financial crises in the 37th and 49th months, that is, the largest negative cumulative cash balances, –$74,100 in the 37th month and –$218,000 in the 49th month
- Possible future competitors

11.5.6.2 Recommendation 2: Financial Crisis

The following financial backups in MMI could be useful:

- MMI has an ending cash balance of $100,000 at present.
- The partner, Cheng Kung Corporation, may wait for the profit share payment in month 4c, so that a cash flow deficit can be avoided by the cash receipt in month 5b.

There would be no risk for PO contract termination by Saito Industries during these 2 years, because the cash flow in the last month of each PO period, such as month 4c or 6c in Figure 11.16, exhibits all positive numbers (positive gross profit).

After completing the 2-year manufacturing contract, MMI can sell the used tape-casting equipment to Cheng Kung at discounted prices, which will create additional cash.

11.5.6.3 Recommendation 3: Possible Future Competitors

This is the destiny of the pioneering firm in piezoelectric actuators. The only solution is to continue with technology development in order to keep the technology leader position.

11.6 Business Strategy Case Study II: MMI's Restructuring

Barb Shay, vice president of MMI, sat in her office chair on a Friday evening, watching the snow fall outside the window. It was the beginning of MMI's third year, and she knew that during the 3 months ahead she would have to make difficult decisions regarding the future of her firm. Although she enthusiastically started MMI, the sales have not been increasing as she had anticipated in these past 2 years, and the largest SBIR program of MMI has been terminated unexpectedly.

11.6.1 Troubled Company

If your start-up company grows smoothly, you may not need to read this section [14]. Statistically, however, many venture companies will experience several serious problems—sometimes inevitable socioeconomic troubles, sometimes internal bankruptcy troubles.

Most causes of troubles or failure can be found within company management. Strategic issues include the following:

- Misunderstanding the market niche
- Mismanaging relationships with suppliers and customers
- Diversifying into an unrelated business area
- Cash flow problems due to a big project
- Lack of contingency plans

Management issues include the following:

- Lack of management skills, experience, and know-how
- Weak financial function
- High turnover in key personnel
- Accounting problems influenced by a big company

11.6.2 MMI Situation Analysis

The cash flow problem occurred at the end of second year. According to Barb's analysis, the main reasons include the following:

- The first active salesperson quit at the end of first year, and the replacement sales engineer was not very active in promoting sales. This sales engineer reduced the total sales in the second year by 30% in comparison with the first year sales amount.

- The biggest SBIR group program ($300,000 to MMI for 18 months) was terminated unexpectedly at the 6-month point for overspending by the program's principal investigator (PI) company. Two companies, including MMI, were eliminated from the program unexpectedly.
- When MMI was awarded the big SBIR program, they hired two research engineers, which increased the R&D division costs.

These reasons correspond to the following explanations:

- *Cash flow problem due to a big project:* MMI overspent the R&D costs before the real money came in.
- *Lack of contingency plans:* Without considering program funding termination, MMI hired two engineers.
- *Lack of management skills, experience, and know-how:* Barb's 25% contribution to MMI may not be sufficient for the full operation.
- *Weak finance function:* Because MMI already borrowed $300,000 from Cheng Kung, no further loans will be possible from them.
- *Turnover in key personnel:* The first active sales person quit unexpectedly.
- *Accounting problem influenced by a big company:* The SBIR PI company terminated money distribution to MMI.

If Barb had continued the same company operation, the company would have been bankrupt 2½ months from the second week of December. Immediate actions were taken:

- Salaries were immediately cut for Barb and another corporate partner, Tom Meyer.
- Three engineers were immediately laid off, including the unproductive sales engineer.

These actions, however, could delay bankruptcy for only 2 to 3 months. Thus, Barb needed to implement a drastic restructuring strategy within 6 months.

11.6.3 Restructuring Strategy

11.6.3.1 Introduction of New Capital

Barb Shay considered introducing another sponsor company to MMI, this time from Europe, because Cheng Kung is an Asian company with a mass production-oriented strategy, but without a strong marketing strategy in Europe. If MMI could find a European partner, there would be no conflict between Cheng Kung and the new partner. Further, MMI must be a strong navigator into the U.S. market for the European partner. Again, Barb searched for a candidate from her recent consulting firms, and found PiezoDynamics in Germany. The president, Mr. Karl Schultz, is a young, active guy with whom Barb chatted occasionally at the international meeting. Since Euro currency has become very strong against U.S. currency recently, the amount of money requested from PiezoDynamics seemed to be less to the European firm. Barb proposed the following plans to PiezoDyanamics: MMI would introduce the capital $100,000 (154 shares) from PiezoDynamics, Germany; that is, MMI would be a PiezoDynamics company in charge of North America. PiezoDynamics is interested in expanding its sales in the United States, as well as promoting its R&D. MMI will join the PiezoDynamics family and also keep the Cheng Kung partnership position. The expected benefits of this PiezoDynamics capital introduction will be as follows:

- New cash inflow of $100,000 (stock share) from PiezoDynamics to MMI.
- New loan allowance up to $200,000 from PiezoDynamics to MMI (5% interest rate).
- Transfer of $500,000 on PiezoDynamics' existing U.S. sales portfolio to MMI with a 15% margin profit to MMI. This represents a profit of $75,000 in the restructured company.
  - Consequently, the total accumulated investment from direct new cash inflow and loans plus existing sales on PiezoDynamics' products elevates to approximately $375,000.
  - This will create a drastic jump in sales to MMI in the third year.

Note that because PiezoDynamics is a small business corporation, with about 240 employees in total, PiezoDynamics' partial ownership as a shareholder will not affect MMI's eligibility as a small business corporation in the United States.

11.6.3.2 Ownership Change of MMI

According to the introduction of PiezoDynamics capital (154 shares), and the stock option to Tom Meyer (40 shares in 2 years), the present ownership will be changed as shown in Figure 11.18.

- PiezoDynamics will become the largest shareholder of MMI (49%). Accordingly, MMI will propose that PiezoDynamics' President Mr. Karl Schultz become chair of the board of directors. PiezoDynamics will also send another board member. Lenny Chu from Cheng Kung will step down as a board member. If Lenny were to keep the board member position, MMI would be controlled by three aliens, which automatically excludes MMI from SBIR program application eligibility. Refer to the proposed board of directors organization chart in Figure 11.19, where MMI has five board members. Compare this chart with the previous structure in Figure 10.1.
- PiezoDynamics is further interested in expanding capital investment in the future. However, from a practical viewpoint, in order to keep U.S. company privileges, foreign partner PiezoDynamics' share should not exceed 49% of all MMI stocks, as the U.S. stock must exceed 51%.
- Forty-five stocks (14%) shared by Cheng Kung Corporation should be transferred to a small U.S. corporation or an individual investor. Again, due to the SBIR program application

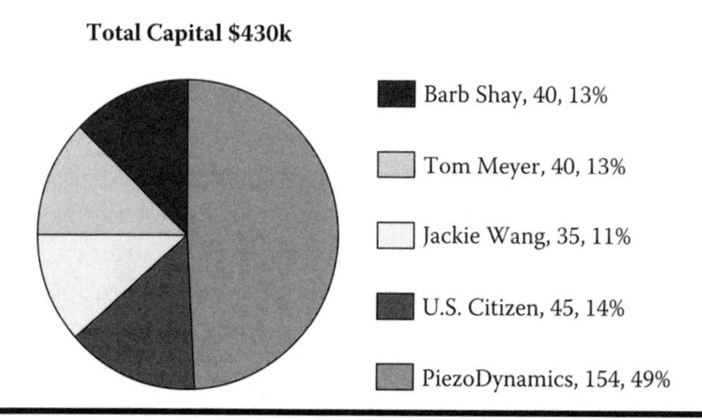

Figure 11.18 Proposed change in the MMI shareholders.

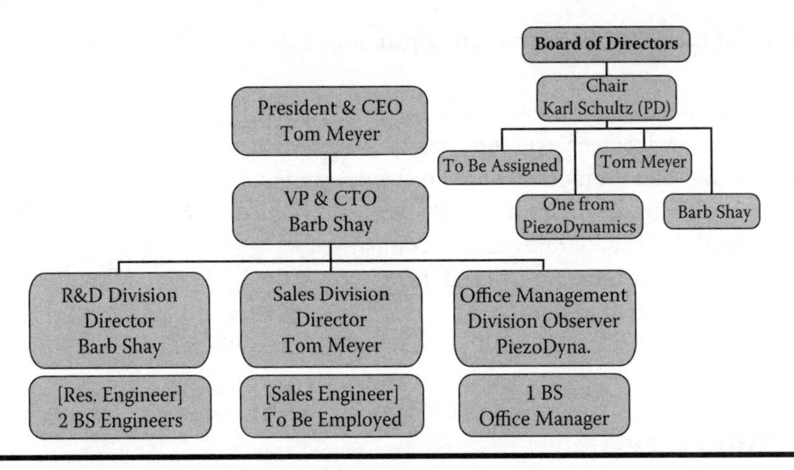

Figure 11.19 Proposed change in the MMI organization chart.

eligibility, MMI should keep a minimum of 51% of U.S. stocks. This can be done after finding an American investor.

11.6.3.3 Introduction of a New President

In order to cover Barb Shay's lack of time to operate MMI, Tom Meyer will be promoted to president and CEO and will operate MMI. Tom was chief engineer in another small business firm for 3 to 4 years and learned how to operate a small venture. Because of his young age and 100% commitment to MMI, the sales may increase drastically. Barb, to the contrary, will be retained as vice president and CTO, but her responsibilities will be fewer, and she will also be R&D director. The new position of president and CEO, Tom Meyer, and his hiring conditions are summarized below.

- President and CEO: Tom Meyer covers the general management of the company, business expansion, in addition to R&D promotion and program management.
- Compensation: Annual wages in the future will be determined by MMI's financial situation; 40 stocks will be shared in the successive 2 years as stock options.
- Ownership of MMI: Meyer will receive 40 shares in total in 2 years, which corresponds to 13% ownership, equivalent to the present vice president and founder, Barb Shay.
- Vice president and CTO: Accordingly, Barb Shay will be vice president and CTO, and take care of the R&D technical management including daily instruction to research engineers.

The organizational position of Tom Meyer is proposed in Figure 11.19.

11.6.3.4 MMI Employee Replacement

MMI will restructure the current number of employees by laying off two current R&D engineers and a sales engineer. Moreover, MMI will hire an experienced sales division engineer, perhaps transferred from PiezoDynamics in Germany. Refer to the proposed organization chart in Figure 11.19 for a probable new team.

A summary of the proposed restructuring plan for MMI is as follows:

1. Introduction of a new president—Tom Meyer
   a. Stronger management structure
   b. Significant sales promotion with a new sales engineer
2. Introduction of PiezoDynamics capital—49% of MMI stocks
   a. MMI will join the PiezoDynamics company
   b. MMI's financial situation will stabilize
   c. MMI's market will expand
   d. MMI will maintain their SBIR program application eligibility

Chapter Summary

11.1 Business strategy includes the following:
- Developing a vision and mission
- Identifying an organization's external opportunities and threats
- Determining internal strengths and weaknesses
- Establishing long-term objectives
- Generating alternative strategies
- Choosing particular strategies to pursue

11.2 SWOT grid: The horizontal axis is a measure of the strengths–weaknesses of your company, while the vertical axis is a measure of the opportunities–threats of the industry (the piezo-actuator industry in this case). The former is called internal environment analysis, and the latter is called external environment analysis.

11.3 External factors: STEP (social/cultural, technological, economic, and political) four forces

11.4 Porter's five-forces model for proximate environment analysis: The nature of competitiveness in a given industry can be viewed as a composite of five forces:
1. Rivalry among competing firms
2. Potential entry of new competitors
3. Potential development of substitute products
4. Bargaining power of suppliers
5. Bargaining power of consumers

11.5 Most causes of troubles or failure can be found within company management.
Strategic issues include the following:
- Misunderstanding market niche
- Mismanaging relationships with suppliers and customers
- Diversifying into an unrelated business area
- Cash flow problem due to a big project
- Lack of contingency plans

Management issues include the following:
- Lack of management skills, experience, and know-how
- Weak finance function
- High turnover in key personnel
- Accounting problems influenced by a big company

Practical Exercise Problem

P11.1 Production Strategy

If your small firm is requested to manufacture one million pieces of the product, how will you handle this inquiry? Make a decision among the following three choices: (1) start mass production in your firm in collaboration with the partner company, (2) decline this manufacturing, or (3) transfer this manufacturing task to a competitive company with some back margin (1% of the sales cost).

References

1. David, F. R. *Strategic management*, 11th ed. Upper Saddle River, New Jersey: Pearson Prentice Hall, 2007.
2. Hirashima, Y. *Product planning in feeling consumer era*. Tokyo: Jitsumu-Kyoiku Publ., 1986.
3. Fusakul, M. *Proceedings of 5th International Conference on Intelligent Material (ICIM 2003)*, State College, PA, June 14–17, 2003.
4. Minolta Camera, Product Catalog "Mac Dual I, II," 1989.
5. Fujii, A. *Proceedings of JTTAS*, Meeting on December 2, Tokyo, 2005.
6. Saito, Y. Measurement system for electric field-induced strain by use of displacement magnification technique. *Jpn J Appl Phys* 1996; 35:5168–5173.
7. Porter, M. *Competitive strategy: techniques for analyzing industries and competitors*. New York: Free Press, 1980.
8. Kurashima, N. *Proceedings of Machine Technical Institute Seminar, MITI*, Tsukuba, Japan, 1999.
9. Boecking, F., and B. Sugg. *Proceedings of New Actuator 2006*. Bremen, June 14–16, A5.0, p. 171, 2006.
10. Uchino, K. *Micromechatronics*, 463–468. New York: CRC/Dekker, 2003.
11. Paul Peter, J., and J. H. Donnelly Jr. *Marketing management*. New York: McGraw-Hill Irwin, 2007.
12. Warwick, D. P. *A theory of public bureaucracy*, Cambridge, MA: Harvard University Press, 1954.
13. McIlnay, D. P. *Guide to a SWOT analysis*, MBA 550 class handout. Loretto, PA: Saint Francis University, 2008.
14. Timmons, J. A., and S. Spinelli. *New venture creation*. New York: McGraw-Hill Irwin, 2007.

Chapter 12

Corporate Ethics—Keep it in Mind!

When we hear the term "corporate ethics," we are reminded of WorldCom and Enron, companies whose corporate executives made fictitious financial reporting to line their own pockets. This chapter begins with a discussion of general issues on ethics, law, religion, and education, then follows with a discussion on the ethics in society and culture, in corporate human relationships, and in research and development (R&D) by comparing U.S. and Japanese corporations.

12.1 Ethics and Morals

According to Wikipedia [1], *ethics* and *morals* are respectively akin to theory and practice. Ethics denotes the theory of right action and the greater good, while morals indicates their practice. The antonym is *immoral*, referring to actions that violate ethical principles, which is related to our theme—corporate ethics. Personal ethics signifies a moral code applicable to individuals, while social ethics means moral theory applied to groups. Social ethics can be synonymous with social and political philosophy, inasmuch as it is the foundation of a good society or state. Ethics is not limited to specific acts and defined moral codes, but encompasses the whole of moral ideals and behaviors—it is a person's philosophy on life.

The formulation of ethical standards can be difficult. The perception of ethics is both subjective and qualitative. Ethical business managers are responsible to numerous stakeholders [2]. These groups include the following:

- Employees (individually and collectively in the form of unions)
- Customers
- Suppliers
- Competitors
- Stockholders
- Government and community entities (local, national, and global)

12.2 Ethics, Law, Religion, and Education

Ethics are cultivated between the ages of 2 and 12, when a human is educated, typically by parents, teachers, priests, and others. The concept is based on both social culture and religion. Some of the most important ethics that people and businesses may neglect are enforced as laws. Generating harmful products such as air pollution and toxic chemicals is an example that should be strictly regulated by the law.

12.2.1 Darwin's Evolution Theory

Darwin's evolution theory is one of the most famous theories in biology, and is widely accepted by most countries. However, I've learned that some schools in mid-western states in the United States do not teach this theory because it is against their religious beliefs. This seems to go against educational ethics. Science should be taught equally to all students, separately from religious education. Totally accepting the scientific theories or criticizing the science is the individual's choice (according to his or her religion or culture). However, a teachers' ethical obligation is to teach all publicized theories to all the students. It seems ridiculous for the state, city, or community to restrict the teaching of well-known science to the students.

12.2.2 Production Regulation

The twenty-first century has been called the "century of environmental management." We are facing serious global problems such as the accumulation of toxic wastes, the greenhouse effect, contamination of rivers and seas, lack of energy sources such as oil and natural gas, and so on. Externalities such as pollution and resource depletion are perhaps the best-known examples of market failures. Sometimes the costs of solving such problems are borne not by the direct sellers and buyers, but by third parties such as people downwind from the pollution and future generations.

12.2.2.1 MMI Example

In 2006, the European community implemented the Restriction on Hazardous Substances Directive (RoHS), which explicitly limits the usage of lead in electronic equipment. Therefore, we may need to regulate the usage of lead zirconate titanate (PZT), the most widely used piezoelectric ceramic material, in the future. Japanese and European communities may experience governmental regulation on PZT usage within the next 10 years. Pb-free piezo-ceramics started to be developed in 1999. RoHS seems to be a significant threat to the piezoelectric industry, which so far uses primarily PZT piezo-ceramics. However, this is an opportunity for companies such as MMI and Toyota, which are preparing alternative piezo-ceramics in order to gain more of the piezoelectric device market share.

12.2.2.2 Gun Control

Most people do not like guns, and a majority of countries regulate guns strictly. I had never seen a real gun before immigrating to the United States. Many international tourists are very surprised at seeing rifles and guns sold even in regular supermarkets in the United States, and can understand how the massacre happened at Virginia Tech University, even in an academic institute. Although ethically most people know that guns should be regulated, the U.S. Constitution may not change the rules so easily. The Bill of Rights, Amendment II, explicitly mentions that "A well-regulated Militia, being necessary to the security of a free State, and the right of the people to keep and bear Arms, shall not be infringed."

It is therefore unfortunately difficult to regulate guns in the United States. This is an example of *deregulation* of unethical products.

12.2.2.3 Tobacco and Food Control

Despite the slow regulation on guns, tobacco control occurred incredibly quickly in the United States. Tobacco and alcohol, which are still popular in most other countries, are used according to the individual's liking or taste. Abstaining from smoking in a public area is a matter of etiquette or simple ethics. However, Americans were quick to legally regulate it.

I heard recently that a ridiculous law pertaining to fast-food restaurants was proposed in San Francisco, because of Americans' obesity problem. The reason for Americans' obesity is merely big portions of fatty foods. Most international tourists are surprised with the huge portions of food served in regular restaurants, and actually lose their appetites. If the portions were reduced by half, which is actually a standard portion for an adult in most parts of the world, it would be possible to lose 100 pounds. I know this because I lost 70 pounds in a 3-month period several years ago, merely by controlling food portion and sugar, with some exercise. Before banning McDonald's, why don't we regulate ice cream? Obesity is the individual's responsibility. If a person is anxious about his or her weight and hates fast food, that person does not need to visit McDonald's. It is the individual's choice, and there should not be a legal restriction for food suppliers. Do you think there is something strange about the social ethics on this point?

For your information, I will introduce an interesting regulation in Singapore, on chewing gum. Chewing gum is illegal in Singapore. No shop sells it. If you are chewing gum in a public place, you may be arrested by police. This law was put into effect because unethical people spat out gum on the public roads, which made the city dirty. What do you think about the chewing gum regulation?

12.3 Business Ethics

12.3.1 Conflict of Interest

A conflict of interest arises when an individual enriches himself or herself at the expense of his or her employer or client. If you are a university faculty member, when you start a company, how can you avoid this conflict of interest? I strongly recommended that you exchange an Employment Agreement Appendix with your department on the Conflict of Interest Disclosure with your university, which includes the following contents:

- Define your working time in your company.
- Do not hire your students directly.

Defining the time you will work is most important to avoid accusation of illegal practice. If you are hired by the university 100%, you cannot be a principal investigator (PI) of federal contracts research in your company. Even if you can work on Saturday or Sunday, federal contracts do not allow assignment of your working time out of regular 40-hour work week. If you are hired by the university 100%, your intellectual properties, based on your professional expertise, belong to the university.

You should not hire your students directly. However, hiring a student from a different department, with whom you have no relationship as an advisor or thesis committee member, is allowed, including summer interns.

12.3.2 Confidentiality

It is unethical for organizations to release confidential information about their employees or customers to third parties without express permission. Likewise, it is unethical for the employee to release confidential information about his or her employer to third parties. One of the most important things in the high-tech firm is trade secret maintenance, which is related to the employee's job change or termination. American engineers tend to change jobs every several years, resulting in the inevitable transfer of trade secrets, even among competitive firms (and sometimes leading to head-hunting). Accordingly, we sometimes face a serious conflict with the company in which our former employee found his new position. Typical general conflicts include the following:

- Market research data and R&D and marketing strategies
- Similar product lines in the new company
- Know-how in product-manufacturing processes
- Research proposal ideas
- Customer list

In order to prevent this sort of problem, the firm needs to legally regulate disloyal behavior through use of the Employment Agreement. (Refer to Chapter 9, Section 2.3 for the Agreement example.)

12.3.3 Executive Compensation

Figure 12.1 shows the ratio of average pay for corporate chief executive officers (CEOs) to average pay earned by factory workers [3]. The original 12 in 1960 increased exponentially to 500 in 2000. Among the highest paid CEOs in 2002 was the CEO of Wells Fargo, who earned $8 million in salary and bonus, and was granted an additional $64 million in stock options. The CEO of Wal-Mart earned $2.9 million in salary and bonus, and $50 million in stock options. For

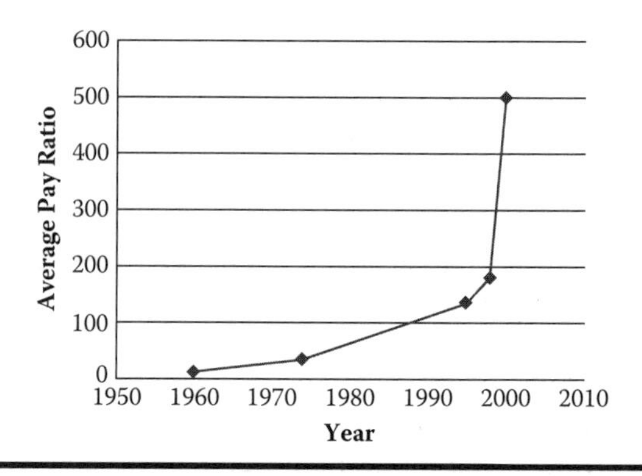

Figure 12.1 The ratio of the average CEO payment over the average factory worker payment.

your information, the average wage for Wal-Mart sales associates for 2001 was $8.23 per hour, or $15,800 per year [3].

Many of these large pay packages are the result of stock options that corporate boards grant to their executives. The portion of CEO pay derived from stock options increased from 27% in 1992 to 60% in 2000. Stock options are the right to purchase the company's stock at a predetermined price any time within a set period, often a 5- or 10-year time frame. For example, an executive might be granted the right to purchase 1 million shares at a price of $10 per share any time within 10 years. If the stock price rises to $20 per share, the executive is able to cash in these options and make a significant profit, equal to $10 million. Many observers believe that such options were much to blame for accounting scandals in recent years. Stock options create a strong incentive for executives to increase the company's share value by whatever means possible. The data for 1996 showed that the top executives of firms that had laid off more than 3000 workers in the previous year received an average 67% increase in their total compensation package for the year. In 1996 the average gap between CEO pay and the wages for the lowest paid worker for the top 12 job-cutting companies was 178:1 [3]. (Refer to the Japanese executive's attitude for the companies' recession in the next section.)

In theory, this ties the executive's pay to increases in shareholder wealth, but in practice it can often lead to short-term, unsustainable increases in share value, as the Enron case demonstrated.

12.3.4 Production Ethics

12.3.4.1 Product Liability

Product liability is the area of law in which manufacturers, distributors, suppliers, retailers, and others who make products available to the public are held responsible for the injuries those products cause. Examples include the restriction on lead (Pb) inclusion in cosmetics and toy coating, or the warning "Do not use this electric hair dryer while bathing" in the hair dryer manual. This is a reasonable, ethical, and customer-oriented rule.

However, some of the warnings required of manufacturers seem to be ridiculous. A woman swallowed a mobile phone and choked. It is obviously a crazy action, which no normal person will try. However, this woman initiated a lawsuit against the mobile phone company because the cell phone manual did not mention "Do not swallow this phone." After that, the manufacturer was required to add this unnecessary and strange sentence to the manual.

12.3.4.2 Quality Control

I was a longtime customer of General Motors' cars. However, I experienced multiple repairs for all five previous cars in the first 3 to 6 month period after the purchase. Thus, at my wife's suggestion, I purchased a Honda. I have been surprised with its no-repair situation over these 2 years (except for my making a scratch by careless parking). I believe that this difference occurred from the quality control (QC) practice in both auto manufacturers. Japanese industries like the idea of Six Sigma management, that is, only 3.4 defects per million products. In contrast, most factories in the United States are operating at three to four sigma quality levels [4]. Product defects take time and effort to repair as well as creating unhappy customers. QC is one of the most important production ethics in manufacturers.

12.3.5 Truth in Advertising

If an advertisement is deceptive, it is unethical, regardless of its intent. The law supports this position, but enforcement may be lacking in stringency.

12.3.6 Discrimination/Sexual Harassment

As we learned in Chapter 10, bias against individuals on the basis of race, ethnicity, creed, age, gender, or sexual orientation is unethical and illegal. Aside from being morally reprehensible, it is also irrational. For example, a manager who fails to promote an individual on the basis of prejudice is ruining the firm's human capital.

12.3.7 Firing Employees

From a legal standpoint, whether a termination is rightful may depend on whether it is based on *at will* (employer's discretion or whim) or *just cause* (employer's justification) considerations. Generally, discrimination cannot justify termination. From an ethical standpoint, however, employers should not fire at will, and employees are entitled to due process and the opportunity to state their case and be judged fairly (i.e., overruling a biased or unfair immediate boss). If an employee is terminated, it is the employer's ethical if not legal obligation to obviate the damage inflicted (e.g., via severance or outplacement services).

12.4 Comparison of Corporate Ethics between the United States and Japan

There are large differences in corporate ethics between the United States and Japan, though there is some consistency in general. Most high-tech entrepreneurs will need to compete or collaborate with worldwide corporations, thus we need to learn about global business ethics in addition to nation-limited legal regulations. As we already discussed in Chapter 10, corporate management styles can be symbolized by a regatta in the United States and *mikoshi* in Japan. Employee's productivity is evaluated by a *differential method* in the United States, while an *integral method* is used in Japan; this encourages job transfer in the United States and permanent employment in Japan, leading to significant differences in managerial ethics. In this chapter, other differences between the two countries in production line QC schemes and in research topics will be introduced. For example, Japanese industries have already created the Six Sigma QC, while U.S. corporations are still struggling around three- to four-sigma levels. This originates from the difference in the employee's loyalties to the corporation. I discussed how these corporate management ethical differences originate from differences in culture and lifestyle. The United States is an individual-based society while Japan is a group-based one. U.S. technical education styles focus on "why," while Japan focuses on "how-to." In any case, American ethics are not the global standard, and we need to seriously understand and respect other people's and countries' thinking styles and business ethics when we consider the global business relationship.

12.4.1 Background

As previously mentioned, I had joint appointments for 18 years as a university professor and a company executive in Japan. In 1991, through a recommendation by the Office of Naval Research, I was brought to Pennsylvania State University to be the founding director of the International Center for Actuators and Transducers (ICAT) and to transfer technologies I had developed in Japan. This is because I was known worldwide as one of the pioneers in the field of piezoelectric actuators.

Table 12.1 Comparison between the United States and Japan in Business Atmosphere

	United States	*Japan*
Living philosophy R&D style	Individual	Group
Education R&D	Why	How-to
Industry type R&D Performance appraisal	Differential	Integral
Management "Big science"	Regatta	*Mikoshi* (portable shrine)
Needs quality control	Reliability	Newness

To tell the truth, I hesitated initially, because my R&D development products would be utilized for military applications. My ethical background and upbringing in Japan prohibited me from developing warfare weapons. Following assurances from the Navy Program Officer that my technology would not be directly used for weapons, but for defense activities, I reluctantly accepted the Penn State position.

Typical Americans who grew up in the United States may not understand my ethics. They join the military service voluntarily or more enthusiastically, and are proud to work for weapons development. The United States is the only country that has a voluntary soldier enlistment system. To the contrary, because of the miserable end of World War II, and the shock of the historical atomic bombs in Hiroshima and Nagasaki, most Japanese are very sensitive about being involved in wars, weapons, or nuclear industries. Further, they are trained to consider that it is a sort of sin to be involved in warfare. As demonstrated in this example, a person's ethics depend on their individual history, education, and experiences, and differ significantly from country to country. American ethics are not the global standard, but are biased due to their historical living style, especially corporate ethics.

Based on my lifelong experiences in both the United States and Japan, as well as both in universities and industries, I will here compare and contrast the corporate ethics of the United States and Japan. Table 12.1 summarizes keywords for understanding these differences. Compared to Table 10.2 in Chapter 10, "why" vs. "how-to" and "reliability" vs. "newness" are added.

This section is based on a chapter of one of my previous books, "The Difference between Japan and the United States in Research and Development Policy," published in 1987 [5], and on my 2000 textbook, *Ferroelectric Devices*, Section 11.3, "Development of Bestseller Devices." [6]

12.4.2 Ethics in Society and Culture

12.4.2.1 Living Philosophy and Religion

Most Asians, including the Japanese, were farming races. Farming required laborious teamwork in a long horizontal time frame such as a half year. From ancient times, the farming races have been trained to work as a group, by restricting individual liking. To the contrary, Americans (European races) mostly originated from hunting races, a society in which just a small number of individuals were regarded as heroes by hunting large animals such as deer. This thousands-of-years period in history created differences in societal principles, culture, and ethics.

The Japanese created Shintoism and Buddhism (actually, a mixed religion at present), respecting natural forces, such as wind, trees, fire, earth, and water/rivers as Gods. The Japanese have hundreds of Gods without considering one unique heroic God. To the contrary, Americans (and Europeans) prefer to accept Jesus Christ, originally a human, as the one unique God. These religious backgrounds can also be understood from the original racial differences.

12.4.2.2 Corporation and Individual Ethics

Japanese industries still use the basic concept of permanent employment, provided the employee is loyal to the company. "Industrial warriors" are still highly respected in Japanese industrial society. Their lifestyles are arranged around the company schedule, and are based on a group decision. Even though Japanese employees have more than 2 weeks of paid holidays per year, in practice, it is difficult for a worker to use more than a couple of days continuously because of pressure in the work environment, so taking a long holiday is against his or her working ethics. This societal atmosphere can be understood as having originated from the farming race. However, for example, several Japanese friends of mine have unfortunately passed away due to stress-related illnesses caused by their managerial positions. This sort of loyal attitude to a corporation, shown by sacrificing the individual self, is highly respected from an ethical viewpoint in Japan.

In contrast, the American lifestyle centers on the individual, and originates from the ancient hunting race. Even directors in American companies can easily take off more than a week during the summer and Christmas. In an extreme case, one sales engineer of my affiliating company did not go to a tradeshow, because that day coincided with her daughter's birthday. Putting her priority on her private or home affairs over her obligatory tasks for the company did not conflict with her working ethics.

12.4.2.3 Education Principles

Many readers will remember the movie *The Karate Kid*, where Mr. Miyagi, the teacher, tries to teach the kid how to "wax on and wax off," but the kid did not understand why this was essential to learning karate. This encapsulates the different educational philosophies of Japan and the United States.

"Repeat and memorize" is the major focus of Japanese educational curriculum. When I was young, enforced rote memorization without coherence was how I was taught. For example, I learned that Borneo, Sumatra, Java, Luzon, Mindanao, Celebes, Bali, Lombok, Sumbawa, Sumba, Flores, Timor, Jolo, Seram, Bangka, and Belitung were the names of 16 islands the Japanese Imperial Army invaded successively in World War II. Though this is an extreme case (after 60 years, I still remember!), Japanese teachers do try to force students to remember as many proper nouns and years as possible. For example, in social studies, "What is the longest river in the Kanto Plain?" and "What year did the American Civil War begin?" A high percentage of Japanese students can correctly answer 1863, while many Americans cannot. However, most Japanese university students cannot correctly answer the questions "Why did Tokugawa Shogunate collapse?" or "Why did the American Civil War happen?" because they have not learned this sort of reason or logic in the school system.

Understanding is the major focus in the American educational system. When I started teaching electroceramics at Pennsylvania State University, I was embarrassed with so many questions of "Why is it?" because I never experienced that in my years teaching in Japan. American students are curious to learn why. For example, Penn State offers a graduate course called "Individual Study." A student focuses on a special topic, performs a literature survey, and writes a research report. The course requires the student to make a short presentation in front of the faculty advisor to receive credit. We find it useful for evaluating students applying for graduate fellowships. That cannot be found in a Japanese graduate curriculum, because this sort of self-motivated training concentrating on deep understanding is rare in Japan. Japanese professors are usually surprised with American graduate student's confidence in explaining their research, when they visit a U.S. laboratory.

"Repeat and memorize" is a very important way to improve basic living skills. The unemployment rate in the United States is roughly double that in Japan. This is not merely due to the economic situation, but may also be caused by the lower abilities of workers in the United States. Statistics say that only 40% of Americans can correctly solve simple math problems such as 60 divided by 12, which should have been learned in his or her elementary school period. More than 70% of Japanese junior high school students can correctly answer it [5]. It may be difficult for those with poor math skills to find a position even as a cashier at supermarkets in Japan, even though barcode readers are popularly used now, because they may be excluded through a qualification exam conducted during the job interview.

There was an intriguing experiment at Sumiton Elementary School in Alabama [5]. In order to improve basic living calculus skills, this small school employed the Japanese Kumon math training program. Emphasizing just simple "repeat and memorize" and "can do" rather than "can understand," this school suddenly appeared among the top ranking schools in the Nationwide Scholastic Assessment Math Achievement Test.

Debate is a course unique to the United States. Students are divided into two groups, pro and con, independent of their individual preferences, for a particular theme, such as "Is this war right or wrong?" They must defend their side with evidence and proof. There is no debate training in Japan. As an example, during a lunch break in my laboratory at Sophia University in Japan, in 1990, there was an interesting debate on the Gulf War. An exchange graduate student from Pennsylvania State University, whose boyfriend was serving in the Army, started the debate with criticism against Japanese government policy: "Why will the Japanese government pay only money but not sweat?" The first response came from a Korean student: "The Korean government is against this war. However, just because we do not want political conflict with the U.S. government, we are not taking an obvious opposite side. We are not sending our troops, nor paying money. I agree with this government policy." The five Japanese students would not respond at all. Even under my encouragement, their replies were simply "I do not know," and "I am not interested in it." Only one student replied with his opinion: "I do not like war." It was immediately shut out by the instigator's strong offensive words, "The war is now going on! I want to know how we should save the present situation." Only a silence followed from the Japanese students. A similar defeat can be occasionally seen in international academic societies and workshops. Even though a Japanese engineer presents a beautiful piece of research, he is defeated in the discussion or debate on his originality or uniqueness by American researchers.

In summary, American students "can understand, but cannot do," while Japanese students "cannot understand, but can do." The former attitude usually enhances a person's creativity, but also may increase the unemployment rate, in practice. To the contrary, the latter suppresses creativity, but keeps the unemployment rate low. Both systems have their merits and should be combined.

For example, it is obvious that keeping the repeat and memorize method up to the university level in Japan suppresses self-motivation and creative study. I was really surprised when I heard from a Japanese graduate student, "Professor Uchino, where is the textbook which explains this research topic? Without a textbook, or without detailed instruction, I cannot work on that topic!" This is a typical Japanese student's response. It is time for both countries to take the best elements from the other's system.

It is important to note how the educational system reflects to the business relationship. The Japanese system based on long-run perseverance will create a strong human relationship with steady ethics, while the American system based on short-run logic will generate a legal relationship with complete agreement documents.

"When a stone is thrown, it will hit an attorney in the U.S." The attorney density is more than 10 times higher in the United States than in Japan. Since attorney fees are very expensive in Japan, the Japanese prefer not to use attorneys in routine business agreements. Therefore, formalizing business agreements can be a big problem for the U.S. companies dealing with them. Among my company's 11 Japanese partner companies, only six set formal trading/distribution agreements. Typically, in the initial 6 months or so, the product distribution of the partners must be made under our company's sole risk. The Japanese partners are just watching our manufacturing or sales performance until they gain trust in us. According to the trading division's comments in these companies, U.S. companies will try to set a rigid business agreement from the start. If problems occur, the U.S. company readily starts a lawsuit to obtain compensation. Whether true or not, this is the Japanese impression of U.S. companies. They prefer to build trust through performance before signing agreements.

In conclusion, the Japanese business relationship respects the ethics of the partner company, while the American company cannot believe the partner's ethics, and sets solid legal protection.

I still remember a ridiculous lawsuit against McDonald's several years ago. In most countries, serving coffee or tea as hot as possible is the best service to the customers. However, one customer at McDonald's accidentally spilled her coffee on her lap in her car. From the majority's viewpoint, this accident was merely due to her careless mistake. However, probably enticed by her attorney, she started a lawsuit against McDonald's claiming that she was burned because her coffee was too hot. This claim is totally against social ethics. However, because of her win, we must now suffer low-temperature coffee and tea.

12.4.2.4 Industry Type

The ratio of manufacturing industries to total industries is 24% in the United States and 34% in Japan, and the ratio of science and technology students to total university students is 5% for the United States and 20% for Japan. In other words, Japan has four times more engineers (the university student numbers are almost the same) and more manufacturing industries, while the United States has fewer engineers and many more financial, insurance, and legal corporations. We can also describe American industries as a differential type, where the profit is created from the change in the price. The profit from the stock market does not depend on the absolute stock price, but on the time derivative of the stock price. In contrast, manufacturers in Japan are the integral type, where the total investment (or accumulated) price is valued.

Most MBA courses teach "other people's money," or OPM—borrowing other people's money, investing it in options, derivatives, etc., returning the loan after a short-run period, and finally obtaining the remaining money. This is a sort of magic; we can create cash from "zero." If this is successfully done, it is admired in the United States. However, this sort of business strategy is not respected in Japan, and is even against its business ethics, based on the original farming culture. Japanese feudalism includes a hierarchy of occupations and industries, that is, from the top, warrior, farmer, manufacturer, and finally merchant. The Japanese normally respect making actual products, rather than making money from other people's money.

12.4.3 Ethics in Management

12.4.3.1 Office Atmosphere

Office structure and atmosphere also reflect the differences between the U.S. individual emphasis and the Japanese group emphasis. As already shown in Figure 10.4, individual cubicles are popular in U.S. industries for the privacy of employees, while Japanese industries do not allow private cubicles. Instead, everyone is in one room. Different from the U.S. custom, the manager's desk is situated at the front of this big room (there is no privacy even for a manager), so that he can observe employee's behavior all the time. This system encourages the slowest workers to keep up with the group.

12.4.3.2 Management Structure

We already discussed the management structures of American and Japanese industries in Section 1.1.3, where we learned the American and Japanese structures resemble regatta and *mikoshi*, respectively. Further, American managers seem to prefer larger power in a bigger company, which may be visualized as a "whale" or "brontosaurus" type (see Figures 1.4 and 1.5). In contrast, Japanese managers prefer a "sardine"-type structure. Loose coupling by medium- and small-sized companies creates a power similar to a big whale. But, unlike a whale, a group of sardines can change shape adaptively according to an enemy's presence. Based on each sardine's synchronized intention, these companies can make a *keiretsu* (industry family tree).

A similar discussion was made by using the analogy of *kabuki* and musical theaters [7] (see Figure 10.7). The *kabuki* attracts the audience with one or two key actors (there are no actresses in *kabuki*), which resembles American industries, such as Mr. Iacocca when at Chrysler. In contrast, the musical is an assembly of many minor actors and actresses, which is closer to Japanese industry situations. I would like to point out an interesting exchange in arts in both countries. In artistic expression, Japanese prefer one heroic person's story (*kabuki*), while Americans prefer uniform teamwork (musical). People sometimes respect or prefer an attitude in the arts that is opposite to the actual situation.

12.4.3.3 Management Culture

It is interesting to compare my management structure description with the concepts of Theory X and Theory Y by D. McGregor [8]. Theory X is a set of propositions of the conventional view of management's task in harnessing human energy to organizational requirements, indicative of an autocratic management style, while Theory Y is based on more adequate assumptions about human nature and human motivation, and therefore has broader dimensions, indicative of an egalitarian management style. As we discussed in Chapter 10, the American management style is basically Theory X-based, while the Japanese management style is basically Theory Y-based. Again this seems to have originated from the historical and social hunting and farming differences.

F. W. Taylor proposed the Four Principles of Scientific Management in the late nineteenth century [9]:

1. The deliberate gathering together of the great mass of traditional knowledge by the means of time and motion study.
2. The scientific selection of the workers and then their progressive development.
3. The bringing together of this science and the trained worker, by offering some incentive to the worker.
4. A complete re-division of the work of the establishment, to bring about democracy and cooperation between the management and the workers.

Japanese management closely follows Taylor's principles. Japanese managers guarantee employment to workers under the supposition of their loyalty to the company. Productivity increases are in the common interest of both managers and workers because profits are more equally divided between the parties. In Japan, the salary ratio between the president and the lowest ranking worker is typically less than 20:1. For example, the president of NEC earns $500,000, while a McDonald's counterperson earns $25,000. In the United States the ratio is more than 300:1, as we already learned from Figure 12.1 in Section 12.3.3 [3]. There is a more trusting relationship between the employee and employer in Japan. Rarely do Japanese employees read their employment agreements, because they trust the company. Often, the company does not have an actual employment agreement signed by the employee. I do not recall whether I signed an employment agreement when I was hired by Sophia University in Japan—there was not an official agreement form! The employee does not negotiate salary individually because it is primarily determined by his or her age, with a large exponential salary increase after spending more than 10 years of service with the company.

In the United States, employment is totally different. American employees normally negotiate for higher salaries when applying for a position, without regard for the company's financial situation at that time. These sorts of negotiations reinforce the concept of the employee working for his or her best interest, with the company a secondary consideration. Managers interpret this as lack of loyalty, and know the employee can change companies at any time. F. W. Taylor's theory of scientific management was meant to address this issue. In Japan, it is accepted as common practice in corporate ethics by both managers and workers.

Ford Motors recently laid off 20,000 workers, and laying off United Airlines employees created tens of millions of dollars in bonuses for the executives. These actions are not illegal, but aren't they against corporate ethics in the United States? Of course, Japanese industries make inevitable minimum layoffs sometimes, but the layoff is not very popular. My friend, a Taiheiyo Cement Corporation executive, told me about his company's policy: in the beginning of the twenty-first century, when Japan's economy faced a serious recession, it reduced the salaries of the managers higher than the assistant manager level by 20% uniformly, and distributed this amount to the lower-level workers without executing layoffs. This is another example of scientific management, which American managers do not want to follow (refer to Section 12.3.3).

12.4.3.4 Employment and Evaluation Criteria

Most Japanese employees put a priority on employment stability, aiming for permanent employment. Next they are interested in the pension system after retirement. These are discussed when they interview for their first job. In contrast, Americans seek better pay and tend to change jobs and companies often.

These employee attitude differences originate from the employer's evaluation criterion. The Japanese salary system is based on a way to invest for the future capability of the worker based on permanent employment (hardware-like). Thus, "fresh-employee's education" programs are very popular in Japanese companies. My research center at Pennsylvania State University has multiple Japanese graduate students who are completely financially supported by their companies, under the supposition that these employees will return to their home companies after finishing their degrees and contribute more than this initial investment. On the other hand, even if very capable, a young employee cannot expect a high salary. It will take 10 years to get a significant salary increase. In other words, the accumulation of work done for a long working period in a company (integral) determines the salary. This salary system guarantees employment to old age, but suppresses the working motivation of young engineers. This situation is depicted in Figure 10.5.

In contrast, the American salary system is based on a way to purchase the present capability of the worker based on the employee's transfer in the future (software-like). Independent of age or the time worked at the company, the salary is determined by the profit delivered by this worker for 1 year (or an equivalently short period), that is, the derivative/differential of the capability in terms of years (refer to Figure 10.5). This salary system motivates young engineers to work harder in the short term. However, with age and slowing of fresh ideas, there is a big risk of salary reduction or unemployment. This salary system will not create long-term loyalty to the company. The difference in philosophy between hardware- and software-like resources was discussed in Section 10.4.3.

12.4.4 Ethics in Research and Development

12.4.4.1 Research and Development Attitude

American researchers have lone wolf-type personalities. Because research originality is highly prized, they tend not to disclose their research content even to their colleagues, to protect their intellectual property. I heard a ridiculous story from a friend in an IBM laboratory that he first learned of his neighbor engineer's research topic through a newspaper that reported the neighbor's award from an academic society. This working environment provides highly creative research, but it may not be suitable for product development that requires a large team effort. The American researcher prefers the "home run" approach to research.

In contrast, Japanese industries highly respect group efforts for developing products, rather than lone wolf-type research, Researchers should work at a uniform pace with inconspicuous individuality. In other words, "single hit and squeeze play" is the Japanese way, rather than "home run." The personality differences between Americans and Japanese may have originated from the fundamental historical differences, that is, the hunting race and the farming race. There was one hero who shot a big animal in the hunting race, while teamwork was essential in the farming race.

The home-run approach corresponds to winning the Nobel Prize, while the "single hit and squeeze play" approach corresponds to continuous product developments. I collected interesting data for new model products for facsimile and copy machines in Japan in 1988 (see Table 12.2). Ten manufacturers created 115 new facsimile models in a year, which corresponds to one product

Table 12.2 New Model Products for Facsimile and Copy Machines Commercialized in Japan in 1988

	Number of Manufacturers	*Number of New Models Commercialized in 1998*
Facsimile	10	115
Copy	14	96

in just 3 days, or one manufacturer developing one new product every month (the statistics are similar for copy machines). Severe competition motivated this new product development race.

I stayed in the same apartment house in State College, Pennsylvania, two times—1979 and 1991—and found the same model refrigerator (GE 750W) in the apartment house. This means that General Electric has been continuously manufacturing the same model from 1979 till 1991—although GE has developed a new model, the customer has not been motivated to renew the fridge for at least 12 years. This is significantly different from Japanese electric companies, which have been continuously remodeling the refrigerator and improving the energy consumption by one-third over the past 12 years. Accordingly, Japanese customers are motivated to exchange the fridge every 5 years. Note that the refrigerator is the most power-consuming appliance in the home. American customers can choose a cheaper GE refrigerator for half the price of a Japanese refrigerator, or an energy efficient Japanese type with an ecological or environmental fit; this also reflects consumer characteristics in the United States and Japan.

12.4.4.2 Big Science Projects

Let us consider why "big science" is so advanced in the United States as compared to Japan. I will discuss this issue from a management structure viewpoint. In 1978 while at Pennsylvania State University, I worked on a National Space and Aeronautics Administration (NASA) contract researching electrostrictive actuators for precise positioning of optical lenses and mirrors. After 1 year, I was invited to the NASA workshop on the Adaptive Mirror Project (it was named *Hubble* telescope when launched on the space shuttle). Surprisingly, the same research topic was also being done independently at an American company, then called Honeywell Laboratory. Through the annual workshop, the better result was determined, and that set a new starting point for the next year's race. During the year, no contact was permitted between Honeywell and Penn State. Each group needed to put forth the maximum effort to win the next year's competition. U.S. space technology was far behind that of the former Soviet Union in the 1970s. I learned that this type of development method, called the NASA System, accelerated the space development, leading to the successful launch of the space shuttle.

Japanese "big projects" are lead by the Ministry of International Trading and Industry or the Ministry of Education. The project members are chosen from similar industrial group members and have close communication with each other, leading to synchronous research speed with each other and no competition. The NASA system seems to be advantageous from the viewpoint of creativity and research speed, but the Japanese style makes for a good team with humanistic harmony.

Figure 12.2 depicts the difference in the structure of "big science projects." Note that there is only a vertical command line between the military program officer and the contracting researchers in the U.S. system, while the horizontal interaction is strong among the same level researchers in the Japanese system. The former structure seems to be suitable to long-term big science, while the latter structure fits to short-term small development.

The statistics for R&D investment comparing the United States and Japan in 1980 are collected in Table 12.3. In the United States, 50% of the R&D funds come from the military: federal institutes such as the Defense Advanced Research Project Agency (DARPA), Army, Navy and Air Force, etc.; among the 50% nonmilitary R&D funds, a large portion is by nonmilitary federal institutes such as NASA, the National Institutes of Health (NIH), National Science Foundation (NSF), etc., and contributions by private industries are rather small. In addition, the military share is significantly increasing. This investment situation accelerates the R&D structure initiated and controlled by the government, leading to the vertical command line as shown in Figure 12.2a. In

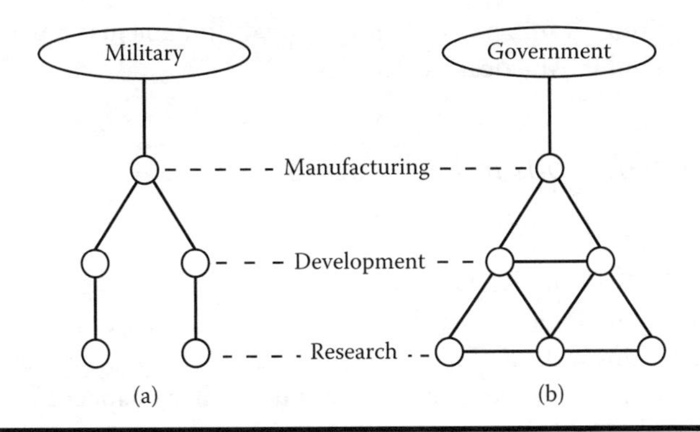

Figure 12.2 Difference in the structure of "big science projects" between (a) the United States and (b) Japan.

Table 12.3 Research and Development Investment Comparison

USA		*Japan*	
Military	50% (10% annual increase)	Government initiative	30% (0.1% annual increase)
Nonmilitary	50%	Industry	70% (15% annual increase)

contrast, the Japanese government funds less than 30% of R&D investment, with a slight annual increase, and more than 70% of R&D is funded by private industries. This amount increases significantly every year. Thus, Japanese projects are initiated or flagged by the government, but the project is propelled by private industries' money, and the friendly horizontal interaction shown in Figure 12.2b becomes reasonable.

12.4.4.3 Research and Development Style

A general manager at the Mitsubishi Electric Research and Development Center described the R&D policy of Mitsubishi, using traditional Japanese proverbs:

- "A tall tree catches much wind." All researchers need to keep the same development speed (group/team uniform pace).
- "Do not do a task today which can be postponed until tomorrow." This was modified from the original proverb "Do a task today which you can do today," which seems to be similar to the "go slow" policy of the regular worker, as pointed out by F. W. Taylor [9].

Table 10.3 illustrated the difference in style between U.S. and Japanese R&D. Japanese industries respect group/team development by keeping individual work levels uniform. This concept is suitable to product development with high cooperation among the team members. Because an individual's role is not separately evaluated, salary is primarily based on the researcher's age, not on his or her individual contribution.

Contrast this with American researchers, who do not necessarily cooperate with each other. Because salary is awarded based on the individual's contribution, a researcher will work at his or her own pace. This attitude is suitable to highly original fundamental research. Lone wolf-type researchers are often found in American companies.

12.4.4.4 Ethical Restrictions in R&D Topics

12.4.4.4.1 Military versus Civilian Applications

Let us start from the discussion on R&D topics for military or civilian applications, so as to expand upon the reasons for my hesitation when I was recruited to Pennsylvania State University (refer to Section 11.4). The shipwreck of *Titanic* happened at 11:45 p.m. on April 10, 1912. If the ultrasonic sonar system had been developed, it would not have happened. This tragic incident motivated ultrasonic technology development. To tell the truth, however, this civilian tragedy did not increase the research money. In 1914, World War I raised the research investment dramatically to accelerate this technology in order to search for German U-boats in the deep sea. Dr. Langevin, a professor at the Industrial College of Physics and Chemistry in Paris, started the experiment on ultrasonic signal transmission into the sea, in collaboration with the French Navy. Using multiple small single-crystal quartz pellets, sandwiched by metal plates (the original Langevin-type transducer), he succeeded in receiving the returned ultrasonic signal from the deep sea in 1917. This was the beginning of piezoelectric applications.

As I mentioned, Japanese researchers usually hesitate to work on military applications. Because Japan does not possess military structure or power, but has only the Ministry of Defense written into their Constitution after World War II, they are trained to believe that military power is ethically bad. To the contrary, American people are proud to join the military without enforced conscription. Moreover, because more than 50% of U.S. R&D money is provided by military-related projects, it is inevitable that it is put toward military applications, as illustrated in Table 12.3. It is totally up to the researchers whether to choose a peaceful application, such as a fish-finder development, or a military application, like a submarine sonar development, i.e., a large fish finder, using the same underwater piezoelectric transducer technology (refer to Figure 12.3).

Figure 12.3 compares other examples using compact piezo-ultrasonic motors. Seiko Instruments developed a compact helicopter, possibly usable for a surveillance camera on the battlefield, while ICAT/Penn State developed automatic zoom/focus mechanisms for cellular phone applications, in collaboration with Samsung EM. However, the same helicopter is useful for peaceful volcano surveillance, and the world's smallest camera module may be utilized in a spy camera. Again, how the product is used depends on the developer's ethics, or is determined by the users.

12.4.4.4.2 Medical Ethics

Not long ago, I met an enthusiastic, middle-aged researcher in Tokyo. He was working on a piezoelectric actuator system for artificial insemination and proudly declared that their system could impregnate cows precisely and quickly, at the rate of some 200 per hour (see Figure 12.3). One question that came to mind in considering such a system was "What are the moral/ethical implications of this sort of mass production of cows?" On the other hand, this equipment could solve starvation in Africa. American people are interestingly very conservative with this artificial fertilization technology.

Ultrasonic motors have been utilized in prosthetic arms, because they are inaudible under operation, which is liked by the customers of prosthetic arms. Also, piezoelectric pumps are very

Figure 12.3 Ethical problems in R&D topics, reflecting the differences between the United States and Japan.

close to being commercialized for artificial hearts. Ethically, how much would you accept as artificial organs? Everything except your brain?

12.4.4.4.3 Ecological Ethics

To overcome the greenhouse effect and other air pollution problems, the Japanese government took a strategy to eliminate diesel engine vehicles, because diesel engines create toxic exhaust gas. To the contrary, German government took a strategy to increase diesel engine vehicles, because low-grade diesel oil does not require large electrical energy for distillation. Which logic will be better accepted? If we calculate the total energy required from the mine to the wheel for both diesel and gasoline, the gasoline exhibits better efficiency for "tank to wheel," but requires much larger electric energy for "mine to tank." Thus, in a global sense, diesel engine vehicles are expanding drastically. I consulted multiple European and Japanese automobile industries with piezoelectric diesel injection valve developments, as shown in Figure 12.3. The American government is rather behind in the ecological ethics viewpoint. Until recently, they have not encouraged diesel engine vehicles; major American automobile companies have also been behind in fuel-cell vehicle development, chasing Japanese automobile companies.

12.4.4.5 Quality Control

Product development is occasionally supported by the government for military applications in the United States. The researcher must understand the various differences in philosophy behind

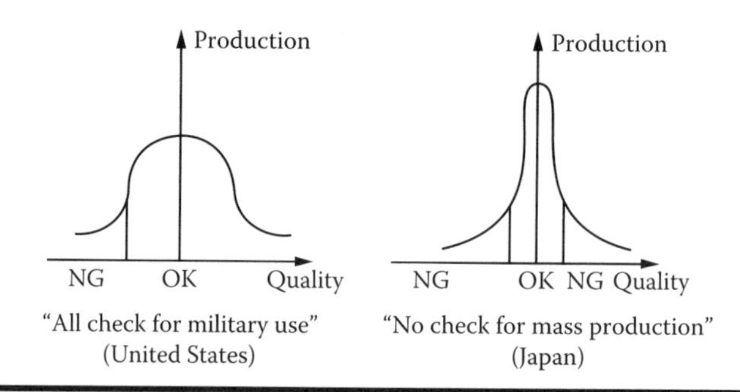

Figure 12.4 Differences between basic trends in quality control for military use (United States) and mass-consumer products (Japan).

military (United States) and civilian commercialization (Japan). For military applications, because the production quantity is relatively small (several hundred to thousands of pieces), manual fabrication processes are generally used, leading to high prices for the products. The strategy of targeting military customers may be adopted by a small venture company as it is starting up.

The difference between required specifications and QC is also very interesting. Figure 12.4 shows the basic trends in QC for military use and mass-consumer products. Due to the use of manual production, the production quality distribution is wider for military use. However, all military products will be checked. 100% inspection is too costly for mass-consumer products. Therefore, the standard deviation of the production quality must be very small. Also, you will likely realize that too high quality of the products is also not good (NG), as explained later in the example of the Toshiba light bulb.

It is no surprise that the significant success of Japan's manufacturing industry is attributed to its total quality management (TQM) efforts. It is ironic, however, in light of America's relatively lackluster performance, that TQM as we know it was introduced to the Japanese many years ago by an American, W. Edwards Deming. In fact, Deming is so revered in Japan that the country's most prestigious award for QC, the Deming Prize, was established in his honor more than 50 years ago by the Japanese Union of Scientists and Engineers. Japanese industries like the idea of Six Sigma management, that is, only 3.4 defects per million products. In contrast, most factories in the United States are operating at three- to four-sigma quality levels [4]. That means they could be losing up to 25% of their total revenue due to processes that deliver too many defects—defects that take time and effort to repair as well as creating unhappy customers. TQM is another example which demonstrates American's "can understand, but cannot do" philosophy, and the Japanese's "cannot understand, but can do" philosophy.

Let us consider the example of Toshiba light bulbs. Toshiba is one of the largest light bulb suppliers in Japan. The light bulbs typically have an average lifetime of around 2000 hours. Their QC curve has a standard deviation of ± 10% (1800–2200). If some of the production lots happen to be of a little better quality lifetime of 2400 hours, what will happen? A company executive might mention bankruptcy of the division. For this kind of mature industrial field, the total number of sales is almost saturated, and this 10% longer lifetime translates directly to a 10% decrease in annual income. Therefore, "too high quality" must be strictly eliminated from mass-consumer products. Of course, Toshiba has the technological ability to extend a bulb's lifetime. If you have a chance to visit Japan and to look for light bulbs, 2400-hour-lifetime bulbs can be found in shops. You should not be surprised, however, to find the price exactly 10% higher than the usual 2000-hour bulbs.

A final comment: Sometimes even famous Japanese consumer-product companies may contribute to military or governmental programs such as NASA's space shuttle. The main reason is to obtain a certificate of high quality for that company's product, leading to very effective advertisement even though the development effort will not bring in significant profit.

12.4.5 Concluding Remarks

I have taken two MBA courses—one in Japan and the other in the United States. The American MBA class emphasized the humanistic (Theory Y) approach, starting from the mechanistic (Theory X) approach. In contrast, the Japanese MBA class tried to teach a more mechanistic approach, because the managers are already very humanistic. This difference reflects the business relationship between an American and a Japanese company; the American company pushes to set a business agreement first (legal attitude), while the Japanese company hesitates to set a legal agreement before establishing mutual trust, or knowing the partner company's business ethics (ethical attitude). I do not believe that there is one best way for management and corporate ethics. A contingency or conditional model should be taken into account, particularly to consider management in racially or culturally different environments. The compromise between management and ethical standards in the American and Japanese industries is a choice common to both countries. In any case, American ethics is not the global standard, and we need to seriously understand and respect other people's and countries' thinking style and business ethics when we consider the global business relationship.

Chapter Summary

12.1 Ethical business managers are responsible to numerous stakeholders:
- Employees (individually and collectively in the form of unions)
- Customers
- Suppliers
- Competitors
- Stockholders
- Government and community entities (local, national, and global)

12.2 Business ethics:
1. Conflict of interest: If you are university faculty, when you start the company:
   a. Define your working time in your company.
   b. Do not hire your students directly.
2. Confidentiality: Typical general conflicts for employee confidentiality include
   a. Market research data and R&D and marketing strategies
   b. Similar product lines in the new company
   c. Know-how in product-manufacturing processes
   d. Research proposal ideas
   e. Customer list
3. Executive compensation
4. Production ethics
   a. Product liability
   b. Quality control
5. Advertising
6. Discrimination/sexual harassment
7. Firing employees

12.3 Comparison between the United States and Japan in business atmosphere:

	United States	*Japan*
Living philosophy, R&D style	Individual	Group
Education, R&D	Why	How-to
Industry type, R&D performance appraisal	Differential	Integral
Management, big science	Regatta	*Mikoshi* (portable shrine)
Needs, quality control	Reliability	Newness

12.4 Ethical restrictions in R&D topics:
- Military versus civilian applications
- Medical ethics
- Ecological ethics

Practical Exercise Problems

P12.1 Business Ethics Questions

When Japanese currency, the yen, was strengthened in 1980, a Japanese industry, Murata Manufacturing, closed a 220-employee Toyama factory in Japan, and opened a 3200-employee factory in Chiang Mai, Thailand. The motivation for this action is, of course, to use the cheap labor in Thailand. You can compare the employee numbers, 220 and 3200, which cost almost the same direct labor payment. However, the Murata vice president mentioned explicitly that Murata was not interested in transferring Murata's technology to Thailand employees, because once transferred, they would start to manufacture the products for other competitive companies in the world. What do you think of his logic? Almost all company executives in the United States and Japan will agree with Murata's factory relocation decision and with the vice president's technology protection policy, or, further, try to imitate this success story. However, I have slightly different opinions from my personal ethical viewpoint:

- 220 Japanese employees were dismissed, except 10 who were transferred to Thailand as executives. Is this dismissal against the business ethics or corporate responsibility to society?
- Since Murata moved into Thailand, is Murata obligated to transfer the technology to local Thai people? It is not the old colonial period, but the new global investment period.
- Is it not possible to believe in the societal ethics of the local Thai employees, and to educate them to have loyalty to the corporation, not to betray? Note: Thailand is one of the most serious intellectual property pirate countries pointed out by the U.S. Government [10].

I am very much interested in obtaining feedback on these final questions from many company researchers in the world with different educational or cultural backgrounds.

P12.2 Technology Dilemma

Question: "A manager in my company wants me to spy on a colleague. I will be using a spying computer program that is 100% hidden, does screen captures, etc. Is there a document that I can have management sign to limit my liability? I want signatures from all managers stating that they are authorizing me to spy. Thoughts? I have done something similar before, but this is the first time that the company has asked me to compile data against a user for possible use in court."

What would you suggest this individual do in response to the managers? Consider the following:

- What are the key facts relevant to your response?
- What are the ethical issues involved in peer spying in the workplace?
- Who are the stakeholders?
- What alternatives would you suggest to this individual, and what alternatives exist for employers who wish to gather information about employees surreptitiously?
- How do the alternatives compare? How do the alternatives affect the stakeholders?

References

1. Wikipedia, "Ethics," http://en.wikipedia.org/wiki/Ethics (accessed on January 10th, 2009).
2. Sobel, M. *The 12-hour MBA program*. New Jersey: Prentice Hall, 1994.
3. DesJardins, J. *Business ethics*. New York: McGraw-Hill High Education, 2009.
4. Brue, G. *Six sigma for managers*, a briefcase book series. New York: McGraw-Hill, 2002.
5. Uchino, K. Difference between Japan and the United States in Research and development policy. *Sophia* 1992;41(2):213–224.
6. Uchino, K. *Ferroelectric devices*. New York: Dekker/CRC, 2000.
7. Makino, N. *The fourth economic crisis*. Tokyo: Hiraku Publication Company,1997.
8. McGregor, D. M. The human side of enterprise. *Management Review*, November 1957. New York: American Management Association.
9. Taylor, F. W. The principles of scientific management. *Bulletin of the Taylor Society* 1916:13–23.
10. Patent Hub, "Patent Pirates," http://www.patentshub.com/pirate/(accessed on January 10th, 2009)

Chapter 13

Now It's Your Turn— The Future of Your Company

You have already learned most of the basic knowledge you need in order to start up your own high-tech company. This chapter will first review some key points. Once you have completed the review, and thus finished the book, it is your turn to aggressively move forward. Wishing you "bon voyage," I dedicate to you a special article "Manual for Communications Success in Japanese Business Culture ," in Section 13.3. If your company expands and finds customers in Japan, you will need to visit Japan to promote your product. This article will help you successfully navigate the cultural and business practice differences between the United States and Japan, and hopefully accelerate your company's business globalization.

13.1 Review of Key Points

Typical MBA courses include the following curricula: economics, accounting, finance/investment, marketing, and human resources, with courses like Managerial Communications and Quantitative Business Analysis (Applied Mathematics), and finally Strategic Management and Business Ethics. As shown in Table 13.1, this textbook provides you with almost all the knowledge necessary for operating a corporation, above and beyond what you may find in an MBA program.

In particular, the topics in Chapters 2 and 9 cannot typically be found in most MBA coursework, but are most essential to high-tech entrepreneurs. Because of this, I took the liberty of writing this textbook for engineer entrepreneurs in particular. You should have also mastered how to write successful business plans (Chapter 4) and research and development (R&D) proposals (Chapter 5). Without these skills, your company will not be successful for very long. The article in Section 13.3 may be particularly intriguing once your company grows in a couple of years.

Mastering accounting and financial management (Chapter 6) seems to be the hardest for engineers, and therefore I strongly encourage you to work on these items continuously.

Now it is time for you to fly. Bon voyage!

Table 13.1 The Relationship of the Book Chapters to Regular MBA Course Curriculum Contents

This Book Chapter		*MBA Course Curriculum*
Chapter 1	Industrial Evolution	Entrepreneurship
Chapter 2	Best Selling Devices	N/A
Chapter 3	Corporation Start-up	Entrepreneurship
Chapter 4	Business Plans	Strategic Management Entrepreneurship
Chapter 5	Corporate Capital and Funds	Entrepreneurship
Chapter 6	Corporate Operation	Management Accounting Financial Management Managerial Economics Investment Analysis
Chapter 7	Quantitative Business Analysis	Quantitative Business Analysis
Chapter 8	Marketing Strategy	Marketing Management
Chapter 9	Intellectual Properties	N/A
Chapter 10	Human Resources	Human Resource Management
Chapter 11	Business Strategy	Strategic Management Policy Analysis
Chapter 12	Corporate Ethics	Business and Society Ethics in Management
Chapter 13	Now It's Your Turn	Perspectives on Management Managerial Communications

13.2 Business Globalization

I am indebted to *Managing Human Resources* by Bohlander and Snell [1] for describing the general contents.

13.2.1 International Corporations

International business operations can take several different forms. Table 13.2 shows four basic forms of international organizations.

First, the *international corporation* is essentially a domestic firm that builds on its existing capabilities in order to penetrate overseas markets. This is the most fundamental structure to expand into foreign markets.

Second, a *multinational corporation* is a firm that usually has fully autonomous units in operation in multiple countries. The companies in this category traditionally give their foreign subsidiaries a good deal of flexibility to address local issues such as consumer preferences, political pressures, and economic trends in the subsidiary's region of the world. Frequently these subsidiaries are run as independent companies, without much integration.

Table 13.2 Types of International Organizations

		Local Responsiveness	
		Low	*High*
Global Efficiency	High	*Global:* views the world as a single market; operations are controlled centrally from the corporate office.	*Transnational:* specialized facilities permit local responsiveness; complex coordination mechanisms provide global integration.
	Low	*International:* uses existing capabilities to expand into foreign markets.	*Multinational:* several subsidiaries operating as stand-alone business units in multiple countries.

Third, the *global corporation* can be viewed as a multinational firm that maintains control of operations back in its home office. Japanese companies, such as NEC and Panasonic, tend to treat the world market as a unified whole and try to combine marketing activities in each country to maximize efficiency on a global scale. These companies operate much like a domestic firm, except that they view the whole world as their market place.

Fourth, a *transnational corporation* attempts to achieve the local responsiveness of a multinational corporation, while also achieving the efficiency of a global corporation. To balance this "global" and "local" dilemma, a transnational corporation uses a network structure that coordinates specialized facilities positioned around the world. By using this flexible structure, a transnational corporation not only provides autonomy to independent country operations, but also brings these separate activities together into an integrated whole.

13.2.2 Trading Practices

13.2.2.1 Import/Export Restrictions

I will remind you here that U.S. government organizations have a list of permitted countries for production bases. The Commerce Control List Overview and the Country Chart on export administration regulations can be found at http://www.access.gpo.gov/bis/ear/pdf/738.pdf. Similarly, there is a list of permitted countries for the export of high-tech products. Further information can be found at http://www.access.gpo.gov/bis/ear.

Your high-tech firm should remember the following regulations:

Section 734.11: Government-Sponsored Research Covered by Contract Controls

(a) If research is funded by the U.S. government, and specific national security controls are agreed upon to protect information resulting from the research, [Section] 734.3(b)(3) of this part will not apply to any export or re-export of such information in violation of such controls. However any export or re-export of information resulting from the research that is consistent with the specific controls may nonetheless be made under this provision.

(b) Examples of "specific national security controls" include requirements for prepublication review by the Government, with the right to withhold permission for publication; restrictions on prepublication dissemination of information to non-U.S. citizens or other categories of persons; or restrictions on participation of non-U.S. citizens or other categories of persons in the research. A general reference to one or more export control laws or regulations or a general reminder that the Government retains the right to classify is not a "specific national security control." [Please refer also to Chapter 10, Human Resources, for more information.]

Tariffs, import quotas, and other types of import restrictions hinder global business. These are usually established to promote self-sufficiency and can be a huge roadblock for the multinational firm. For example, a number of countries, including South Korea, Taiwan, Thailand, and Japan, have placed import restrictions on various goods produced in the United States, including telecommunications equipment, automobiles, rice, and wood products.

13.2.2.2 Cultural Misunderstandings

Differences in the cultures of foreign countries may be misunderstood or not even recognized because of the tendency for marketing managers to use their own cultural values and priorities as a frame of reference. Refer to the article in Section 13.3 to enhance your knowledge on this point.

13.2.2.3 Political Uncertainty

Governments are unstable in many countries, and social unrest and even armed conflict must sometimes be dealt with. This is an important item to consider when seeking an international business partner. I had a very scary experience in this area. One of my former affiliate companies, jointly sponsored with a local electronic company, set up a manufacturing factory in Sri Lanka at the end of 1980s. However, in the early 1990s, there was a coup near the capital, Colombo, and our factory was bombed and destroyed. The corporate president was captured by a terrorist group, though fortunately he was released after paying an expensive ransom. We lost all of our investment.

13.2.2.4 Economic Conditions

The differences in economic conditions between the United States and your partner's country directly affects your sales revenue and profit. Let us take the example of MMI: MMI purchased a product from France and distributed it in the United States. We consider the profit calculation process below:

1. MMI ordered €1000 of finite element method (FEM) simulation software from a French firm named ABC on September 4, 2007. The price can be calculated by €1000/0.735 (currency exchange rate on September 4) = $1360.
2. MMI sent a quote to a customer, by adding 20% profit margin. The quote was $1360 × 1.20 = $1633.
3. MMI received the product from ABC, and sent it to the customer on October 2.
4. MMI received payment from the customer on November 25, because the accounting process is slow due to the amount of paperwork in this big company.
5. MMI wire-transferred money to ABC on November 28 for paying €1000. MMI needed to send: €1000/0.673 (currency exchange rate on November 28) = $1486
6. The gross income obtained was $16333 – $1486 = $147.

You need to understand that the originally expected gross income of $273 (= $1633 – $1360) shrank drastically to only $147 during a period of less than 3 months, because U.S. currency was weakening during that period. (Refer to Figure 13.8 and Practical Exercise Problem P13.2.) If we convert the number into the currency exchange rate change per year, it is a 35% discount rate! It may not be a good time for a U.S. firm to start an import business.

13.3 Case Study: Product Promotion in Japan

Fifty-five percent of the American people recognize that Japan is their most important business partner in Asia, with supporting data that 93% of Americans think the current U.S.–Japan relationship is "good" or "very good" [2]. Under these circumstances, we can expect continuous business growth with Japan in the high-tech industry for the foreseeable future.

This section summarizes the business communication tactics you will need to successfully promote U.S. products in the Japanese market. We adopt here a particular scenario: A female vice president of an electronic components development company will visit a large Japanese electronics company to promote her company's new components.

Japanese business culture and style are important to know before visiting Japan. Issues include Japan's male-dominated society and its unique decision-making process. The *nemawashi*, or prior underwater negotiation, is the key to business negotiations, followed by a face-to-face presentation as a ceremonial event. Japanese industries look for the highest quality components, rather than merely the lowest price, as in Korean and Taiwanese business culture. Therefore, promotion should focus on quality. Quality includes the delivery date, which should be negotiated and kept. What will this vice president need to prepare?

13.3.1 Background of the Japanese Business Atmosphere

Even though Japanese economic power has been reduced over the last 10 years since the recession in the 1990s. Japan is still one of the United States' strongest business partners.

According to the 1998 Gallup Census [2], the American people believe

- Japan can be relied on (60%)
- Japan is the United States' most important partner in Asia (55%)
- Relations between the United States and Japan are "very good" or "good" (93%)

Based on these statistics, we can expect continuous business growth between the United States and Japan for at least the first decade of the twenty-first century.

There are three significant differences in the business atmospheres of Japan and the United States: Japan has a male-dominant society, a double standard, and moderate majority.

In Japanese industries, 98% of managers and general managers are males. Females are typically assistants, who do not work for a long period of time. Female workers typically consider their jobs temporary, until marriage. The following statistics are from my experience as a professor of physics at one of the top 10 Japanese universities. Among 16 female students who found an industrial job in Japan after finishing a bachelor of science degree, 12 resigned in less than 3 years because of marriage, and the remaining four resigned by age 30 because of social and company pressure. Japanese society respects women as professional homemakers, and expects women to marry by the age of 26–28. A single, working female who is over 30 years old will feel a strong social pressure to quit her job all the time. On the contrary, male workers usually consider the job permanent, and generally, in my generation, would not resign until age 55–58. This now goes up to age 60–62. There is a strict age rule for retirement in Japan.

There is a famous Japanese proverb, "A longer stake must be hit," which means that a person who tries to surpass his colleagues will be punished. In Japanese culture, every worker needs to keep the same pace as the other workers. He (or she, rarely) must not work too hard, nor too poorly. Work must be *chuyo*, or moderate, which is the most important factor for the best team effort. From this perspective, a person who occasionally says "I did it," or "I successfully made it," rather than "We did it as a team," or "We made it," will not get along with his or her colleagues. This *chuyo*-ism is sometimes excessive in Japan: A CNN news broadcast in June 2008 focused on one company's extreme control of the employees' weight. The company's personnel department staff regularly records all employees' weight and waist size to calculate their body mass. The company then fines employees with a number higher than a certain threshold. The employee must reduce his or her body mass immediately. This measure is not enforced at present for foreign employees, who are specially treated as *gaijin*: an American employee of the company criticized this control as "very difficult to imagine in the U.S."

One of the most difficult Japanese business-culture aspects for Americans to understand is the double standard of real intention (*honne*) and theory (*tatemae*). There is usually a difference between a person's words and actual intentions. Without understanding a person's mind, the words expressed sometimes do not have any real meaning. The Japanese saying "Reading the mind" came from a religious background, and is related to another Japanese proverb, "Silence is golden." The Japanese often respect silence over a wordy presentation or busy conversation, while Americans often find silence hard to endure (a kind of torture!). I was very surprised to hear that a U.S. elementary school teacher misuses the label of "autism" very often for a quiet student. In this sense, maybe 90% of Japanese students should be categorized as autistic. In order to know the actual intentions deep in a person's mind, a *nemawashi*, or underwater negotiation, is essential. This process will be discussed in Section 13.3.2.

This report summarizes the business communication tactics which will successfully promote U.S. products in the Japanese market. We will adopt here a particular scenario as described below (refer also to Figure 13.1): A female vice president (Barb Shay) at MMI Corporation, which is developing electronic components, will visit Saito Industries, a large Japanese electronics company, and present MMI's new components to Saito general manager Toru Nakamura and manager Kenichi Suzuki for use in their devices.

Although the scenario is a bit extreme, it is also probable, and you can enjoy and learn about the perception gap and unique business atmosphere in Japan.

13.3.2 Before Arrival—Preliminary Contact

As described in Section 13.2.1, a business agreement with a Japanese company is usually made through *nemawashi,* prior underwater negotiation. Without securing a preliminary agreement during this period, visiting Japan is meaningless. This section details this process.

13.3.2.1 Communication Methods

Barb Shay must find a suitable contact person, a so-called *gatekeeper*, at Saito Industries. For this purpose, she will target either the manager, Kenichi Suzuki, in Saito's sales division or the general manager, Toru Nakamura, in the R&D division.

Figure 13.1 Characters in this section article.

The first contact should be made via mail (registered air mail, UPS, or FedEx) and be addressed to the higher-ranking person, that is, Mr. Toru Nakamura. Even if the practical contact point will be manager Kenichi Suzuki as the negotiation proceeds, all correspondence should be addressed to Mr. Nakamura with a copy to Mr. Suzuki. This is standard Japanese business practice. Never skip the higher-ranked person without a special request from Mr. Nakamura himself.

Because 98% of Japanese industry people do not have a PhD (based on the 3000 business cards I have collected), Barb should refrain from using her academic title (e.g., Barb Shay, PhD) on her letter. It will be interpreted as arrogance. Follow the *moderate majority principle* or, when in Rome, do as the Romans do! Also, Barb should explicitly mention that she is a female. It is as difficult for Japanese people to recognize Barb as a female name as it is for Americans to recognize that Toru is a male name. In the Japanese male-dominated society, if Barb does not initially disclose her gender as female, her Japanese counterparts will be embarrassed once her gender is revealed, possibly creating conflict in the business relationship. Therefore, Barb should disclose her gender from the start.

In order to accelerate the process, Barb can start to use fax and e-mail after the first contact. Note the U.S. custom of computer e-mail is not very popular in the Japanese business community. A cellular phone e-mail system is much more popular in Japan. Some Japanese engineers read their computer e-mails only once a week or so. In addition, some companies set a high firewall for incoming e-mail, which automatically erases all English e-mails as virus e-mails. Even when computer e-mail can be used, 10–20% of the e-mails sent from the United States may bounce back regularly due to these firewalls. Since the computer e-mail system is not reliable, I strongly recommend that the e-mailed content also be faxed if a letter is important. A fax sent to Saito Industries will definitely be put on Mr. Nakamura's desk, even if he does not read computer e-mail regularly. Direct telephone contact from Barb to Mr. Nakamura or Mr. Suzuki should be avoided except for real emergencies. A phone call can create an embarrassment to Saito Industries due to the language barrier.

Before visiting Saito Industries' Tokyo office, Barb needs to disclose and explain all of the negotiation items to Mr. Nakamura and Mr. Suzuki (or sometimes to higher officials in the company), and receive an unofficial oral agreement with respect to most of the parts. This is the *nemawashi*, and is extremely important to Japanese business discussions. Japanese executives do not like surprise proposals at the face-to-face meeting. Even if Barb presents a beautiful and attractive presentation at Saito's Tokyo office, if she has forgotten to notify them of some of the discussion items, the business discussion will be automatically terminated, because the discussion item is new to Saito's executives. Barb may lose any opportunity to discuss her proposal further. Remember that there should be no surprise proposals. The presentation at the Tokyo office is a ceremony to formalize the agreement, which has been already roughly agreed upon during the prior *nemawashi* period.

13.3.2.2 Forms of Address

Barb should never call Toru Nakamura by his first name. Barb should call him "Mr. Nakamura" at all times. Even his wife does not usually call him by his first name. A wife usually calls her husband just "husband" (*shujin*) in public, and "you" (*anata*) on face-to-face occasions. In basic Japanese business protocol, Mr. Nakamura calls Mr. Suzuki "Suzuki-kun" or "Suzuki manager" (*kacho*), and Mr. Suzuki calls Mr. Nakamura "Nakamura-san" or "Nakamura general manager" (*bucho*). "Kun" means Mr. and is used for a lower-level or younger person. "San" is used for a higher-level or older person. In Japanese business interactions, title, rank or seniority should always be explicitly mentioned.

For your reference, Nakamura is called by his close male friends just simply "Nakamura," without adding "san" or "kun," while his female friends will call him "Nakamura-san." Some of his friends actually may not know his first name. Only his parents and brothers/sisters call him by his first name, using "Toru-chan." "Chan" is used for very intimate young relatives.

If Barb calls Mr. Nakamura by his first name, the others could assume that Barb is Mr. Nakamura's concubine, which is not rare for a high-ranking person. This is not a joke. Be cautious when using a first name in Japan!

13.3.2.3 Airport Pick-Up and Hotel Reservations

Because women are so rare in Japanese businesses, once Barb discloses she is female, the attitude of Mr. Nakamura and Mr. Suzuki will change drastically; if she is relatively young, they will become very chivalrous. Mr. Nakamura or Mr. Suzuki will meet her at Tokyo's Narita International Airport when she arrives, and take her to her reserved hotel. During the initial arrangements, Barb can request that Mr. Suzuki make a reservation at a convenient hotel near Saito Industries. Using the female advantage is another tactic for succeeding in Japan. This seems like a contradiction in a male-dominated society, but it is true!

If necessary, Barb can request that Mr. Suzuki take her back to the airport when she departs. The Japanese will be happy to take care of this sort of arrangement, as long as Barb requests them in a polite way.

For your reference, if Barb is male, another scenario may happen as follows: the possibility for Mr. Suzuki to meet Barb at the airport will be 50%, and driving Barb back to the airport at departure probably will not happen.

13.3.3 In Japan

13.3.3.1 Cellular Phone Rental

When Barb departs from the United States, or arrives at Narita Airport, she must rent a cellular phone at the airport. As advised in Section 13.2.2, since Japan's computer Internet system is very poor, communication with Mr. Nakamura or Mr. Suzuki (one of whom will already be in the airport to meet Barb) should be made via cellular phone. Barb must learn to send e-mail via her cellular phone. This is particularly important because talking on cellular phones is prohibited on trains or buses. Therefore, Barb will need to communicate with Mr. Suzuki via cellular phone e-mail.

13.3.3.2 Cash Kingdom

Japan is a "cash kingdom." Most expenses should be paid in cash, except for the hotel. Barb needs to pay for rail, airport shuttle, and taxi fees all by cash, even though they will total more than JY20,000 (Japanese yen; almost US$200). Barb needs to keep at least US$400 in her purse at any time. An airport bank is the best place to exchange currency. It is recommended to initially exchange US$1000 (JY100,000). Few restaurants accept credit cards, though dinner is typically more than US$100 per person.

Credit cards can be used only in hotels and department stores. Remember that Barb needs to pay cash for transportation, restaurant meals, and most shops. Traveler's checks are not accepted except in hotels. Most likely, Mr. Suzuki will use his own car to take Barb from Narita Airport to her hotel. He will pay for the expressway fee at the gate, which is very expensive: a 40-minute ride on the highway typically costs US$50 or higher. Barb should offer to pay for the transportation expenses, even though she is a guest. It will be interpreted as a show of politeness since transportation costs are so high. For example, filling a gas tank generally costs US$150 or higher.

13.3.3.3 Smoking Kingdom

When Barb arrives at Narita Airport, she will notice a strong smell of cigarettes. Even though there are fewer smokers in the younger generation, senior managers such as Mr. Nakamura and Mr. Suzuki will definitely smoke. Do not complain or criticize it. Smoking is permitted even during business meetings, and is encouraged in the after-5-o'clock session. Barb may even be invited to smoke Mr. Nakamura's favorite cigar or cigarette. Barb may decline this offer politely, but never criticize Mr. Nakamura's smoking. Drinking and smoking are believed to be the best communication lubricants in Japan. Without these, smooth business discussions cannot be expected.

13.3.3.4 Hotel Conditions

International travelers are initially surprised by the small room and bed in business hotels in Tokyo (Figure 13.2a). For someone who is more than 6 feet tall, and/or weighs more than 200 pounds, the bed will be uncomfortably small. Another problem could occur when Barb picks up computer e-mail sent from the MMI U.S. headquarters. None of the business hotels in Tokyo, which are one rank lower and cheaper than regular hotels, has a satisfactory Internet system. No Internet connection can be expected in the hotel room, unless Barb requests that Mr. Suzuki find her an expensive high-rank hotel or a special American-subsidiary hotel such as Holiday Inn,

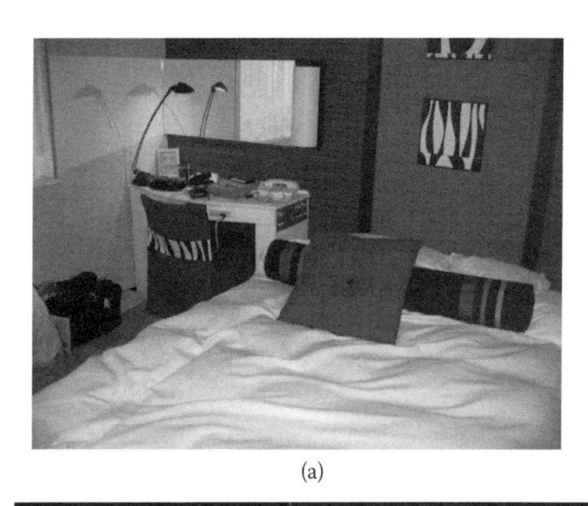

(a)

(b)

Figure 13.2 (a) Typical room in a business hotel, and (b) the Internet connection space beside the hotel lobby at New Ohtani Inn, downtown Tokyo, 2006.

Hilton, or Marriot, which are usually extraordinarily expensive. I offer the following tips for finding Internet service:

Designated hotel internet connection place: Since there will be a long line for the connection after lunch time, I recommend Barb work early in the morning, around 6 a.m. Figure 13.2b was taken at 6 a.m. in a business hotel. Do not expect a comfortable working space. It is just for short connections to pick up and send e-mails. Many people will be waiting to use the facilities.

Telephone connections from the hotel rooms: If Barb can get the phone number of the local Internet provider, a telephone connection is possible. However, phone connections are extremely slow in Japan (only 5–6 kilobites per second, or kbps). I sometimes call my U.S. university computer center via an international call to get their Internet connection. A typical e-mail with a 200 kb file can take 40–50 minutes to send via the international telephone, resulting in a telephone charge around JY4000 (US$40).

Internet café: There are many coffee shops that provide wireless Internet connections in downtown Tokyo. This is a comfortable way to work, if Barb has a lot of free time during the day. Coffee shops are typically open from 9 a.m. until 9 p.m. Unfortunately, this time period overlaps with business meetings and the after-5-o'clock session. Also, the shops are very expensive; a cup of coffee costs JY1000 (US$10).

Thus, for local communications, e-mail via the rental cellular phone is essential. Since talking on cellular phones is prohibited on commuter trains, Mr. Suzuki will not be able to receive a typical phone call when coming to meet Barb at the hotel or airport. Phone e-mail is essential for personal communication. [Note: The Internet situation has improved drastically in big cities since 2008. However, rural Japan still encounters problems.]

13.3.3.5 Business Meetings

It is a strict custom in Japan to wear formal suits in business meetings. A necktie is mandatory for a businessman in a meeting. Women such as Barb should wear a skirt suit. A pantsuit is sometimes allowed for women, but a skirt is preferred in Japan. Sometimes, employees are allowed to come to work without a tie under special orders from the company. Such orders are made for limited hot and humid summer periods, as shown in Figure 13.3, which reports the order by the Japanese government in 2006.

Figure 13.3 The Japanese government orders the bureaucrats to wear cool summer clothes (no tie) in the summer of 2006. (From *Yomiuri* newspaper, July 25, 2006. With permission.)

After Mr. Suzuki's arrival at the hotel, Barb will start the business meeting. Barb should introduce herself in Japanese:

> "Hajime mashite. Watashi wa Barb Shay desu. Dozo yoroshiku."
> Meaning: "How do you do? I am Barb Shay. Very glad to meet you."

Mr. Suzuki, if he is a typical Japanese businessperson, will be very nervous to meet Barb, because he will need to speak English. This sort of simple conversation in Japanese with Mr. Suzuki will totally change his attitude toward Barb, and definitely bring a smile to his face. This is the first step for Barb's communication success.

When Barb arrives at Saito Industries with Mr. Suzuki, she is brought to a VIP meeting room to wait for Mr. Nakamura. Barb's second challenge is to consider where she should sit. As shown in Figure 13.4, in Japan, the highest rank seat is farthest from the entrance door, or nearest to the window. Barb should sit initially in seat F in this room configuration. Seats B and A should be left for the Saito Industries executives. Incorrect selection of a seat position is sometimes interpreted as an arrogant attitude that neglects the Japanese hierarchical system. Typically, the top executive (Mr. Nakamura) will invite Barb to move to a better seat, such as seat E, if he has a good first impression of Barb, or a good impression from the prior negotiations.

The third challenge is the informal initial conversation. After learning that this is Barb's first visit to Japan, the most common question would be "What do you think of Japan?" This is actually a ridiculous question to a newly arrived person. However, Barb should prepare a humorous answer beforehand to successfully pass this third examination step. A sample answer may be "I like the small room arrangement in my hotel very much. Without moving from the bed, I can reach

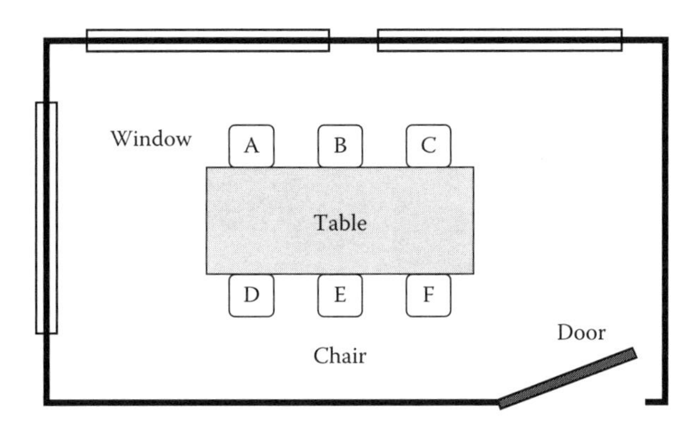

Figure 13.4 VIP meeting room arrangement in Saito Industries. The seat Barb should take is the second key to her business communication success.

anything; the TV, tea pot, etc. It is very convenient!" If spoken very slowly, this sort of reply may cause a big laugh to Saito Industries members, which is a large success for Barb. Even if Barb was uncomfortable the previous night due to the small bed size, Barb should not complain about it, because Mr. Suzuki kindly reserved the hotel. Barb should not criticize explicitly the Japanese business male-dominated or hierarchical society. These topics are really "taboos."

Again, Barb should not call Mr. Nakamura or Mr. Suzuki by their first names. The best way is for her to call them "Nakamura-san" and "Suzuki-san" all the time.

13.3.3.6 Product Sales Promotion

When the discussion topic shifts to the main topic, Barb will promote MMI's new products after a brief introduction to MMI. Though she obtained a favorable reply from Mr. Suzuki and Mr. Nakamura during the *nemawashi* period, this is the first official presentation about MMI products.

Barb must

- Carefully choose easy-to-understand English words (Japanese cannot typically distinguish b and v, r and l, th, s and sh)
- Speak very slowly (typically half speed)
- Prepare presentation slides with Japanese words for the key points

Mr. Nakamura and Mr. Suzuki will be delighted to see some Japanese words on the slides (refer to Figure 13.5), which will help make Barb's presentation successful.

The presentation should initially stress the following two points:

1. Quality over price
2. Teaming

Exemplified by Sony Corporation's policy, most Japanese industries respect product quality over inexpensive price. This is totally different from the business atmosphere in Taiwan and Korea. Barb needs to explain primarily how MMI products are of high quality, reliable, and differentiated from competitor's products. Quality does not mean just the long lifetime, but also the "newness." Since typical Japanese update their cars and electronic devices (TV, cellular phone, etc.) every

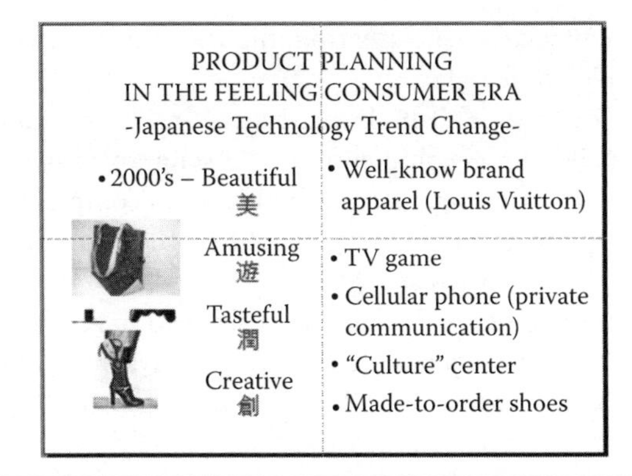

Figure 13.5 Presentation slide example with some Japanese characters.

3 years or so, long lifetime or durability, like that found in Samsonite suitcases is not the most attractive sales point. But, like Nintendo computer games, newness or innovation is key.

For the twenty-first century, wide market research has shown that Japanese appreciate the keywords and key concepts "beautiful," "amusing," "tasteful," and "creative" (refer to Figure 13.5). Barb needs to include these in her presentation. A good example is Game Boy by Nintendo, which is a video game system for TVs, and has become popular worldwide among children. Nintendo used to supply Japanese traditional playing cards. At the beginning of the 1970s, when most Japanese electronic industries were chasing U.S. technologies in semiconductor devices, a major semiconductor company had a large number of failed 8-bit chips (the Japanese technology at that time had a low quality level). Nintendo was able to purchase them at a very low price. Since most of the basic functions worked, they could be used to develop computer-aided toys. The prototype Game Boy did not utilize any advanced technologies, but utilized the cheap 8-bit chips with well-known technologies. The key to this big hit was its ability to fit a social trend, hit on the key word "amusing," and to firmly attract the kids' attention.

Barb should not use "I" in her presentation, which is against the Japanese business policies of "moderate majority" and "teaming." She should use "we" all the time. Even if she has the authority to decide some agreements, it is recommended she defer, and answer "I think it seems to be a good offer, but could you wait until I can get permission from MMI executives?" Since the team decision is respected in Japan, the decisive reply should come later, even if Barb has authority as the vice president. If Barb says "I can decide it now," she will be thought of as very arrogant, rather than very capable. Remember the moderate majority principle.

Perhaps Barb needs to add a production partner with facilities in Taiwan or South America to keep down the costs and stay competitive with the mainland Chinese companies. She should add a strong remark about how punctually the products will be delivered to Saito Industries, without any delay and not too early, so storage fees will be unnecessary on Saito's side. Barb will use a presentation slide to explain how this teaming is structured and how she manages the total team.

During the business meeting, the following four points should also be noticed:

1. *Smoking is allowed in the meeting.* Barb should not complain about this situation. Drinking and smoking are necessary communication lubricants in the Japanese business community.
2. *Japan is a male-dominated and hierarchical society.* Barb should not begin the conversation. Usually, the highest boss will begin (in this scenario, Mr. Nakamura). Then, Mr. Suzuki may request that she answer Mr. Nakamura's questions or comments. Then it is time for

Barb to speak. The sequence is generally from higher to lower ranked persons, from elder to younger, and from male to female.

3. *Don't say "No." Use "Yes, but ..."* If Barb interrupts Mr. Nakamura's speech, or immediately replies "No" to Mr. Nakamura's comments, it will be interpreted as impolite by the Japanese businesspeople. Even if Barb is against Mr. Nakamura's comments, she needs to accept his thought first, then explain an alternative thought on the issue. An appropriate response to a question by Mr. Nakamura such as, "Ms. Shay, I am wondering about the quality control of the MMI manufacturing line. The products seem to have a relatively large variation of performance," may be, "Yes, Nakamura-san, you are truly an observant engineer (praise Mr. Nakamura, first). We had some problems in manufacturing (admit the error, second). But, we solved this issue last year (now, strike back politely)."
4. *Be aware of the double standard between real intention and theory.* When Mr. Nakamura speaks English (sometimes via an English interpreter), he is speaking theory (*tatemae*). However, if he starts to discuss with Mr. Suzuki in Japanese, they are discussing real intention (*honne*) That Japanese conversation is most essential, but Barb cannot understand their secret conversation. Barb should not neglect the Japanese double standard system—if she were to scream "Could you speak in English?" she would be thought of as an invader of their privacy. This *honne* portion can be easily disclosed by Mr. Nakamura in the after-5-o'clock session with the aid of some alcohol. Barb must endure until then. Even if this *tatemae* discussion seems to be a little negative, without asking about the *honne*, Barb need not be disappointed with the discussion. Typically in Japan, *tatemae* is strict in comparison to *honne*, which is discussed with the aid of alcohol, off the record.

13.3.3.7 Business Agreements

"When a stone is thrown, it will hit an attorney in the United States." The attorney density is more than 10 times higher in the United States than in Japan. Since attorney fees are very expensive in Japan, the Japanese prefer not to use attorneys in routine business agreements. Therefore, formalizing business agreements can be a big problem for U.S. companies dealing with Japanese companies.

Among the 11 Japanese partner companies with Barb's company, only six have set formal trading/distribution agreements. Typically, in the initial half year or so, distribution must be made under MMI's sole risk. The Japanese partners are just watching MMI's manufacturing or sales performance, until they gain trust in MMI. Based on the trading division's comments in the company, U.S. companies will try to set a rigid business agreement from the start. If problems occur, the U.S. company readily starts a lawsuit to obtain compensation. Whether this is true or not, this is the Japanese impression of U.S. companies. Japanese companies prefer to build trust through performance before signing agreements.

In this sense, Mr. Nakamura may hesitate to agree to a business agreement (*tatemae*) offered by Barb at this first meeting. However, once Saito Industries recognizes MMI as a reliable partner, continuing the component supply for a half year or so without a strict written agreement, MMI will establish a good working relationship between the companies for years to come. Establishing a good *honne* relation, without a legal constraints, is the most important issue. This business attitude is strongly related to *keiretsu*, family-like subsidiary companies, in Japan. Once MMI is recognized as one of the *keiretsu* companies of Saito, trading will not be easily terminated. But, Barb needs to endure at least a half year without an official agreement; that is, at MMI's own risk.

At this stage, Barb should not rush or push for a final answer of "Yes" or "No" from Mr. Nakamura. The Japanese prefer to answer in the *tatemae* meeting in a rather ambiguous way (*chuyo*, or moderate), such as "I would like to consider this issue positively." In order to identify whether this answer means "Yes" or "No," Barb needs to wait until the *honne* after-5-o'clock meeting.

13.3.3.8 After 5 O'Clock Session

Public relations fees are divided into two major categories: guest treatment, such as eating and drinking, and advertisement, such as newspaper, TV, booklets, etc. U.S. and Japanese industries spend differently. The U.S. public-relation fees are typically 2.4% of the GDP, with guest treatment less than 10% of that. However, in Japan, public relations fees are 1% of the GDP, with guest treatment roughly 50% of that. In the monetary amount, the Japanese guest treatment fee is US$53 billion per year (in 1998), while the U.S. treatment fee is US$14 billion [14]. Japanese companies tend to spend lavishly on wining and dining their business partners.

Barb is supposed to be invited to dinner by Saito Industries after finishing the official business meeting, including Mr. Nakamura and Mr. Suzuki. Sometimes, a couple of young women will also be invited, to take Barb's gender into account.

Usually a female guest should not choose her favorite meal, but accept what the highest ranking male recommends. If Mr. Nakamura asks whether Barb can eat raw fish *(sashimi)* or not, she may reply, "Yes, I can. But, I like fried fish *(tempura)* much better" (see Figure 13.6). Never decline or outright say "No." For your reference, a male guest has a little privilege to choose meals, in particular alcohol (beer, sake, wine, etc.)

Figure 13.6 Japanese delicacy: shrimp and vegetable tempura.

When the hors d'oeuvre is served, Barb initially needs to follow the Japanese hierarchical alcohol pouring sequence. Mr. Nakamura can first pour Barb's beer. Immediately after this, Barb needs to pour beer into Mr. Nakamura's glass, then Mr. Suzuki's glass. The other persons will pour beer for each other. Remember to never pour alcohol by yourself in Japan even when your glass is empty. It is considered very impolite (for both males and females). Also, even if Barb does not drink alcohol much, she should not decline the initial toast (*kanpai*).

Since a typical dinner costs more than US$150 per person, it is polite for Barb to bring some gifts to Mr. Nakamura, Mr. Suzuki, and others. American whiskey, bitter chocolate, beef jerky (but be aware that customs at the airport sometimes regulates agricultural products depending on the situation, such as an outbreak of "mad cow" disease), specially blended coffee, and famous brand accessories (belts, ties, or cuff links from duty-free shops) are recommended gifts. If Barb can bring it in (i.e., if there are specially inspected ones sold in the airport), American fruits are welcome. Gifts that are not recommended include T-shirts (too cheap, and not many senior persons wear T-shirts in Japan), city-, company-, or university-logo included goods (also seems to be cheap), and colorful candies and other sweets (which are too sweet for the Japanese).

After a few drinks, Mr. Nakamura usually makes a frank comment about the previous business discussion. This is off the record, but will be the most important information received during Barb's travel. She needs to listen to the conversation much more carefully than the official meeting, and additionally ask about the details for the future actual relationship. Asking about the details is not impolite in the after-5-o'clock meeting in Japan. She needs to report this *honne* to the MMI headquarters, in addition to any written *tatemae* memorandum.

During dinner, Mr. Nakamura and Mr. Suzuki will be particularly interested in Barb's motivation for taking her present job, because it is very rare in Japan for a female to have a responsible managerial position. Barb can disclose her academic and industrial backgrounds briefly. However, it is taboo in Japan to mention family matters, religious issues, and politics. Showing family pictures is not popular, and is actually hated among business colleagues. Also, since most of the Japanese are not very religious, they do not discuss religion.

Barb's trip to Japan is simply not complete without a visit to a karaoke bar. Karaoke is one of the main social activities of Japanese adults, especially businessmen or "salarymen." These men visit "lounge" or "snack" bars after work, drink for hours, and divulge their worries and concerns to the women at hand. In between, they sing their favorite tunes at which the women clap enthusiastically, regardless of talent (Figure 13.7). But karaoke is much more than just businessmen loosening their ties. The bars are popular with people from all walks of life. It is a staple form of entertainment. There are even professional karaoke coaches for those who are so concerned with "losing face" or appearing talentless in front of their friends, colleagues or boss.

Barb should not decline to sing a song. Before visiting Japan, she may need to train in an American karaoke bar to learn a couple of popular (not new, not very professional) songs. The songs which the Japanese sing include Elvis Presley's "Love Me Tender," The Beatles' "Hey Jude," or Celine Dion's "My Heart Will Go On." Also, remember that songs by the Carpenters are very popular in Japan even recently. Once Barb receives big applause from the Saito Industries people, she is accepted as a family member, and her visit to Japan will be an almost certain success.

13.3.4 Follow Up

As soon as Barb returns to the U.S. MMI office, before sending a business letter, she should send a thank you letter or card with pictures taken at the dinner and karaoke bars, with a sincere appreciation note. This is the Japanese custom, and will show that Barb enjoyed these experiences,

Then, the business wrap-up should be mailed, e-mailed, or faxed, which will summarize the *tatemae* meeting. Barb also needs to prepare a report to MMI headquarters on the *honne* meeting, in her own words (this may require assistance from Mr. Suzuki, who became rather close with Barb). Barb should not rush to set up an official, strict trading agreement with Saito Industries at this early stage. Relying on Mr. Nakamura and Mr. Suzuki's oral support, MMI can start the preliminary product supply to Saito Industries without a legal document.

13.3.5 Epilogue

Because Japanese industry does not like legal *tatemae* agreements, we may sometimes start the business relationship with just an oral *honne* agreement. This is a drastically different business convention than in the United States. Regardless, as described previously, it takes a long time to receive recognition as a family partner from a Japanese company. It is a laborious process requiring endurance and perseverance on Barb's part. However, once the business relationship starts, as the statistics suggest, Japanese companies are reliable and will not betray their partners, even without legally binding written agreements.

Finally, "The 10 Business Commandments (Not to Do) in Japan" are summarized in Table 13.3. These are to be internalized before communicating with Japanese businessmen. I hope that this short article helps American businesspersons achieve success in the Japanese business culture.

Figure 13.7 Karaoke singing bar.

Table 13.3 The 10 Business Commandments in Japan

1. Don't say "No." Use "Yes, but ..."
2. Don't ask for a "Yes or No" answer outright; they prefer a moderate reply.
3. Don't make a surprise offer, because consistency with the *nemawashi* prior negotiation is essential.

(*Continued*)

Table 13.3 The 10 Business Commandments in Japan *(Continued)*

4. Don't say "I." Use "we" in business discussions.
5. Don't use officials' first names, but call them "Mr. or Mrs. last name."
6. Don't believe just the expressed word (*honne*), but imagine the unexpressed meaning (*tatemae*).
7. Don't rush or be aggressive in business. Be patient and persistent.
8. Don't push a legal agreement; they prefer an oral agreement.
9. Don't decline their hospitality (eating and drinking), and reciprocate with a gift.
10. Don't explicitly criticize the male-dominated society or hierarchical system in Japan.

Chapter Summary

13.1 The relationship of the book chapters with regular MBA course curriculum contents.

This Book Chapter		*MBA Course Curriculum*
Chapter 1	Industrial Evolution	Entrepreneurship
Chapter 2	Best-Selling Devices	N/A
Chapter 3	Corporation Start-Up	Entrepreneurship
Chapter 4	Business Plan	Strategic Management; Entrepreneurship
Chapter 5	Corporate Capital and Funds	Entrepreneurship
Chapter 6	Corporate Operation	Management Accounting; Financial Management
		Managerial Economics; Investment Analysis
Chapter 7	Quantitative Business Analysis	Quantitative Business Analysis
Chapter 8	Marketing Strategy	Marketing Management
Chapter 9	Intellectual Properties	N/A
Chapter 10	Human Resources	Human Resource Management
Chapter 11	Business Strategy	Strategic Management; Policy Analysis
Chapter 12	Corporate Ethics	Business and Society; Ethics in Management
Chapter 13	Now It's Your Turn	Perspectives on Management; Managerial Communications

13.2 Key words to remember in business practice in Japan: Underwater negotiation (*nemawashi*), moderate majority (*chuyo*), teaming, double standard (real intention and theory, *honne* and *tatemae*), family-like subsidiary companies (*keiretsu*).

13.3 The 10 Business Commandments in Japan.
1. Don't say "No." Use "Yes, but …"
2. Don't ask for a "Yes or No" answer outright; they prefer a moderate reply.
3. Don't make a surprise offer. Consistency with the prior negotiation (*nemawashi*) is essential.
4. Don't say "I." Use "we," in business discussions.
5. Don't use officials' first names. Call them "Mr. or Mrs. last name."
6. Don't believe just the expressed word (*honne*). Imagine the unexpressed meaning (*tatemae*).
7. Don't rush or be aggressive in business. Be patient and persistent.
8. Don't push a legal agreement, because Japanese businesses prefer an oral agreement.
9. Don't decline their hospitality (eating and drinking), and reciprocate with a gift.
10. Don't criticize the male-dominated society or hierarchical structure in Japan.

Practical Exercise Problems

P13.1 Japanese Language Essentials

Remember the most essential expressions in conferences/meetings in Japanese.

Examples

1. How do you do?	Hajime mashite
2. Good morning.	Ohayo gozaimasu
3. I am Barb Shay.	Watashi wa Baabu Shay desu
4. I am from the United States	Watashi wa America kara kimashita
5. May I have your name?	Anata no namae wa?
6. Very glad to meet you.	Dozo Yoroshiku
7. Good-bye!	Sayoonara!
8. See you soon!	Jah, mata!
9. Thank you (very much).	Arigato (gozaimasu)
10. You are welcome.	Doo (pronounce like Doe) Itashimashite
11. Excuse me!	Sumimasen!
12. I have a question.	Shitsumon ga Arimasu
13. Tell me about your research.	Anata no Kenkyu wo Oshiete Kudasai
14. I am studying piezoelectrics.	Watashi wa "piezoelectrics" Atsuden Zairyo wo Kenkyu Shiteimasu
15. I will start. (Meal-starting phrase)	Itadaki masu!
16. I am finished. (Meal-finishing phrase)	Gochiso sama!
17. Delicious!	Oishii!
18. What is this?	Kore wa Nan Desuka?
19. How much (in price)?	Ikura Desuka?
20. I will have (or purchase) this.	Kore wo Moraimasu

P13.2 Foreign Currency Training

Determine the U.S. dollar amount corresponding to the following foreign currency amount.

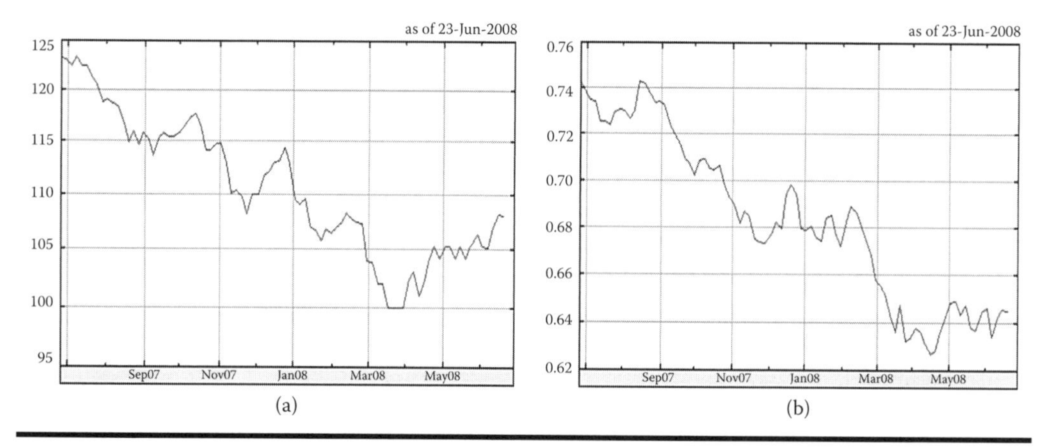

Figure 13.8 Currency exchange rate change with time for (a) Japanese yen and (b) € (euro) from mid-2007 to mid-2008. (From Yahoo! Finance USD/JPY Currency Conversion Chart, http://finance.yahoo.com. With permission.)

Questions

1. JY10,000 for purchasing a *shinkansen* (bullet train) ticket in Tokyo, as of June 2008
2. €17 for purchasing an airport shuttle ticket in Hamburg (as of June 2008)
3. Taiwan$1200 for purchasing a Taiwan bullet train ticket from Taipei to Kaohsiung (as of June 2008)

Answers

1. US$1 = JY107. A bullet train ticket is about US$100 for one way.
2. US$1 = € 0.635. An airport shuttle ticket in Germany costs US$28
3. US$1 = Taiwan$30.4. A bullet train ticket is about US$40 for one way.

The U.S. dollar is getting weaker as of June 2008, with a decreasing rate of 15% per year in the currency exchange rate against any other countries' currency. Refer to Figure 13.8 to check the trend for JY and €. Americans traveling abroad feel that everything is expensive in Japan and Europe recently, and similarly, importing materials from foreign countries is quite expensive. To the contrary, exporting MMI's product to Saito Industries seems to be rather easy just from the price viewpoint.

References

1. Bohlander, G., and S. Snell. *Managing human resources, 14th ed.* Mason, OH: Thomson, 2007.
2. 1998 Gallup Census, published in *Yomiuri* newspaper.
3. *Yomiuri* newspaper, July 25, 2006.
4. Takagi, T. Business culture differences between Japan and America. Soshisha: Tokyo, 1999.
5. Yahoo! Finance, USD/JPY Currency Conversion Chart, **http://finance.yahoo.com.**

Index

C

D

E